Handbook of Agricultural Biotechnology

Scrivener Publishing
100 Cummings Center, Suite 541J
Beverly, MA 01915-6106

Publishers at Scrivener
Martin Scrivener (martin@scrivenerpublishing.com)
Phillip Carmical (pcarmical@scrivenerpublishing.com)

Handbook of Agricultural Biotechnology

Volume I Nanopesticides

Edited by

Charles Oluwaseun Adetunji

and

Julius Kola Oloke

Scrivener Publishing

WILEY

This edition first published 2024 by John Wiley & Sons, Inc., 111 River Street, Hoboken, NJ 07030, USA and Scrivener Publishing LLC, 100 Cummings Center, Suite 541J, Beverly, MA 01915, USA

For more information about Scrivener publications please visit www.scrivenerpublishing.com.

Wiley Global Headquarters
111 River Street, Hoboken, NJ 07030, USA

For details of our global editorial offices, customer services, and more information about Wiley products visit us at www.wiley.com.

Library of Congress Cataloging-in-Publication Data

ISBN 978-1-119-83614-8

Cover image: Pixabay.Com
Cover design by Russell Richardson

Set in size of 11pt and Minion Pro by Manila Typesetting Company, Makati, Philippines

Printed in the USA

10 9 8 7 6 5 4 3 2 1

Contents

Preface

The application of biopesticides is an alternative option that could ensure adequate agro-ecosystem functions and biodiversity, thereby enhancing agricultural production, the effective management of biotic and abiotic stressors, and the maintenance of animals, humans, and a healthy planet. Their effectiveness is due to their availability in larger quantities, eco-friendliness, and their ability to preserve beneficial microorganism and soil activity.

Biopesticides have been identified as a sustainable and permanent replacement to synthetic chemicals. Their application will go a long way toward preventing major challenges that confront sustainable agriculture, the actualization of global food production, and food security, helping to feed an ever-increasing population that is predicted to increase to nine billion by 2050. An interdisciplinary collaboration among policymakers, private sector, researchers, civil society, farmers, consumers, and environmentalists will foster an innovative pathway toward future sustainable agriculture and food systems that could ensure resilience, food security, and a healthy environment.

This book emphasizes the application of microencapsulation and nanoformulations technology that could enhance agricultural production while increasing stability and residual action of nanobiopesticides products, as well as improving their environmental sustainability and field application. A heavy emphasis is placed on the registration of nanobiopesticides with relevant stakeholders, plus the role of scientists and farmers, government, research institutes, industries, universities, agro-industries, and researchers. Additionally, relevant information is provided on the application of nanobiopesticides that could serve as an antibacterial antinematode, antimollusca, antimite, antirodent, and antiviral agent.

The chapters herein outline techniques for the characterization and structural elucidation of the biologically active constitutes that are found in plant materials and could be utilized in the fabrication of nanobiopesticdes. Relevant information is collected on some nanobiopesticides that could be applied as fumigants, which are used in the production of gas or vapor for effective destruction of pests in buildings or soil. Additionally, how to apply nanobiopesticides as biocides, disinfectants, and sanitizers is explained, too.

The book explains the application of some nanobiopesticides as ovicides that could kill eggs of insects and mites, as well as slimicides that could destroy slime-producing microorganisms, such as algae, bacteria, fungi, and slime molds. Other highlights include: a discussion on the application of nanobiopesticides for the rejuvenation of heavily contaminated environments (as well as their role in the mitigation of several abiotic stress); a demonstration of how nanobiopesticides derived from plants could be applied for effective management of pests and diseases in animal husbandry and fishery; and a collection of relevant information on patents, the commercialization of relevant plant-derived nanobiopesticides, and their social economic and industrial relevance.

This book is a useful resource for a diverse audience, including global leaders, industrialists, food industry professionals, agriculturists, agricultural microbiologists, plant pathologists, botanists, agricultural experts, microbiologists, biotechnologists, nanotechnologists, environmental microbiologists and microbial biotechnologists, investors, innovators, farmers, policy makers, extension workers, educators, researchers, and many in other interdisciplinary fields of science. It also serves as an educational resource manual and a project guide for undergraduate and postgraduate students, as well as for educational institutions that seek to carry out research in the field of agriculture and nanotechnology.

I offer my deepest appreciation to all the contributors who dedicated their time and efforts to make this book a success. Furthermore, I want thank my co-editors for their effort and dedication during this project. Moreover, I wish to gratefully acknowledge the suggestions, help, and support of Martin Scrivener and the Scrivener Publishing team.

Professor Charles Oluwaseun Adetunji (Ph.D, AAS affiliate, FRSB (UK) FNYA; FBSN; FNSM, MNBGN)
Director of Research and Innovation, Edo State University, Uzairue, Nigeria
December, 2023

1

Application Nanobiopesticides Derived From Plants

Charles Oluwaseun Adetunji[1]* and Abel Inobeme[2]

[1]*Applied Microbiology, Biotechnology and Nanotechnology Laboratory, Department of Microbiology, Edo University Iyamho, Auchi, Edo State, Nigeria*
[2]*Department of Chemistry, Edo University Iyamho, Okpella, Edo, Nigeria*

Abstract
The utilization of biopesticides for the management of agricultural pest and diseases has been carried for a long time, the incorporation of the vast knowledge gained from nanotechnology only commenced recently. The integration of nanomaterials in the development of biopesticides will, therefore, help in the mitigation of some of the challenges associated with the use of biopesticides alone, such as enhancing the overall efficiency, improved stability, effective delivery while limiting some of the negative effects. Nanotechnology has, therefore, provided the adoption of responsive, biodegradable, biocompatible, and intelligent materials for the fabrication of safe, green, and efficient pesticides. It has been indicated that the integration of nanotechnology to the existing biopesticides has the potential of ensuring the design and formulation of more reliable nanobiopesticides.

***Keywords*:** Nanobiopesticides, nanotechnology, biodegradable, biocompatible, biopesticides

1.1 Introduction

The role of nanotechnology in various areas of human endeavors, such as agriculture, medicine, industries, biotechnology among others cannot be overemphasized. The global population is rapidly increasing and the quest for quality food is increasing continually. The place of agriculture

**Corresponding author*: adetunjicharles@gmail.com

Charles Oluwaseun Adetunji and Julius Kola Oloke (eds.) Handbook of Agricultural Biotechnology: Volume I Nanopesticides, (1–6) © 2024 Scrivener Publishing LLC

for the feeding of this population and environmental sustainability therefore becomes paramount. On yearly basis, it is reported that 25% of agricultural crops on global scale are destroyed by pests and diseases [1]. On this premise, the control and management of pest and diseases is necessary for improving agricultural yield and ensuring food safety and security. More recently, various approaches and pesticides formulations have been employed for the management of pesticides with a view to ensuring food security. Some of these include synthetic, chemical and biological measures. Basically, pesticides are grouped into different classes on the basis of their structures, compositions and targeted organisms into rodenticides, herbicides, bactericides, insecticides, fungicides, among others. The use of chemical pesticides has also been reported to posses it inherent limitation. These substances posses high toxicity to both pests and humans [2]. On global scale it has been documented that over 200,000 persons die on annual basis as a result of poisoning emerging from the use of chemical and synthetic agents. Some of the existing chemicals that have been employed in this regards have also showed some limitations as a result of challenges with efficient delivery mechanism, poor stability within the environment, low specificity, less biodegradability, as well as the high cost of formulation. The chemicals also contaminate the environment since they all pass through it polluting soils, water bodies and the atmosphere [3]. The workers in industries where these pesticides are produced are also prone to occupational hazards due to exposure to these toxic substances. Exposure to these pesticides occurs through the intake of food and water. In order to overcome the aforementioned limitations nanotechnology has shown to have promising results and prospects [4]. At present, nanotechnology has found relevance in various sectors and fields of humans in problem solving. The incorporation of nanopesticide to the area of agriculture has the potential of increasing the production of crop and improvement the mortality of pest. The use of nanobiopesticides gained remarkable attention in their usage against pests as a result of their minute sizes (1–100 nm), enhanced stability, large surface area, ease of application and cost effectiveness [5–7].

This book presents the synthesis and applications of nanobiopesticides in management of various agricultural pests and diseases causing microorganisms. Some of the areas of application covered in this word include their utilization as antirodents, antinematodes, antimites, antimollusca, antiviral agents, biocides, fumigants, ovicides, slimicides, disinfectants as well as their applications in the mitigation of various stresses associated with abiotic factors and the rejuvenation of various environment heavily polluted by numerous emerging contaminants.

1.2 Concept of Biopesticides

Biopesticides are pesticides of biological origin that are obtained majorly from natural substances or microorganisms. They are basically classified into three major groups which are the botanical biopesticides, the microbial based biopesticides and plants derived biopesticides. Biopesticides have gotten remarkable attention as alternative to the conventional agropesticides as a result of their unique features such as specificity in targeting, relatively fewer negative effects, ease of breaking down quickly and higher efficiency. There are several substances, which have been evaluated as potential biopesticides in recent times, which include oxymatrine (an alkaloid), stilbenes found in grape cane, olive oil mill, extracts of *Clitoria ternatea* and stains of *Talaromyces flavus*. The application of biopesticides does not have side effects of concern on the environment due to its eco-friendliness. There are specific products, which have been licensed as biopesticides, although they are still being investigated for any possible health effect [8].

1.3 Uniqueness of Nanobiopesticides

According to various scientific evidences, the emergence of nanotechnology has proven to be a highly reliable tool for the formulation of novel nanocomposites suitable for curbing pest and diseases in agriculture. The nanobiopesticides have outstanding superiority over the biopesticides as well as the other traditional approaches to pest mitigation. They posses unique biodegradability, environmental friendliness, rapid action, good result, easy discharge to plants, gentle release from the vector and remarkably high efficiency. The small sizes of the nanobiopesticides also make them useful carriers [9]. More recently, various nanobiopesticides have been documented to have high efficacies against different pests, such as *Bacillus sphacricus*, *Bacillus firmus*, *Trichoderma harzianum*, *Beauveria bassiana*, and *Bacillus thringicsis*. These groups of pesticide have also been reported not to have negative effects on the population of microorganisms in the environment. The nanobiopesticides are prepared through various routes, with the two major being: extraction of the bioactive compound with pesticidal effect and the blending it with nanomaterials prior to inserting it into a suitable polymeric materials which act as a base support, while in the other, the active pesticidal agent produces the metallic salt with the binded nanoparticles which then hemolysis and merges into the compatible vectors including liposomes, polymer, nanosphere, nanofibers, and micelles [1].

1.4 *In Vitro* Nanobiopesticides Assay

Nanobiopesticides could be evaluated against specific groups of pest so as to assess their efficacies prior to their large scale applications in farmlands. The nanobiopesticides could be produce through various active pesticide based compounds together with mixing of various nanoparticles such as silver oxide, zinc oxide and oxides of aluminum. The overall toxicity of the nanobiopesticide is checked through the use of minimum inhibitory contents which employs the diffusion method based on agar well. It involves the coating of filter paper with the outer layer of nanobiopesticides as well as oral introduction to the targeted pest. Final measurement is done on the concentration of the pest that are dead and those that are alive within an interval of 40 days [2].

1.5 *In Vivo* Treatment with Nanobiofungicides

Nanobiofungicides can be introduced into plants in the right doses so as to ensure their protection from seasonal infections. The use of LC50 and LC90 help in the detection of the specific insects, larvae and attacks by bee. The nanobiopesticides are also used in changing environments like humidity, temperature and stresses from the environment. Under such conditions, the nanobiopesticides are applied directly as sprays in protecting the plants from attack by pests. The application of nanobiopesticides has therefore become the most efficient approach for the control of attacks by pests that transmit diseases [9].

1.6 Conclusion

There is increasing destruction of a large component of global agricultural yield due to the activities of pests. Recent studies revealed the various side effects associated with the use of chemical pesticides such neurotoxicity, Parkinson's disease, endocrine disruption, malignancies and obesity [10–19]. The emergence of nanobiopesticides has been identified as one of the most remarkable breakthrough in the field of nanotechnology. Nanobiopesticides are ecofriendly, benign with unique biodegradation potential.

References

1. Abdollahdokht, D., Gao, Y., Faramarz, S. *et al.*, Conventional agrochemicals towards nano-biopesticides: An overview on recent advances. *Chem. Biol. Technol. Agric.*, 9, 13, 2022, https://doi.org/10.1186/s40538-021-00281-0.
2. Kumar, J., Ramlal, A., Mallick, D., Mishra, V., An overview of some biopesticides and their importance in plant protection for commercial acceptance. *Plants*, 10, 1185, 2021, https://doi.org/10.3390/plants10061185.
3. Boedeker, W., Watts, M., Clausing, P. *et al.*, The global distribution of acute unintentional pesticide poisoning: Estimations based on a systematic review. *BMC Public Health*, 20, 1875, 2020, https://doi.org/10.1186/s12889-020-09939-0.
4. Chaud, M., Souto, E., Zielinska, A., Severino, P., Alves, T., Nanopesticides in agriculture: Benefits and challenge in agricultural productivity, toxicological risks to human health and environment. *Toxics*, 9, 6, 131, 2021, https://doi.org/10.3390/toxics9060131.
5. Hyde, K.D., Xu, J., Rapior, S. *et al.*, The amazing potential of fungi: 50 ways we can exploit fungi industrially. *Fungal Divers.*, 97, 1–136, 2019, https://doi.org/10.1007/s13225-019-00430-9.
6. Shekhar, S., Sharma, S., Kumar, A., Taneja, A., Sharma, B., The framework of nanopesticides: A paradigm in biodiversity. *Mater. Adv.*, 2, 6569–658, 2021.
7. Akutse, K.S., Subramanian, S., Maniania, N.K., Dubois, T., Ekesi, S., Biopesticide research and product development in Africa for sustainable agriculture and food security - experiences from the International Centre of Insect Physiology and Ecology (icipe). *Front. Sustain. Food Syst.*, 30, 2020, https://doi.org/10.3389/fsufs.2020.563016.
8. El-Saadony, M., Abd El-Hack, M.E., Taha, A.E., Fouda, M.M.G., Ajarem, J.S., Maodaa, S.N., Allam, A.A., Elshaer, N., Ecofriendly synthesis and insecticidal application of copper nanoparticles against the storage pest *tribolium-castaneum*. *Nanomater. (Basel)*, 10, 3, 587, 2020.
9. Melanie, M., Miranti, M., Kasmara, H., Joni, M., Nanotechnology-based bioactive antifeedant for plant protection. *Nanomater.*, 12, 630, 2022.
10. Islam, S., Thangadurai, D., Adetunji, C.O., Micheal, O., Nwankwo, W., Kadiri, O., Anani, O., Makinde, S., Adetunji, J.B., Nanostructured biomaterials for targeted drug delivery, in: *Biogenic Nanomaterials*, pp. 83–108, Apple Academic Press, FL, USA, 2022.
11. Adetunji, C.O., Olaniyan, O.T., Okeke, N.E., Production of next-generation biodiesel from high yielding strains of microorganisms: Recent advances, in: *Sustainable Green Chemical Processes and their Allied Applications*. Nanotechnology in the Life Sciences. Inamuddin, A. Asiri (Ed.), Springer, Cham, 2020, https://doi.org/10.1007/978-3-030-42284-4_2.

12. Jhala, Y.K., Panpatte, D.G., Adetunji, C.O., Vyas, R.V., Shelat, H.N., Management of biotic and abiotic stress affecting agricultural productivity using beneficial microorganisms isolated from higher altitude Agro-ecosystems: A remedy for sustainable agriculture, in: *Microbiological Advancements for Higher Altitude Agro-Ecosystems & Sustainability*, R. Goel, R. Soni, D. Suyal (Eds.), Rhizosphere Biology. Springer, Singapore, 2020, https://doi.org/10.1007/978-981-15-1902-4_7.
13. Adetunji, C.O., Samuel, M.O., Adetunji, J.B., Oluranti, O.D., Corn silk and health benefits, in: *Medical Biotechnology, Biopharmaceutics, Forensic Science and Bioinformatics*, pp. 167–178, CRC Press, FL, USA, 2022.
14. Olaniyan, O.T., Adetunji, C.O., Adetunji, J.B., Oloke, J.K., Olaleye, O.O., Hefft, D.O., Ubi, B.E., Soy-based food products consumed in Africa: A panacea to mitigate food insecurity and health challenges, in: *Fermentation and Algal Biotechnologies for the Food, Beverage and Other Bioproduct Industries*, pp. 87–103, CRC Press, FL, USA, 2022.
15. Adetunji, C.O. and Oyeyemi, O.T., Antiprotozoal activity of some medicinal plants against entamoebahistolytica, the causative agent of amoebiasis, in: *Medical Biotechnology, Biopharmaceutics, Forensic Science and Bioinformatics*, pp. 341–358, CRC Press, FL, USA, 2022.
16. Adetunji, C.O., Panpatte, D.G., Jhala, Y., *Microbial rejuvenation of polluted environment*, vol. 3, Springer Nature, Singapore, 2021.
17. Adetunji, C.O., Michael, O.S., Nwankwo, W., Anani, O.A., Adetunji, J.B., Olayinka, A.S., Akram, M., Based drugs derived from medicinal plants, in: *Green Synthesis in Nanomedicine and Human Health*, vol. 103, 2021.
18. Dauda, W.P. and Ubi, B., Nutritional and health benefits of nutraceutical beverages derived from cocoa and other caffeine products: A comprehensive review, in: *Fermentation and Algal Biotechnologies for the Food, Beverage and Other Bioproduct Industries*, pp. 105–118, CRC Press, FL, USA, 2022.
19. Adetunji, C.O., Olaniyan, O., Adetunji, J.B., Osemwegie, O., Ubi, B.E., African mushrooms as functional foods and nutraceuticals, in: *Fermentation and Algal Biotechnologies for the Food, Beverage and Other Bioproduct Industries*, pp. 233–251, CRC Press, FL, USA, 2022.

2

Application of Plant-Based Nanobiopesticides That Could Be Applied as Fumigants

Kehinde Abraham Odelade[1*], Babatunde Oluwafemi Adetuyi[1], Oluwakemi Semiloore Omowumi[1], Nafisat Adeola Moshood[1], Dorcas Anuoluwapo Adeleke[1], Grace Onuwabhagbe Odine[1] and Charles Oluwaseun Adetunji[2]

[1]*Department of Natural Sciences, Faculty of Pure and Applied Sciences, Precious Cornerstone University, Ibadan, Nigeria*
[2]*Applied Microbiology, Biotechnology and Nanotechnology Laboratory, Department of Microbiology, Edo University Iyamho, Auchi, Edo State, Nigeria*

Abstract
Production of crops and farm products has reduced drastically over the years. This has been traced down to the infestation of pests and insects on the crops right on the field. Most of these pests are called plant pathogens. However, several methods have been utilized to eradicate or kill these pests or insects on the field; this is done so as to improve the production of crop. One of those methods includes the utilization of synthetic pesticides. The use of chemical pesticides was however banned in 2015, despite the fact that it shows good pesticidal activity. Chemical pesticides became illicit as a result of the negative impact they exert on both the environment and human health. Production of pesticides from natural substances or microorganisms is called biopesticides. The use biopesticides then replace chemical pesticides due to its minimal adverse effects, high efficacy and potential target specificity. Moreover, nanotechnology was integrated to the production of biopesticides, which give rise to nanobiopesticides. Nanobiopesticides are appealing for usage because of their properties, such as greater solubility, steadiness, huge surface-area-to-volume proportion, mobility, enhanced efficiency, decreased toxicity, and small size. Because of their low toxicity, they are highly recommended. These nanobiopesticides can be synthesized through various ways and can also be

**Corresponding author*: kennybramm@gmail.com

Charles Oluwaseun Adetunji and Julius Kola Oloke (eds.) Handbook of Agricultural Biotechnology: Volume I Nanopesticides, (7–36) © 2024 Scrivener Publishing LLC

applied in pest management. The full knowledge of the application of plant-based nanobiopesticides as fumigant has not been fully known, but few of the applications are discussed in this study.

Keywords: Nanobiopesticides, fumigants, nanotechnology, pesticides, biopesticides, biodegradability, pests, ecofriendly

2.1 Introduction

Lack of foods has been a major concerned in the world today, as the population is growing steadily to about 7.7 billion. This occurred as a result of the natural challenges (both living and nonliving), which include pests, infections, and weeds [1]. There are over 65,000 types of pests recorded in time past. These pests are also referred to as plant pathogens and they include fungi, weeds and arthropods [2]. It has been estimated that pest poses a lot of losses on crops. These losses are estimated as thus; 10%, 20%, 25%, 30%, 35%, and 50% losses in wheat crops, sugar, rice, pulses, oil seeds and cotton respectively. About USD 2000 billion has however been estimated to the crops loss annually which originated from pests and diseases [1]. However, this problem has paved way to a challenge in science, which is to improve the crops against pests without affecting the crop yields. Synthetic pesticides have been used to enhance the inhibition of pests for a good crop production [3] and they have been of great interest due to their large scope of usefulness in the control of insects and their potential to eliminate the presence of pests in agricultural ecosystem. Biopesticides are derived from plants and the secondary metabolites produced by plants, which include glycoalkaloids, organic acids, and alkaloids. All these have been identified as a potential origin of plant effecting pests [4–15].

Pesticides are divided into different groups and they include pesticides produced from chemical compounds, pesticides produced from biological activities (biopesticides), synthetic pesticide, pesticides synthesized from microbes, plant-incorporated pesticides, and biochemical pesticides. Figure 2.1 shows the significance of nanotechnology in the production of nano-based biopesticides [35]. Different kinds of seeds and weeds are treated and controlled through the use of chemical pesticides. They can be delivered to plants directly or indirectly. Indirect delivery is done by spraying chemicals on plants. The utilization of synthetic pesticides was, however, banned in 2015, despite the fact that it shows good pesticidal activity. Chemical pesticides became illicit as a result of the negative impacts they exert on both the environment and human health. A good example is methyl bromide. Methyl bromide is a good chemical pesticide that has been used for over 4 decades to treat many crops such as strawberry, melon seeds, all varieties of peppers and tomatoes against pests, nematodes, and soil pathogens. However, methyl bromide contributes negativity to the depletion of ozone layer and this prompted to its ban according to the regulation given by Protocol by Montreal.

Likewise, different synthetic compounds utilized as pesticides, for example, dazomet and chloropicrin have additionally been suspended; this is because of the antagonistic impact on food and human well-being [16]. Insecticides produced from natural substances or microorganisms are named biological pesticides or biopesticides. Three different types of biopesticides exist, and they include plant-incorporated protectants, microbial biopesticides, and botanical pesticides [17]. Characteristics of biopesticides include minimal adverse effects, capacity to disintegrate fast, high efficacy, and potential target specificity. These characteristics, however, paved way for biopesticides to serve as an alternative to conventional insecticidal methods. A lot of substances that have been investigated as biopesticides include the fractions isolated from *Clitoria ternatea*, an

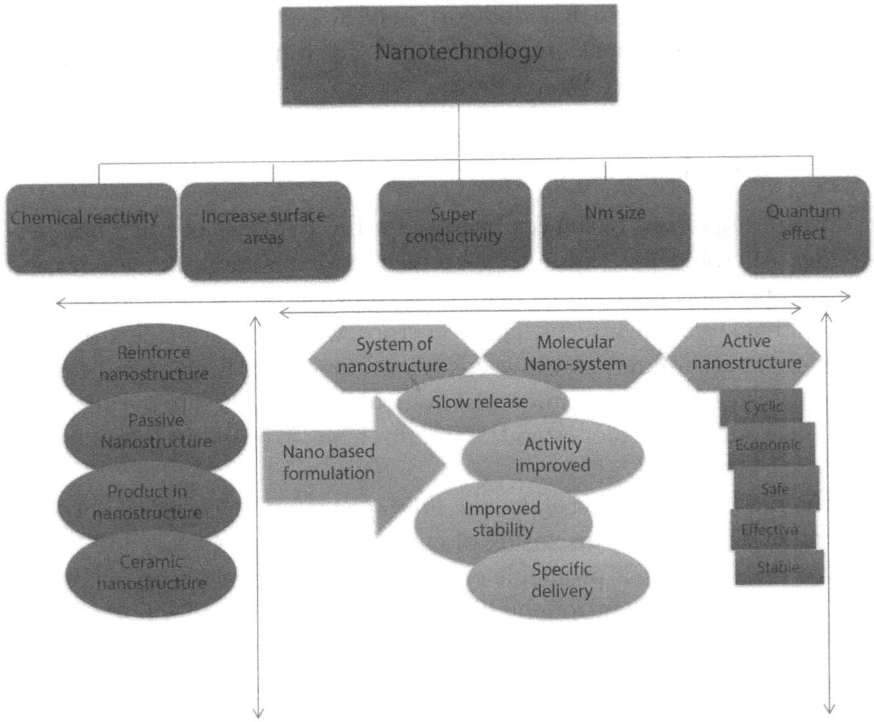

Figure 2.1 Significance of nanotechnology in the production of nano-based biopesticides. Source: Lade *et al.* [35].

alkaloid component called oxymatrine, stilbenes in grape cane, olive mill oil, and *T. flavus* [18]. Due to the negative influence biopesticides have on the human health status, it is advisable for its administration to be done with caution [19].

Recently, a novel field called nanotechnology has gained special attention due to its application in multidisciplines. These include parasitology discipline, agriculture field, pest management, and pharmacology field [20]. Nanoparticles (NPs), a rapidly expanding area of nanotechnology, provide solutions to lots of natural challenges as a result of their cost effectiveness, tiny size (1–100 nm), wide surface area, and cost effectiveness. It has been established that novel nanocomposites can be produced using an effective equipment nanotechnology. This is done so as to prevent crops from pest infestation and also to enhance different crops varieties [21].

2.2 Types of Biopesticides

Biopesticides exist in various kinds and are characterized in accordance to the various factors, which include origin of extractions and sorts of atoms utilized for arrangement/preparation [22]. The different types of biopesticides are explained below.

2.2.1 Microbial Pesticides

Different organisms have been utilized for the synthesis of microbial pesticides. Microbial pesticide affects both entomopathogenic nematodes and distinct pest species. Bioinsecticides are known as the compounds or molecules that target the insects that affect crops, while bioherbicides are pesticides that have influence over weeds via microbes, like fungi. A wide range of research carried out on microbial pesticides has resulted to the detection of a huge quantity of biopesticides, and this has also created a great path for their marketability [22].

The effective utilization of *Bacillus thuringiensis* and a few species of other microbes driven to the disclosure of numerous modern species and strains of microbes, and their profitable poisons and destructiveness components that might be a benefit for the biopesticide business, and a few of these have been deciphered into saleable items as well [23]. Significant species of bacterial entomopathogens involve *Pseudomonas* sp., *Chromobacterium* sp., *Yersinia* sp., etc., whereas fungi contain the following *Paecilomyces* sp., *Beauveria* sp., *Lecanicillium* sp., *Metarhizium* sp., *Hirsutella* sp., *Verticillium* sp., and so on [24]. Other crucial makers of pesticides synthesized from microbes include *baculoviruses* which are species explicit and their infectious activity is related with the crystal-clear impediment bodies which are dynamic on pests (*Lepidopteran* caterpillars) [24].

2.2.2 Biochemical Pesticides

The normal happening products, which are utilized to manage pests via nonharmful instrument are called biochemical pesticides, although pesticides produced from chemical synthesis utilize engineered atoms that specifically destroy insects and pests. The characterization of biochemical pesticides into distinctive types depends on their function in the management of invasions of pests and insects by exploiting pheromones (semichemicals), natural pests' growth regulator or plant fractions/oils. Figure 2.2 shows different categories of biopesticide [36].

2.2.3 Insect Pheromones

Insect pheromones are synthetic compounds made by pests and insects, which are mirrored for utility in managing pests within the coordinate's pest management programs. These synthetic compounds are viable in disturbing pests mating to avoid the aim of mating, hence decreasing the quantity of pests offspring. The pests used in this mechanism behave as allocators of pheromones that had the chance to be bewildered because of the existence of semiochemical flumes dispersed in the environment. Insect semiochemicals are not genuine "insecticides" because they do not destroy pests but impact their olfactory structure, which influence their conduct [25]. An accurate record of the method of activity of pheromones was described by Ujváry [23]. Summarily, the receiving wires (antennae) of the seeing pests accumulate pheromones on the surface, which is further dispersed to the inner parts of sensilla via minute openings within the cuticula. When they entered, they are exchanged via the polar sensillum to the sensory receptor layers by semiochemical binding proteins. Hence, the semiochemical binds to a particular protein receptor, which changes the synthetic compounds signs to an increased electric signs by a second courier structure associated with neuronal apparatus [26].

2.2.4 Essential Oils and Plant-Based Extracts

Plant-based fractions and essential oils have developed as promising insect pest management options in contrast to chemical insecticides in recent years. Because they are generated out of vegetations and consist of a variety of active compounds, these insecticides are naturally occurring insecticides [27].

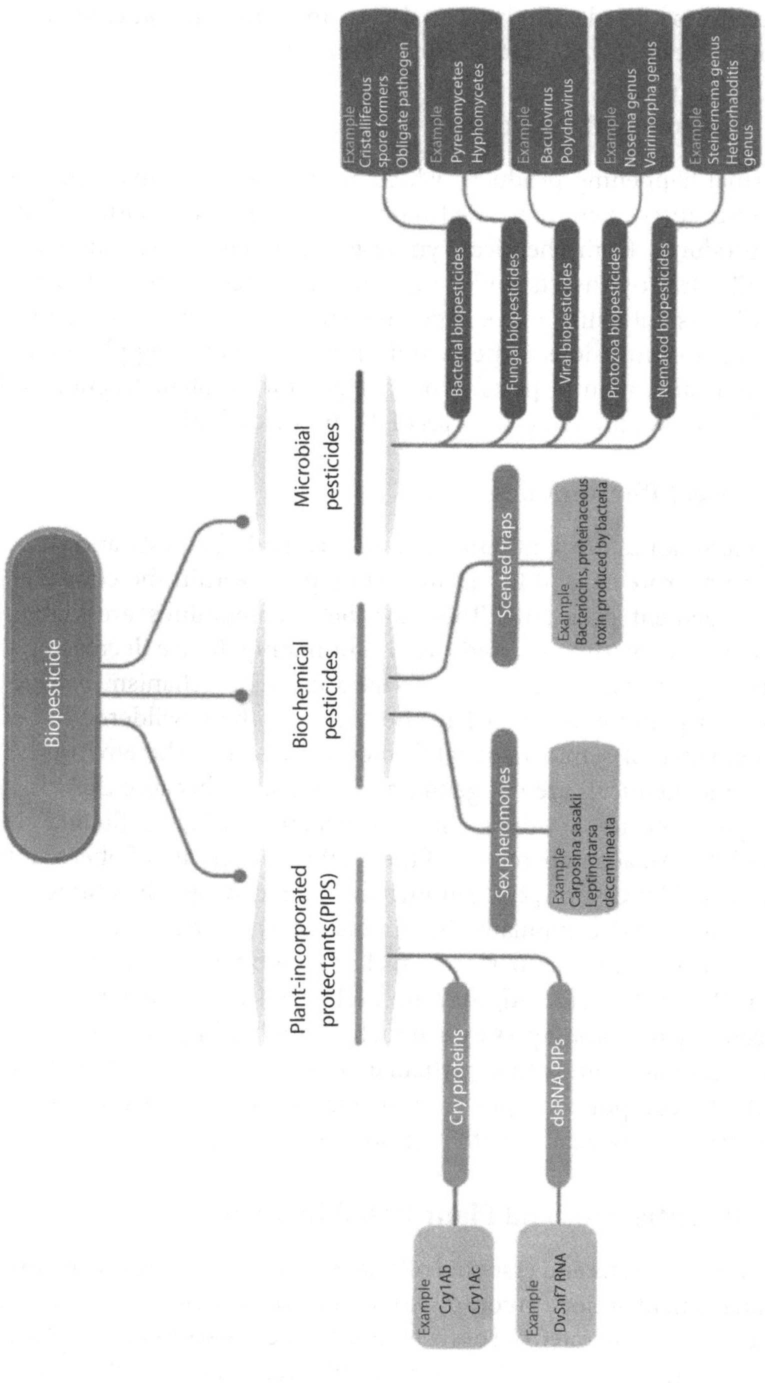

Figure 2.2 Different categories of biopesticide. Source: Abdollahdokht *et al.* [36].

Essential oils and fractions obtained from plants have a variety of activity on insects, and this depends on the physiological qualities of insect spp. as well as the kind of vegetation: they can act as antifeedants, attractants or repellents. The plants extracts and essential oils can also disrupt respiration, hinder insect recognition of plant host, stop the process of laying eggs, and reduce adult development through larvicidal and ovicidal impacts [28–30]. Their make-up differs tremendously. In this context, notable models include lemongrass oil and neem, both of which are widely available in worldwide herbal markets. Some researchers found that the combination of neem oil with entomopathogenic microscopic organisms, such as *Beauveria bassiana*, was extremely effective against on sucking pests [29]. Nonetheless, it is crucial to decide the portion of azadirachtin constituent in neem oil in order not to destroy nonaim microorganisms [31]. For the successful control of the aimed pests without damaging non-target pests, a comparable strategy must be devised for entomopathogenic fungi, which must be backed by supplementary laboratory bioassays, station, and/or field trials [32].

2.2.5 Insect Growth Regulators

Insect growth regulators (IGRs) kill insects by inhibiting key essential processes needed for their survival. Besides, these molecules are extremely specific and less hazardous to microorganisms that are not targeted [26]. IGRs have lately been classified as chitin synthesis inhibitors and chemicals which get involved in the activity of the hormonal system of insects (that is, ecdysteroids and juvenile hormone analogs) based on their method of activity. Even while IGRs are not as lethal to adult insects as mosquitos, cockroaches, and fleas, they can also manage a wide range of insects, which include fleas, mosquitos, and cockroaches. Despite the fact that they are not harmful to people, they forestall proliferation, egg hatching, and shedding starting with one phase then onto the next in young insects, and when combined with different pesticides, they can destroy even the adult insects [33].

2.2.6 Genetically Modified Microorganisms Products

These synthetic compounds are made by genetically engineered microorganisms. The hereditary material is integrated in the plant, which is subsequently utilized to manufacture pesticides, also known as plant-incorporated protectants (PIPs). Insecticidal proteins (or cry proteins) are original insecticidal plant-incorporated protectants brought into genetically modified plants with engineered genes from *Bacillus thuringiensis* soil bacteria [34]. Plant-incorporated protectants likewise necessitate the current status of research for continuous ecological fate evaluations of these particles, especially the RNAi-based PIPs [34], which would be discussed in a separate section.

2.2.7 Plant-Incorporated Protectants

PIPs are chemicals used to control pests, and they are produced by plants as well as the genes expected for the plant to create them [37]. A large number of pesticides being used include low molecular weights biochemical compounds (<0.5 kDa or 500 g·mol^{-1}). The ecological and insightful science of these pesticides and their possible effect on environmental and human well-being have all been well studied. While low molecular weight-chemical pesticides have overwhelmed the market, biopesticides are turning into a greater part of the pesticide business/market [38]. The worldwide market for every biopesticide is presently valued at USD 34 billion, representing generally 6% of the overall pesticide market. Plant incorporated protectants are also biopesticides; they are displayed straightforwardly in the tissues of genetically engineered crops, so as to shield and keep them from pests infestation, such as viruses and insects. This aspect majors on plant-incorporated protectants designed against insects as a result of the broad utilization of insecticidal plant-incorporated protectants all over the planet earth, and also the recent development of novel plant-incorporated protectants targeting insect pests [39]. PIPs are consumed by insect/pests when they feed on transgenic crop tissue. Double-stranded ribonucleic acid (dsRNA) plant-incorporated protectants and cry proteins within the gut of an insect have a variety of effects on insect development and mortality [40]. PIPs and cry protein bind to particular sense organ on epithelial cells present in insect midgut, enter the cell layer, and in the long run create transmembrane pores, resulting in the destruction of cell and pest mortality.

Cry proteins come in a variety of shapes and sizes, each with its own structure and toxicity, which is peculiar to certain insect groups. Cry1 proteins poison *Lepidoptera* (such as corn borer), and Cry3 proteins harm *Coleoptera* (such as the corn rootworm). Cry proteins are the originally produced insecticidal plant-incorporated protectants. PIPs for next-generation dsRNA have as of late been endorsed. The pest insect's dsRNA PIPs are delivered into a target cell after consumption. Inside the cell, double stranded-RNA is separated to 20-nucleotide siRNA (small interfering RNA) atoms, which directs the insect's inner RNAi (RNA interference)

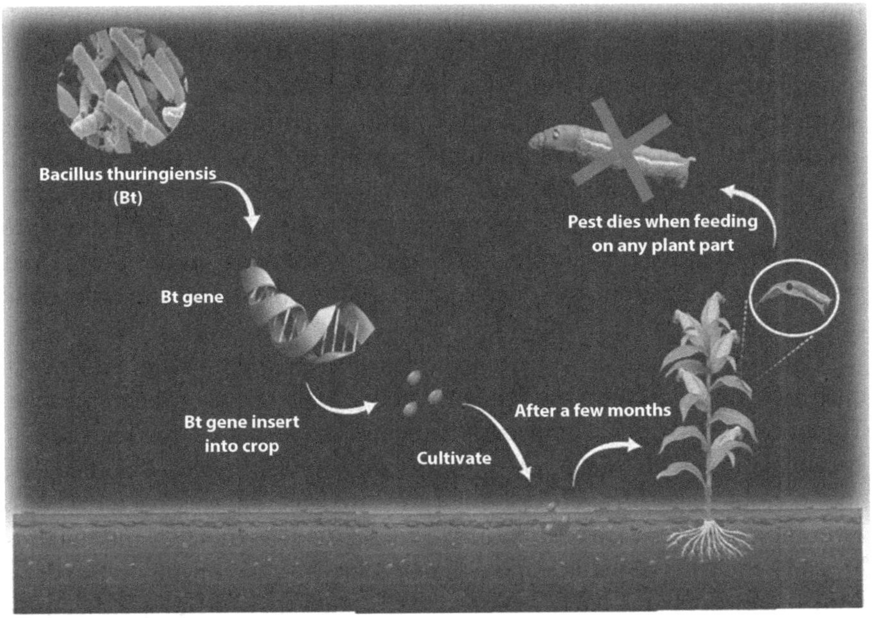

Figure 2.3 An organism that secretes a protein that is poisonous to some insect species. The Bacillus thuringiensis gene, which is responsible for the production of the toxin, can be transported to plants, which increases the resistance of the plants to the matching insect. Source: Abdollahdokht *et al.* [36].

apparatus to break down the corresponding messenger RNA. The aimed mRNA is degraded, preventing it from being converted to fundamental insect proteins, resulting in sublethal consequences (e.g., reduced growth) or high mortality of pests (Figure 2.3). The corn rootworm also known as *Diabrotica virgifera* is aimed by the primary dsRNA PIP endorsed by the FDA. This is done by obstructing the significant vacuolar arranging protein, i.e., formulation of the Snf7 protein [41].

2.3 Nanobiopesticides

Nanobiopesticides are appealing for usage because of their properties, such as greater solubility, steadiness, huge surface-area-to-volume proportion, mobility, enhanced efficiency, decreased toxicity, and small size. Because of their low toxicity, they are highly recommended. When sprayed directly to plants, chemical pesticides pose a significant danger to both the human and environment wellbeing. These threats can be mitigated by using pesticides that contain nanoparticles and nanocomposites in their formulations [42]. Nanobiopesticides are comprised in carriers that allow the precise dispensing of dynamic constituent to achieve the ideal effects in a specific climate. Penetrability, solubility, crystallinity, stiffness, biodegradability, and thermal stability are all improved when nanobiopesticides are added to polymers [43]. Nanobiopesticides have numerous benefits and superiority above biopesticides and conventional techniques. Biodegradability, ease of distribution to plants, ecologically friendly behavior, quick and desired results after utilization, and delayed release from the vector are only a few of the benefits [20].

Nanobiopesticides are an efficient transporter when paired with pesticides which can comfortably penetrate crops as a result of their tiny size. The photo toxicity of Ag-based NPs has been decreased through nanoencapsulation with biological compatible polyvinyl pyrrole molecules, and nanobiopesticides do not pose any toxicity on soil bacteria [44]. Nanobiopesticides have been found to comprise of auxiliary plant metabolites and their regulated metal oxide nanoparticles. According to the data, a lot of research has lately been done on nanobiopesticides; either pest are becoming resistant to synthetic pesticides, or a low quantity of insecticides are expired owing to significant human and ecological concerns. This circumstance, however, necessitates the development of new pesticides synthesized from plant extracts on the nanoscale in order to synthesize nanobiopesticides for the control of pest. *Bacillus firmus* and *Bacillus sphaericus,* which are utilized on diamond black moths, *Trichoderma viride* and *Trichoderma harzianum,* which are utilized on root rots and wilts, neem-based nanobiopesticides against whitefly, and *Bacillus thuringiensis* against *israelensis* are examples of various biopesticides that have been used on pests, yet they do not have the ability to inhibit pests [45].

2.4 Synthesis of Nanobiopesticides

There are different synthetic methodologies that can be utilized in the formulation of nanoparticles. These synthetic methodologies are costly, and the majority of the synthetic substances are made out of dangerous substances. This provokes incited scientists to redirect their attention regarding the utilization of organic strategies like plant extracts and microorganisms for the creation of nanoparticles. These biologically produced NPs are more secure than nanoparticles produced or formulated from chemical compounds [46]. The synthesis of nanobiopesticides is done at room temperature, and its procedure consumes low energy and power. It has been accounted for that the synthesis of nanobiopesticides is economy friendly and can likewise undergo a cycle to restore its usage [47]. A fascinating part of nanotechnology is the utilization of products produced from plants for the synthesis of nanobiopesticides. Polymers, which comprises of sugars or proteins, accommodates the trapped subdued nanoparticles (pesticides and nanoparticles) [48]. The entrapped nanobiopesticides are maintained and freed by the activities of the biodegradable polymers and cross-linking. The active synthetic compounds present in nanobiopesticides are liberated in a sequenced order to guarantee the life span of its pesticidal activity; this is made possible through the help of the trio structure known as cross-linking polymer, nanoparticles and active biopesticides [49]. The capability of nanobiopesticides to destroy pests can be known through the utilization of relative examination of the nanoparticles, biopolymers, plant extract-based metals, crude extract, and the synthesized nanobiopesticides. This can however be done through the evacuation of the active compounds in the crude extract from the nanobiopesticides. Nanobiopesticides can be synthesized with the utilization of silver nitrate salt ($AgNO_3$). The silver nitrate salt can be reduced by the extract obtained from a plant named *Aleo vera,* and incorporated into a particular chitosan (which is a biopolymer) (Figure 2.4) [50].

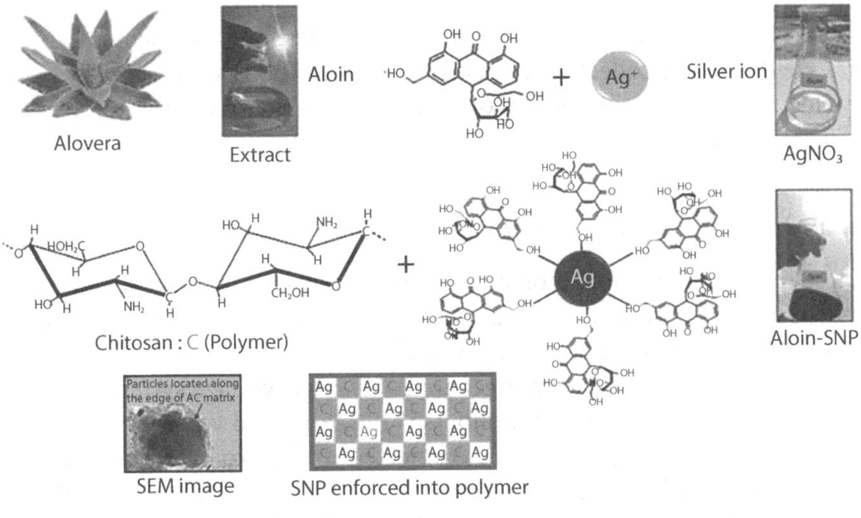

Figure 2.4 Nanobiopesticides synthesis with the use of silver nitrate salt ($AgNO_3$). Source: Lade and Gogle [50].

2.5 Different Methods Used in the Synthesis of Nanobiopesticides

There are lots of strategies accessible for the synthesis/formulations of nanobiopesticides. These techniques incorporate microwave oven, homogenization, high-pressure homogenization, mixing nanoparticles with APC, mixing in magnetic stirrer autoclave, ultra high-pressure homogenization, and so on. These large numbers of strategies are the least complex methods utilized in the synthesis nanobiopesticides. Other high level strategies utilized in nanobiopesticides synthesis include ball milling, electrospinning and grinding. However, only nanobiopesticides formulation and cross-linking in polymers through electro spinning method is examined here [50].

2.5.1 Electrospinning

The application of electrospinning techniques has been utilized in some areas or fields such as farming, pharmaceutical, clinical, and wound dressing.

2.5.1.1 Principle of Electrospinning

The electro spinning generates nanofibers out of the polymer which contains and directs the nanoparticles. The dissolved polymer was passed via syringes to obtained droplets at lower voltage. The melted polymer which was yielded at the collector because of the application of increased voltage by electrostatic force disappears or dissipates until it arrives at the collector to form nanofibers [50].

2.5.1.2 Fundamental Building Blocks of Electrospinning

These include (1) the injection needle siphon: this enables the melted solution move via the spinneret at a directed and consistent level. The spinneret is attached to the injection needle whereby the channeling polymer is processed, this happens in the syringe pump. (2) High-voltage system: the significant power that attracts nanofibers is the expanded voltage DC power. The anode and syringe's needle is connected to the expanded voltage DC power supply, and the negative terminal is appended to a collector (metallic in nature) shielded with aluminum foil: 30 to 50 KV is commonly utilized in here. (3) Collector: this comprises of different conducting materials like

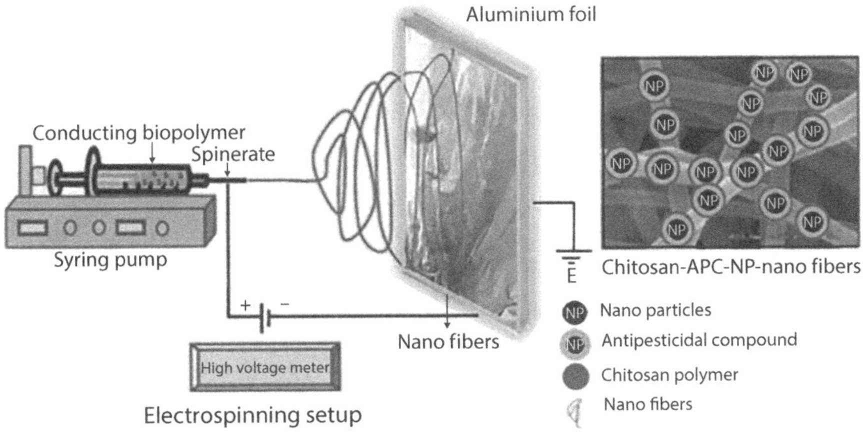

Figure 2.5 Fundamental electrospinning setup and production of nanofiber. Source: Lade and Gogle [50].

aluminum. These conducting materials accept the electric charged fibers and place them on aluminum to constitute a nano fiber web. The collection agents can be revolving in nature or static in nature. Various kinds of collectors include grids, mesh, pin, gridded bars or parallel, wheel liquid bath, rotating rods [50].

2.5.1.3 Methods Used in Electrospinning

The synthesized nanobiopesticides are mixed together in a specific diluent and the conductor polymer is melted in the specific diluent. The nanobiopesticides and the diluent (which contains the polymers) are stirred in a shaker; the diluent is then introduced, electric charge is moved via the instrument and the fibers are collected in the collection agent. The end product is nanofibers of nanobiopesticides and polymers. Figure 2.5 shows the structure of nanofiber [51].

2.6 Possible Nanostructures

Various structures of nanobiopesticides have been synthesized and examined in many advanced countries. Figure 2.6 below shows the common conceivable nanostructures shapes utilized for the synthesis of nanobiopesticides using secondary metabolites obtained from plants. Discussion on these nanostructure-holding nanoparticles and APC is given below.

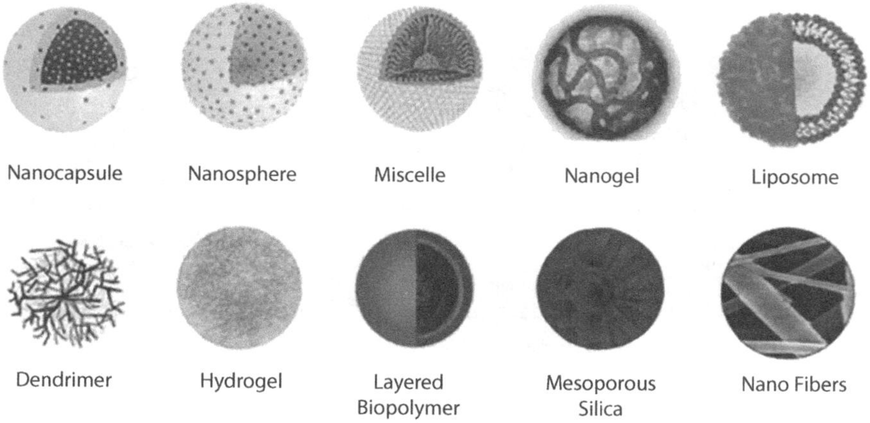

Figure 2.6 Possible nanostructure obtained from different methods. Source: Nuruzzaman *et al.* [52].

Nanospheres: the active constituent is evenly dispensed in the polymeric matrix in this mass aggregate; nanocapsules: in this mass cluster, the active constituent is gathered in one mass (concentrated) close to the middle core and lined by the polymeric matrix; nanogels: a cross-linked polymer that has can possibly integrate increased the quantity of water; Micelles; this totality is composed in aqueous solution by atoms (which have hydrophobic and hydrophilic component) [49]; nanofibers: these are fibers with diameters within nano meter range produced out of various conducting polymers. Other possible nanostructures include layered biopolymers, dendrimer, liposomes, mesoporous silica, and liposomes.

2.7 Mechanisms of Nanobiopesticides

Efficient dispersion is required for nanoeffective synthesis in real-world utilizations. Water, fertilizers, and pesticides may now be used in an ecologically beneficial way because of the utilization of nanosensors and intelligent delivery structure. The utilization of satellite pictures of the field by farm managers helps them to easily locate and identify agricultural pests on the farm. It also helps in the collection of proof of pressure instigated by outrageous intensity of flood, heat and drought. To create a more accurate ecological model, nanomaterials and GPS would be coupled with satellite photos of farmlands. Farmers may now modify agricultural inputs consequently on account of this innovation. The detection of plant viruses and soil nutrients can be done through the utilization of nanosensors in the field, which allows for accurate plant management. Pesticides utility and contamination are limited when slow-release biopesticide present in the nanoparticles is delivered to their target [35].

Utilization of nanobarcode is another alternative used in the delivery mechanism of nanobiopesticides. A novel innovation called nanobarcode is being used to determine the quality/nature of farm produce. Various researchers from Cornell University utilized grocery barcodes to develop rapid, efficient, low-cost and easy approach for the detection of illnesses and diseases. Grocery barcodes were used to create this technique. Self-folding branching DNA structures can be used to scan these nanobarcodes or tiny probes using a microscope. A fluorescent color ratio can be utilized to recognize an illness biomarker on farm products or on the farm. Since nanobarcodes and microbe biomarkers are so viable, they ought to be identified by any fluorescent-based machine that has the potential of discovering illness or infection [53]. Microemulsion and nanoemulsion can improve the kinetic steadiness, optical translucency, solubility, and bioavailability of nanobiopesticides while reducing the emulsion size and viscosity [54]. The potential antimicrobial properties of incorporating Ag NPs into electrospun polyacrylonitrile fibers are fascinating. This approach could be utilized to betrap a dynamic constituent or a nanobiopesticide for utilization in insecticides or pesticides that are applied to the ground. An electrospun nanofibrous mat filled with nanobiopesticides is electrospun into the soil and then discarded to kill the soil-borne insect [55].

The utilization of nanoparticles for the conveyance of biopesticides and pesticides in the field encounter a few difficulties like many ecological perturbations, the huge regions under spray coverage, and expense viability. In the standard spraying system, the entire plant is sprinkled with the synthetic compound for easy utilization including a low-value preparation, high-volume, while nanomaterial-based preparations are supposed to include high-value utilization and low-volume. This monitored nanoparticle delivery structure needs an aimed released method zeroed in on utilizing information of the conduct and life cycle of the pest or pathogenic microbes [56]. The overall utilization of nanobiopesticides to crops for pest management is displayed in Figure 2.7. These nanopesticides can comprise of biological constituents (like polymers) and/or chemical constituents (like metal oxides) in different structures like particles and micelles, nanofibers, nanogels, or nanocapsules [49]. The nanobiopesticides are utilized on crops to manage the insects or pests and also to protect the plants. Notwithstanding, the impacts of these nanobiopesticides on insects and

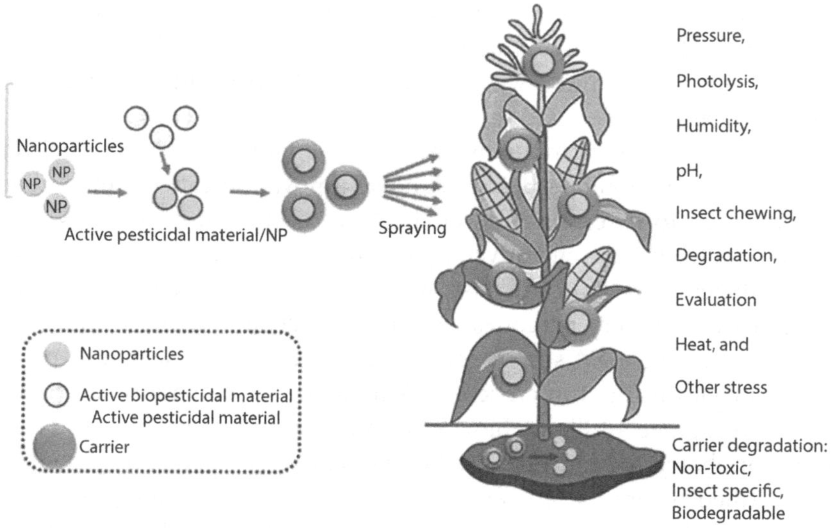

Figure 2.7 Delivery mechanism of nanobiopesticides in crop for pest management. Source: Lade and Gogle [35].

plants, have not been deeply known. It is crucial to decide biopesticide formulation toxicity for nontoxic-limit concentrations.

2.8 Possible Action of Nanobiopesticide

The biopesticides can be combined and mixed with known polymers via encapsulation for successful activity and the slow-release mechanism. When the levels of light and other explicit factors like moisture, temperature, pH or water are achieved, the splitting of the biological polymer (such as poliglusam) and slow delivery of nanoparticles as well as pesticides (like biocides) may happen. This particular circumstance permits dissipation or debasement of the poliglusam polymer and delivers the acting nanobiopesticides in the atmosphere, soil and plants. These nanoformulations are by and large normal to different pesticides which have the benefit of expansion in the obvious solvency of an inadequately dissolvable active constituent and delivering the bioactive constituent in a gradual way, and additionally safeguarding the bioactive constituent against untimely desement [57]. These nanobiopesticides influence pests or insects infesting plants, though the polymers are nonharmful and are degenerated after releasing the nanoparticles.

2.8.1 Safe Target Delivery of Pesticides

Recently, new research on the useful control release of conventional pesticides formulations have emerged due to its limitation. Controlled pesticide release structures are utilized on farmlands to manage a particular pest through the selection of a proper and relevant result channel. The use of a smart innovation that screens and delays pesticide release seeks to reduce pesticide demand in crops and also allows for more efficient use of pesticides over the long run [58].

2.8.2 Nanopesticides

Research has shiwn that engineered nanoparticles possess the capability to be used and deployed as pesticide carriers. Various formulations that may be utilized to enhance the effectiveness of present-day pesticides active ingredients or their ecological security have also been proposed. This range of formulations include nanoencapsulations, nanoemulsions, nanofibers, nanogels, nanovesicles [59].

2.8.3 Nanoemulsion

AIs dispersed in the oil and water produce a biphasic dispersion system is called a nanoemulsion [60]. This structure is kinetically stable and has a translucent/transparent appearance due to the size of the nanometer drops (20 nm–200 nm) [61]. An oil-in-water (O/W) dissipitation of pesticides can melt weakly water-soluble insecticides into minute oil drops, considerably enhancing their bioavailability and adequacy [62]. Moreover, as compared to standard pesticide formulations, nanoemulsion greatly limits the utilization of organic solvents and surfactants, and has obtained a lot of attention from scientists lately [63]. A group of researchers, for instance, gave a nanoemulsion solution, which contains ecological acceptable surface-active agents, esterified fats, and 41% (w/w) glyphosate isopropyl amine herbicide [63]. The characteristics of nanoemulsion, which incorporate little molecule size (< 200nm) and reduced surface tension than the cationic surfactant system (RoundupR), empowers the drops to be disseminated and deposited similarly on the leaves with decreased contact point, which results in great spreading, wetting and saturation. The nanoemulsion formulations had a much lower effective dosage 50 (ED50) than RoundupR, which implies that they were more organically efficient. A group of scientists made an avermectin B1-loaded nanoemulsion with 2% avermectin B1, 5% polyethylene glycol castor oil, and 7.5% of organic compound solvent that met the Food and Agriculture Organization's (FAO) quality markers [61]. There are few benefits nanoemulsion display over microemulsions and marketable oil/water emulsions. These benefits include enhanced insecticidal activity, lowered cytotoxicity, and a reduced contact angle on leaves. Albeit, a nanoemulsion of pesticides that possess exceptional physical steadiness can be achieved rapidly. However, the commonly used strategies, which include the oil stage, emulsifiers and high-energy emulsification, can be harmful [64]. Green nanoemulsions as well as reduced-energy strategies (for example, phase transition, self-emulsification, phase inversion temperature methods, and so on) are likely methodologies for overcoming these difficulties. Another set of researchers created a green nanoemulsion of β-cypermethrin utilizing inexhaustible unsaturated fat methyl ester (methyl laurate) as the oil stage, "green surfactant" alkyl polyglycoside, and the nonionic surfactant polyoxyethylene 3-lauryl ether (C12E3) as the mixed surfactant [65]. The nanoemulsion, including β-cypermethrin, had a nearly monodisperse droplet size dispersion polydispersity index (PDI) less than 0.2 Mw/Mn), that showed great wetting and dispersing characteristics on the hydrophobic surface in contrast to the marketable β-cypermethrin nanoemulsions, proposing a biological compatible technique for the release of pesticides.

2.8.4 Nanoencapsulation

Nanoencapsulation of pesticides is a conveyance methodology where AIs are captured in different nanomaterials and managed in a regulated fashion [66]. AIs are shielded from inauspicious degradation (such as hydrolysis, photolysis, biodegradable, and so on) and a few preventable losses that happen via volatilization and filtering; this is profoundly efficient in practical utilizations contrasted with antiquated pesticide synthesis [67, 68]. Nanomaterials may likewise give the encapsulation system tireless delivery conduct to expand the control time or upgrade responsive delivery highlights to accomplish exact control after the rational design. For nanoencapsulation of pesticides, an assortment of nanomaterials have been utilized, including organic materials (such as manufactured and natural polymers [69], lipids [70], plant-derived nanoparticles [35], and inorganic materials (such as silica-based NPs, carbon-based NPs, calcium-based NPs, and clay-based NPs) [71]. In order to enhance the adhesion, uptake, portability dispersibility, and target release, AIs are frequently entrapped in nanocarriers through covalent bonding, incorporation, electrostatic complexation collaborations. The results of this include enhanced bioavailability and a prolonged proficient span of AIs. Various kinds of polymer based nanoencapsulated pesticide, such as nanocapsules, nanospheres, nanomicelles, nanogels, and nanofibers, have been seriously researched due to their outstanding biodegradability and biocompatibility and their overall of functional groups [72]. Lipid-based pesticides nanoencapsulation improves the penetration and absorption in insect epidermis, which results to the upgrade of the activity of the insecticides and furthermore considers for the successful encapsulation of hydrophilic or hydrophobic AI asides the utilization of organic solvents. The development or dispersibility of Abm, a nematicide in soil is enhanced by the nanoencapsulation of plant based viral nanoparticles pesticides contrasted to the free abamectin, prompting upgraded plant safety against plant–parasitic nematodes present in soil; this was shown in a few research articles for the past few decades [53].

2.9 Application of Plant-Based Nanobiopesticide

Nanobiopesticides enjoy a benefit from their nanosize, large surface area or volume proportion, steadiness, enhanced effectiveness, increased solubility, portability, and reduced poisonousness [73]. Harmful substances or pesticides utilized in nanoformulations goes through some moderation to stay away from nanopesticide run off in atmosphere, soil and aquatic environment; however, the poisons might assemble in the food chain which is a significant climate and wellbeing challenge. To prevent that occurrence, molecules of nanopesticide are encapsulated into a conveyor like biological polymers, composites or micelles, to enhance the insecticidal activity and stopping build up in the climate. These nanobiopesticides are delivered through ecological factors like light, moisture, temperature, and humidity. Like biological polymer, clay nanotubes created as carriers of pesticides are delivered by the natural factor to have collaboration with pests. Likewise, nanofertilizers and nanobiopesticides are included in unambiguous carrier to give controlled delivery to precise effect [74]. The nanobiopesticides are banded to biological polymers to uncover other characteristics like rigidity, vulnerability, transparency, heat stability, solvency, and biodegradability [75]. No known impacts of nanomaterials on soil organisms is seen and this is an additional benefit of nanomaterial [76], and the phototoxicity of the powdered silver NPs was suppressed by encapsulating them with biocompatible polyvinyl pyrrole, which is a nanoencapsulation [44]. Investigating such biological compatible encapsulations to invert the poisonousness of nanomaterials improve the opportunities of utilizing nanomaterials in vegetation growth and development [77] hypothesis which influenced arrival of bioactive vegetation development catalysts and other synthetic substances enfolded in nanocomposites made of layered twofold hydroxides (anionic clays) may be some other choice for organic farming. As of late, nanocontainers, nanoencapsulates, nanoemulsions and nanocages, are accounted for as some methods used in nanopesticide [74] with various functions in crop safety.

In general, biopesticides have a tight target reach and a quite certain method of activity. They are slow acting. They have generally basic times of utilization/application. They eliminate totally instead of suppressing. They have restricted field industriousness and a brief time frame of realistic usability; are more secure to people and the environment than regular pesticides. They do not present any residue issues; do not continue in the climate; pose a reduced risk to advantageous parasites and predators (non-target microorganisms). They are commonly degraded in the environment; are effortlessly utilized by animals getting sub deadly dosages; have a wide range of action; and are not difficult to process and utilize. Thinking about these benefits, biopesticides can be used for the synthesis of nanomaterials and used in agribusiness to improve people and furthermore to our inclination. There is a magnificent concern connecting with the nanobiopesticides material that can possibly apply harmful impacts on people, soil, and climate. The positive effect of nano pesticide could make a few adverse consequences, which expect to be settled by significant examination.

Nanobiopesticides are utilized in pest management to forestall their activity in agriculture sectors. These biopesticides will be useful in focusing on various pests in more efficient ways by lessening the synthetic substances to make advantageous and harmless to the ecosystem production. The research progress of pest management in agriculture is slow because of the muddled subatomic mechanisms and sites of activity. Ongoing investigations indicate that utilizations of nanobiopesticides are efficient in managing pest by restoring the customarily utilized synthetic substances. These nanobiopesticides possess secondary effects when contrasted with straightforwardly applied synthetic compounds. They have incredible worth to deliver active constituents that are useful in keeping up with the various issues related with agricultural systems, like eutrophication. In spite of the fact that nanobiopesticides are broadly utilized in various fields to manage pests, their use in human beings and animals stays indistinct as they have gone to the food chain [78].

As the total populace increases quickly, the feeding of people will sum up to approximately nine billion by year 2050. It will require heaps of nanobiopesticides to destroy the pests and for the capacity of nourishment for an extensive period of time. Nanobiopesticides will be a rising methodology to the control of pests which keeps up with ecological maintainability with less harmful effects on human wellbeing. The application of nanobiopesticides is likewise useful in keeping up with the balanced diet in plants; limit the dangers to food safety, and developing harmful materials. Nanobiopesticides are significantly utilized in farmlands for the control of arthropod infestation or pest attack; however they have synthetic formulations which consist of nanoparticles that cause toxicity challenges and medical problems. Such nanobiopesticides should be normalized globally to lessen their harmful impacts on plants and the food chain. The utilization of nanobiopesticides in farming is encouraging, yet lots of investigation is required to comprehend their harmful activity and screen their utilization time to soils [78]. Biopesticides have numerous advantages over conventional chemical pesticides. They are harmless to the ecosystem, target explicit, and nonharmful to nontarget microorganisms and subsequently powerful enough to replace chemical pesticides for the control of pest. Lately, the utilization of biopesticides is acquiring impulse on the grounds that they can be productively utilized in maintainable farming practices. Biopesticides are profoundly powerful in modest quantities and decay rapidly without a deposit of dangerous residues and thusly can diminish the utilization of traditional pesticides as a fundamental part of IPM programs. Nevertheless, notwithstanding the benefits of utilizing biological pesticides, their utility is not yet circulated as anticipated. The reasons for this are listed below:

1. Huge expenditure of pesticide creation because of the expenses used in the monitoring, development, and obtaining legal liberation for novel biological agents.
2. Little time spans of usability because of the responsiveness of biopesticides to variances in humidity and temperature.
3. Restricted field viability as a result of environmental variations in soil condition, humidity, temperature and so on.
4. Farmers are uninterested in them because of the high specificity of biopesticides, that it they are just viable against aimed

pathogenic microbes and pests. They required the utilization of numerous biotic agents to manage various microbes and pests on farmlands. These agents are befuddling, expensive and lumbering, and are additionally not accessible for all pests or pathogenic microbes.

Nanoparticles are discovered in the environment, yet can likewise be produced through chemical synthesis. Chemical NPs are discovered in the space of science (chemistry field), biotechnology, mechanical engineering, and hardware/electronics. Ag NPs are utilized for their high electrical conductivity, low temperatures and stability. The major applications Ag nanoparticles has been utilized include wound dressings, textiles, antimicrobial coatings and biomedical devices. A broad range of materials can be utilized to produce nanoparticles like ceramics, magnetic materials, silicates, metal oxides, semiconductor quantum dots, biological molecules, dendrimers, and so on. Organic origins like microorganisms are used to separate nanoparticles. Also, in a few organic frameworks, the importance of the utilizations of NPs can be 100 nm and above. The utilization of nanoparticles has been found in the protection and enhancement of crop. NPs are utilized for fertilizers, pesticides degradation, biological material delivery in pesticides and sensors. The improvement of nanocides, the monitored delivery of nanomaterial-assisted genetic material, nanomaterial-assisted fertilizers, release into plants, and nanosensors for the identification of pathogenic microorganisms and pesticides for agrochemical elimination or degradation to further develop soil wellbeing and nutrition are few of the great improvement in application of nanotechnology in farming [79]. Pesticides coating is a helpful choice for security in the farmland. Nanospheres are utilized to evaluate the release of insecticides out of the phenyl pyrazole group [80]. Permeable nanosilica molecules are displayed to safeguard abamectin against ultraviolet deactivation [81]. Likewise, nanoclays are displayed to safeguard agrochemicals [82]. The utilization of polymeric NPs can enhance the release of chlorothalonil and tebuconazole, which are fungicides utilized for the conservation of hard woods [83]. Pesticides release can be increased utilizing nanocapsules. Polymeric nanocapsules are utilized for successful conveyance of the pesticides acetamiprid and ivermectin [84]. Nanosized Al_2O_3, $Fe2O_3$, SiO_2, or TiO_2 are proposed for utilization as pesticide carriers for enhanced biological activity and reduction of residues [85].

2.10 Environmental Sustainability of Nanobiopesticides

Nanobiopesticides are nonharmful to the environment, have biodegradation characteristics, and are shipped to the various components of the crops. Because of their biological availability in the crop structure, they are useful in figuring out the associations and conduct of various pests which infest crops. Dispersion of Ag nanoparticles with mixes of *aloe vera* fractions and $AgNO_3$ is useful to manage the development of pest, like *H. armegera*. Synthesis of biological pesticides via nanotechnology assists to manage the pathogenic microbes and pests that infest crops [86]. Nanobiopesticide utilizations in plants do not only manage pests but also control natural contamination by lessening the dangers of poisonous metals in plants, which are useful in purifying the climate. For instance, the release of Zn-based nanoparticles to rhizospheric parts of plant safeguards the plants as they display high effectiveness contrasted with the poisonous synthetic substances directly applied to plants. Nanobiopesticides are likewise useful in the bioremediation cycle by degrading poisonous compounds [87, 88].

Nanobiopesticides can be degraded biologically and moved to the various tissues of plants. A few investigations have exhibited that soil applications of nanobiopesticides under ideal circumstances are useful for the debasement of harmful metabolites created in plants. These metabolites lead to the building up of poisonous metals. It prompts an increment in the possibilities of mortality of plant tissues. Then again, traditionally utilized synthetic substances likewise enhance the possibilities of mortality of plant tissues because of cellular toxicity in some cells. In this way, the utilization of nanobiopesticides in the environment is dependable than other synthetic constituents [89].

Nanobiopesticides assume a significant part in keeping up with natural supportability by replacing traditionally utilized synthetic substances in the form of sprays. The utilization of nanobiopesticides to manage the pests also keeps up with the environmental chain. Nanobiopesticides for land preservation guarantee the most extreme yields and maintain the agriculture system. Thus, nanobiopesticides are additionally useful for further developing the quality of soil and increasing production of foods under various developments. Diverse utilizations are discovered in crop security by controlling pests and other creatures like birds and honey bees via ecological development [90].

2.11 Conclusion

The gaseous state of plant-based nanobiopesticides can be used to fumigate or disinfect the environment and even plant due to the presence of nanoparticles in the formulation of nanobiopesticides.

References

1. Sinha, K., Ghosh, J., Sil, P.C., New pesticides: A cutting-edge view of contributions from nanotechnology for the development of sustainable agricultural pest control, in: *New Pesticides of Soil Sensers*, pp. 47–79, Elsevier, Amsterdam, 2017.
2. Berini, F., Katz, C., Gruzdev, N., Casartelli, M., Tettamanti, G., Marinelli, F., Microbial and viral chitinases: Attractive biopesticides for integrated pest management. *Bio. Adv.*, 36, 3, 818–838, 2018.
3. Chowański, S., Adamski, Z., Marciniak, P., Rosiński, G., Büyükgüzel, E., Büyükgüzel, K., A review of bioinsecticidal activity of Solanaceae alkaloids. *Tox.*, 8, 3, 60–69, 2016.
4. Ibanez, S., Gallet, C., Després, L., Plant insecticidal toxins in ecological networks. *Tox.*, 4, 4, 228–243, 2012.
5. Adetunji, C.O., Akram, M., Michael, O.S., Shahzad, K., Ayeni, A.E., Hasan, S., Adetunji, J.B., Hasan, S.M., Olaniyan, M., Muhibi, M.A., Polysaccharides derived from natural sources: A panacea to health and nutritional challenges, in: *Polysaccharides: Properties and Applications*, pp. 701–738, John Wiley & Sons, Inc., USA, 2021.
6. Adetunji, C.O., Palnam, W.D., Adetunji, A.J., Ubi, B.E., Ayeni, A.E., Thangadurai, D., Akram, M., Khan, F.S., Single cell sequencing of microbial species, in: *Sequencing Technologies in Microbial Food Safety and Quality*, pp. 349–358, CRC Press, USA, 2021.
7. Adetunji, C.O., Michael, O.S., Anani, O.A., Pharmafoods for body cleansing of toxic exposure to chemical and biological warfare agents, in: *Phytochemistry, the Military and Health*, Amsterdam, pp. 239–255, Elsevier, Amsterdam, 2021.
8. Adetunji, C.O., Anani, O.A., Michael, O.S., Application of nanoceutical technology for fast and efficient control of illness, in: *Phytochemistry, the Military and Health*, pp. 497–508, Elsevier, Amsterdam, 2021.
9. Michael, O.S., Adetunji, O.S., Ayeni, A.E., Akram, M., Adetunji, J.B., Olaniyan, M., Muhibi, M.A., Marine polysaccharides: Properties and applications, in: *Polysaccharides: Properties and Applications*, pp. 423–439, John Wiley & Sons, Inc., USA, 2021.
10. Michael, S., Bamidele, O., Ogheneovo, P., Ariyo, T., Adedayo, L., Oluranti, O.I., Soladoye, E.O., Adetunji, C.O., Awobajo, F., Watermelon rind ethanol

extract exhibits hepato-renal protection against lead induced-impaired antioxidant defenses in male Wistar rats. *Curr. Res. Physiol.*, 4, 252–259, 2021.

11. Adetunji, J.B., Adetunji, C.O., Olaniyan, O.T., African walnuts: A natural depository of nutritional and bioactive compounds essential for food and nutritional security in Africa, in: *Food Security and Safety*, O.O. Babalola (Ed.), Springer, Cham, 2021.
12. Adetunji, C.O. and Fawole, O.B., *In-vitro* effects of crude extracts of *Jatropha curcas* on *Aspergillus niger v.* Tieghem associated with post harvest rot of onion bulb. *Int. J. Phytofuels All. Sci.*, 1, 1, 1–5, 2012.
13. Adetunji, C.O., Fawole, O.B., Olaleye, O.O., Williams, J.O., Arowora, K.A., Adetunji, J.B., Abiodun, A.A., Kazeem, M.O., Antibacterial activity and phytochemical screening of extracts from the stem of Aspila Africana. *Sci. Focus*, 17, 1, 116–121, 2012.
14. Akram, M., Michael, O.S., Saeed, M., Adetunji, C.O., Zahid, R., Adetunji, J.B., Ethnopharmacological properties of Asian medicinal plants during conflict-related blockades, in: *Phytochemistry, the Military and Health*, pp. 53–68, Elsevier, Amsterdam, 2021.
15. Akram, M., Adetunji, C.O., Laila, U., Michael, O.S., Samson, O.E., Overview of the traditional systems of medicine in different continents during post-war recovery, in: *Phytochemistry, the Military and Health*, pp. 37–52, Elsevier, Amsterdam, 2021.
16. Huang, X., Zhao, J., Zhou, X., Han, Y., Zhang, J., Cai, Z., How green alternatives to chemical pesticides are environmentally friendly and more efficient. *Eur. J. Soil Sci.*, 70, 3, 518–529, 2019.
17. Mathew, I.L., Singh, D., Singh, R., Tripathi, C., *Bacillus thuringiensis:* The biocontrol agent in a food web perspective. *Biolife*, 2, 1, 348–362, 2014.
18. Damalas, C.A. and Koutroubas, S.D., Current status and recent developments in biopesticide use. *Agr*, 8, 1, 1–6, 2018.
19. Kumar, A., Dhananjaya, P.S., Tyagi, M., Biotoxic cyanobacterial metabolites exhibiting pesticidal and mosquito larvicidal activities. *J. Microbiol. Biotechnol.*, 13, 1, 50–56, 2003.
20. Sayed, A.M., Kim, S., Behle, R.W., Characterisation of silver nanoparticles synthesised by Bacillus thuringiensis as a nanobiopesticide for insect pest control. *Biocontrol Sci. Technol.*, 27, 11, 1308–1326, 2017.
21. Chaloupka, K., Malam, Y., Seifalian, A.M., Nanosilver as a new generation of nanoproduct in biomedical applications. *Trends Biotechnol.*, 28, 11, 580–588, 2010.
22. Ruiu, L., Microbial biopesticides in agroecosystems. *Agr.*, 8, 235, 2018.
23. Ujváry, I., Chapter 3—Pest control agents from natural products, in: *Handbook of Pesticide Toxicology*, 2nd ed., R.I. Krieger and W.C. Krieger (Eds.), pp. 109–179, Academic Press, San Diego, CA, USA, 2001.
24. Chen, C., Zhang, G., Dai, Z., Xiang, Y., Liu, B., Bian, P., Fabrication of light-responsively controlled-release herbicide using a nanocomposite. *Chem. Eng J.*, 349, 101–110, 2018.

25. Gonzalez-Coloma, A., Reina, M., Diaz, C.E., Fraga, B.M., Santana-Meridas, O., Natural product-based biopesticides for insect control, in: *Module in Chemistry, Molecular Sciences and Chemical Engineering*, Elsevier Inc., Amsterdam, The Netherlands, 2013.
26. Gurr, G.M., Thwaite, W.G., Nicol, H.I., Field evaluation of the effects of the insect growth regulator (tebufenozide) on entomophagous arthropods and pests of apples. *Aust. J. Entomol.*, 38, 135–140, 1999.
27. Magierowicz, K., Górska-Drabik, E., Golan, K., Effects of plant extracts and essential oils on the inc. behavior of Acrobasis advenella (zinck.) caterpillars and females. *J. Plant Dis. Prot.*, 127, 63–71, 2020.
28. Tripathi, A.K., Upadhyay, S., Bhuiyan, M., Bhattacharya, P.R., A review on prospects of essential oils as biopesticide in insect-pest management. *J. Pharmacogn. Phytotherapy*, 1, 52–63, 2009.
29. Halder, J., Rai, A.B., Kodandaram, M.H., Compatibility of neem oil and different entomopathogens for the management of major vegetable sucking pests. *Natl. Acad. Sci. Lett.*, 36, 19–25, 2013.
30. Ali, M.A., Doaa, S.M., El-Sayed, H.S., Asmaa, M.E., Antifeedant activity and some biochemical effects of garlic and lemon essential oils on *Spodoptera littoralis* (Boisduval) (Lepidoptera: Noctuidae). *J. Entomol. Zool.*, 5, 1476–1482, 2017.
31. Mordue, A.J., Morgan, E.D., Nisbet, A.J., Azadirachtin, a natural product in insect control, in: *Comprehensive Molecular Insect Science*, L.I. Gilbert (Ed.), pp. 117–135, Elsevier, Amsterdam, The Netherlands, 2005.
32. Dannon, H.F., Dannon, A.E., Douro-kpindou, O.K., Zinsou, A.V., Houndete, A.T., Toffa Mehinto, J., Elegbede, I.A.T.M., Olou, B.D., Tamo, M., Toward the efficient use of beauveria bassiana in integrated cotton insect pest management. *J. Cotton Res.*, 3, 24, 2020.
33. Gwinn, K.D., Bioactive natural products in plant disease control, in: *Studies in Natural Products Chemistry*, vol. 56, R. Attaur (Ed.), pp. 229–246, Elsevier Inc., Amsterdam, The Netherlands, 2018.
34. Parween, T. and Jan, S., Pesticides and environmental ecology, in: *Ecophysiology of Pesticides*, T. Parween and S. Jan (Eds.), pp. 1–38, Academic Press, Cambridge, MA, USA, 2019.
35. Lade, B.D., Gogle, D.P., Lade, D.B., Moon, G.M., Nandeshwar, S.B., Kumbhare, S.D., Nanobiopesticide formulations: Application strategies today and future perspectives, in: *Nanobiopesticides Today and Future Perspectives*, pp. 179–206, Elsevier, London, United Kingdom, 2019.
37. Gupta, S. and Dikshit, A.K., Biopesticides: An ecofriendly approach for pest control. *J. Biopesticides.*, 3, 186, 2010.
38. Parker, K.M. and Sander, M., Environmental fate of insecticidal plant-incorporated protectants from genetically modified crops: Knowledge gaps and research opportunities. *Environ. Sci. Technol.*, ACS Publications, 51, 12049–12057, 2017.

39. Yang, J., Sun, Q., Zhu-Salzman, K., Qin, M., Feng, H., Kong, X., Host-induced gene silencing of brown planthopper glutathione S-transferase gene enhances rice resistance to sap-sucking insect pests. *J. Pest Sci.*, 94, 1–13, 2020.
40. Cagliari, D., Dias, N.P., Galdeano, D.M., Dos Santos, E.A., Smagghe, G., Zotti, M.J., Management of pest insects and plant diseases by non-transformative RNAi. *Front. Plant Sci.*, 10, 1319, 2019.
41. Zotti, M., Dos Santos, E.A., Cagliari, D., Christiaens, O., Taning, C.N.T., Smagghe, G., RNA interference technology in crop protection against arthropod pests, pathogens and nematodes. *Pest Manag. Sci.*, 74, 6, 1239–50, 2018.
42. Sasson, Y., Levy-Ruso, G., Toledano, O., Ishaaya, I., Nanosuspensions: Emerging novel agrochemical formulations, in: *Insecticides Design Using Advanced Technologies*, pp. 1–39, Springer, Berlin, Heidelberg, 2007.
43. Bouwmeester, H., Dekkers, S., Noordam, M.Y., Hagens, W.I., Bulder, A.S., De Heer, Review of health safety aspects of nanotechnologies in food production. *Regul. Toxicol. Pharmacol.*, 53, 1, 52–62, 2009.
44. Lu, W., Senapati, D., Wang, S., Tovmachenko, O., Singh, A.K., Yu, H., Effect of surface coating on the toxicity of silver nanomaterials on human skin keratinocytes. *Chem. Phys. Lett.*, 487, 1-3, 92–96, 2010.
45. Hazafa, A., Murad, M., Masood, M.U., Bilal, S., Khan, M.N., Farooq, Q., Iqbal, M.O., Shakir, M., Naeem, M., Nano-biopesticides as an emerging technology for pest management, in: *Insect*, pp. 1–31, IntechOpen, UK, 2021.
46. Nguyen, H.M., Hwang, I.C., Park, J.W., Park, H.J., Enhanced payload and photo-protection for pesticides using nanostructured lipid carriers with corn oil as liquid lipid. *J. Microencapsulation*, 29, 596–604, 2012.
47. Benelli, G., Plant-mediated biosynthesis of nanoparticles as an emerging tool against mosquitoes of medical and veterinary importance: A review. *J. Parasitol. Res.*, 115, 23–34, 2016.
48. Atef, M.M.S. and Behle, R.W., Comparing formulations for a mixed-microbial biopesticide with Bacillus thuringiensis var. kurstaki and beauveria bassiana blastospores. *J. Arch. Phytopathol. Plant Prot.*, 50, 15–16, 2017.
49. Ragaei, M. and Sabry, A.H., Nanotechnology for insect pest control. *Int. J. Sci. Environ. Technol.*, 3, 2, 528–545, 2014.
50. Lade, B.D. and Patil, A.S., Silver nano fabrication using leaf disc of *Passiflora foetida Linn. Appl. Nanosci.*, 7, 5, 181–192, 2017.
51. Prasad, K.G., *Advances in application of nanotechnology training module chapter 5, electrospinning for production of nano materials*, pp. 48–53, Sept 24–28, ICAR–CIRCOT, MS, India, 2018.
52. Nuruzzaman, M., Rahman, M.M., Liu, Y., Naidu, R., Nanoencapsulation, nano-guard for pesticides: A new window for safe application. *J. Agric. Food Chem.*, 64, 1447–1483, 2016.
53. Li, Y., Cu, Y.T.H., Luo, D., Multiplexed detection of pathogen DNA with DNA based fluorescence nanobarcodes. *Nat. Biotechnol.*, 23, 7, 885–889, 2005.

54. Xu, L., Liu, Y., Bai, R., Chen, C., Applications and toxicological issues surrounding nanotechnology in the food industry. *Pure Appl. Chem.*, 82, 2, 349–372, 2010.
55. Schiffman, J.D. and Schauer, C.L., A review: Electrospinning of biopolymer nanofibers and their applications. *Polym. Rev.*, 48, 2, 317–352, 2008.
56. Manjunatha, S., Biradar, D., Aladakatti, Y.R., Nanotechnology and its applications in agriculture: A review. *J. Far. Sci.*, 29, 1, 1–13, 2016.
57. De Souza, J.R., Mondego, J.M., Junior, A.L.B., Effects of different formulations of neem oil-based products on control *Zabrotes subfasciatus* (Coleoptera: Bruchidae) on beans. *J. Stored Prod. Res.*, 56, 49–53, 2014.
58. Barzman, M., Barberi, P., Birch, A.N.E., Boonekamp, P., Dachbrodt-Saaydeh, S., Graf, B., Eight principles of integrated pest management. *Agron. Sustain. Dev.*, 35, 4, 1199–1215, 2015.
59. Kookana, R.S., Boxall, A.B., Reeves, P.T., Ashauer, R., Beulke, S., Chaudhry, Q., Nanopesticides: Guiding principles for regulatory evaluation of environmental risks. *J. Agric. Food Chem.*, 62, 19, 4227–4240, 2014.
60. Solans, C., Izquierdo, P., Nolla, J., Azemar, N., Garcia-Celma, M.J., Nano-emulsions. *Curr. Opin. Colloid Interface Sci.*, 10, 3–4, 102–110, 2005.
61. Singh, Y., Meher, J.G., Raval, K., Khan, F.A., Chaurasia, M., Jain, N.K., Nanoemulsion: Concepts, development and applications in drug delivery. *J. Control. Release*, 252, 28–49, 2017.
62. Feng, J., Zhang, Q., Liu, Q., Zhu, Z., McClements, D.J., Jafari, S.M., Application of nanoemulsions in formulation of pesticides, S.M. Jafari, D.J. McClements (Eds.), pp. 379–413, Academic Press, USA, 2017.
63. Jiang, L.C., Basri, M., Omar, D., Rahman, M.B.A., Salleh, A.B., Rahman, R.N., Green nano-emulsion intervention for water-soluble glyphosate isopropylamine (IPA) formulations in controlling Eleusine indica (E. indica). *Pestic. Biochem. Physiol.*, 102, 1, 19–29, 2012.
64. Castro, M.J.L., Ojeda, C., Fernandez, C.A., Advances in surfactants for agrochemicals. *Environ. Chem. Lett.*, 12, 1, 85–95, 2014.
65. Ci, J., Cao, C., Kuga, S., Shen, J., Wu, M., Huang, Y., Improved performance of microbial fuel cell using esterified corncob cellulose nanofibers to fabricate air-cathode gas diffusion layer. *ACS Sustain. Chem. Eng.*, 5, 11, 9614–9618, 2017.
66. Nuruzzaman, M., Ren, J., Liu, Y., Rahman, M.M., Shon, H.K., Naidu, R., Hollow porous silica nanosphere with single large pore opening for pesticide loading and delivery. *ACS Appl. Nano Mater.*, 3, 1, 105– 113, 2020.
67. Zhao, X., Cui, H., Wang, Y., Sun, C., Cui, B., Zeng, Z., Development strategies and prospects of nano based smart pesticide formulation. *J. Agric. Food Chem.*, 66, 26, 6504–6512, 2018.
68. Vurro, M., Miguel-Rojas, C., Perez-de-Luque, A., Safe nanotechnologies for increasing the effectiveness of environmentally friendly natural agrochemicals. *Pest Manag. Sci.*, 75, 9, 2403–2412, 2019.

69. Liang, W., Yu, A., Wang, G., Zheng, F., Hu, P., Jia, J., A novel water-based chitosan-La pesticide nanocarrier enhancing defense responses in rice (*Oryza sativa* L.) growth. *Carbohydr. Polym.*, 199, 437–444, 2018.
70. De Oliveira, J.L., Ramos Campos, E.V., Goncalves da Silva, C.M., Pasquoto, T., Lima, R., Fraceto, L.F., Solid lipid nanoparticles co-loaded with simazine and atrazine: Preparation, characterization, and evaluation of herbicidal activity. *J. Agric. Food Chem.*, 63, 2, 422–432, 2015.
71. Chen, K., Yu, G., He, F., Zhou, Q., Xiao, D., Li, J., Feng, Y., A pH-responsive emulsion stabilized by alginate-grafted anisotropic silica and its application in the controlled release of λ-cyhalothrin. *Carbohydr. Polym.*, 176, 203–213, 2017.
72. Shakiba, S., Astete, C.E., Paudel, S., Sabliov, C.M., Rodrigues, D.F., Louie, S.M., Emerging investigator series: Polymeric nanocarriers for agricultural applications: Synthesis, characterization, and environmental and biological interactions. *Environ. Sci. Nano*, 7, 1, 37–67, 2020.
73. Sasson, Y., Levy-Ruso, G., Toledano, O., Ishaaya, I., Nanosuspensions: Emerging novel agrochemical formulations, in: *Insecticides Design Using Advanced Technologies*, pp. 1–39, Springer, Berlin, Heidelberg, 2007.
74. Bergeson, L.L., Nanosilver pesticide products: What does the future hold? *Environ. Qual. Manag.*, 19, 4, 73–82, 2010.
75. Jayaraj, R., Megha, P., Sreedev, P., Organochlorine pesticides, their toxic effects on living organisms and their fate in the environment. *Interdiscip. Toxicol.*, 9, 90–100, 2015.
76. Sharma, S., Singh, S., Ganguli, A.K., Shanmugam, V., Anti-drift nano-stickers made of graphene oxide for targeted pesticide delivery and crop pest control. *Carbon*, 115, 781–790, 2017.
77. Oancea, S., Padureanu, S., Oancea, A., Growth dynamics of corn plants during anionic clays action. *Lucr. Stiint. Ser. Agron.*, 51, 212–217, 2009.
78. Bipin, D.L., Dayanand, P.G., Sukhdev, B.N., Nano biopesticide to constraint plant destructive pests. *J. Nanomed. Res.*, 6, 1–9, 2017.
79. Chormade, V., Deshpande, M.V., Paknikar, K.M., Perspectives for nano-biotechnology enabled protection and nutrition in plants. *Biotechnol. Adv.*, 29, 792–803, 2011.
80. Boehm, A.L., Martinon, I., Zerrouk, R., Rump, E., Fessi, H., Nanoprecipitation technique for the encapsulation of agrochemical active ingredients. *J. Microencapsul.*, 20, 433–441, 2003.
81. Liu, Y., Wei, F., Wang, Y., Zhu, G., Studies on the formation of bifenthrin oil-in water nano-emulsions prepared with mixed surfactants. *Colloids Surf. A*, 389, 1–3, 90–96, 2011.
82. El-Nahhal, Y., Nir, S., Marguiles, L., Rubin, B., Reduction of photodegradation and volatilization of herbicides in organoclay formulations. *Appl. Clay Sci.*, 14, 105–119, 1999.

83. Liu, Y., Wei, F., Wang, Y., Zhu, G., Studies on the formation of bifenthrin oil-in water nano-emulsions prepared with mixed surfactants. *Colloids Surf. A.*, 389, 1–3, 90–6, 2011.
84. Zhang, W.X., Wang, C.B., Lien, H.L., Treatment of chlorinated organic contaminants with nanoscale bimetallic particles. *Catal. Today*, 40, 387–395, 1998.
85. Wang, L., Dong, J., Chen, J., Eastoe, J., Li, X., Design and optimization of a new self-nanoemulsifying drug delivery system. *J. Colloid Interface Sci.*, 330, 2, 443–448, 2008.
86. Bhagat, D., Samanta, S.K., Bhattacharya, S., Efficient management of fruit pests by pheromone nanogels. *Sci. Rep.*, 3, 1–8, 2013.
87. Srivastava, R., Awasthi, K., Tripathi, D., Nanotechnology towards sustainable agriculture. *Nanotechnology*, 6, 6, 155–158, 2018.
88. Patil, S.S., Shedbalkar, U.U., Truskewycz, A., Chopade, B.A., Ball, A.S., Nanoparticles for environmental clean-up: A review of potential risks and emerging solutions. *Environ. Technnol. Innov.*, 5, 10–21, 2016.
89. Khayatnezhad, M. and Nasehi, F., Industrial pesticides and a methods assessment for the reduction of associated risks: A review. *Adv. Life Sci.*, 8, 2, 202–210, 2021.
90. Grillo, R., Fraceto, L.F., Amorim, M.J., Scott-Fordsmand, J.J., Schoonjans, R., Chaudhry, Q., Ecotoxicological and regulatory aspects of environmental sustainability of nanopesticides. *J. Hazard. Mater.*, 404, 124–148, 2020.

3

Application of Plant-Based Nanobiopesticides as Biocides

Kehinde Abraham Odelade[1*], Babatunde Oluwafemi Adetuyi[1], Goodness Oluchukwu Jonah[1], Oluwaseun Dorcas Opayinka[1], Divine Uwanna Okonimeh[1], Joanna Success Mishael[1], Funmilayo Peace Adebayo[1], Oluwakemi Semiloore Omowumi[1] and Charles Oluwaseun Adetunji[2]

[1]Department of Natural Sciences, Faculty of Pure and Applied Sciences, Precious Cornerstone University, Ibadan, Nigeria
[2]Applied Microbiology, Biotechnology and Nanotechnology Laboratory, Department of Microbiology, Edo University Iyamho, Auchi, Edo State, Nigeria

Abstract
Nanotechnology field is one of the fewest settled areas of research in the present material sciences. This nanotechnology field is evolving daily and it is making a significant impact in every area of human existence. It also creates a developing sensation of satisfaction in the life sciences particularly biotechnology in addition to biomedical equipment. Nanotechnology has been used in the production of different products that kills pests and insects. Some of these include biopesticides, biocides, and so on. An essential requirement needed for various biological systems and crop protection is the application of biocides. However, different characteristics like modified and controlled release are important in the synthesis of sustainable biocide release frameworks. Nanobiopesticides, together with its advantages and disadvantages, were discussed in this review. Also, the different strategies used in the production of nanomaterials on the basis of controlled release formulations (CRFs) for the applications of biocides were also discussed.

***Keywords*:** Nanobiopesticides, biocides, nanotechnology, biopesticides, polymers, nanoparticles, micelles, ecofriendly

**Corresponding author*: kennybramm@gmail.com

Charles Oluwaseun Adetunji and Julius Kola Oloke (eds.) Handbook of Agricultural Biotechnology: Volume I Nanopesticides, (37–62) © 2024 Scrivener Publishing LLC

3.1 Introduction

Unwanted plants, illnesses, and pests wreak havoc on potential agricultural productivity all over the world. Medicinal herbs, forest trees, unwanted plants, crop plants, and other plants provided food for insects. They also eat meals and other stored goods, resulting in massive losses and disintegration in the quality of food. With regard to crop safety, traditionally chemical pesticides of various classes were used widely across international borders. The death rate of people has been assessed by the World Health Organization (WHO) to approximately twenty thousand human beings and this was a result of pesticides toxins in the third world each year. Chemical pesticides derived from the derivatives of petroleum are accounted for to be accumulating in groundwater and soil, posing a threat to ecosystem health. Pest outbreaks, insect resistance to pesticides, health risks to humans, domestic and wild animals and pest resurgence, are all regarded to be significant negative effects of pesticides [1]. Materials with a diameter of ten nanometer or 1 millionth of a meter are used in nanotechnology. The two techniques known as top-up and button-up are followed in nanotechnology utilizing synthetic and actual methodology. However, numerous researchers have proposed lots of adverse effect associated to the use of these techniques. A lot of individuals in industries have been involving these particles in a pragmatic way. Additionally, agriculturists are really involving this innovation (nanotechnology) for exceptional plant development inducement [2], plants infection determination, genetic enhancement of plants, food biotechnology and post-harvest control [3, 4], infections, as well as resistance to stress [5], control of vegetation infection [6], and conveyance of synthetic substances and DNA in vegetation [7]. The seeds of tomato grown in carbon nanotubes (CNT) displayed that improved growth as a result of increased water take-up prompted by the entrance of carbon nanotubes as a conveyor to release atoms like nutrients or biocides in the seedlings at the time of germination. Many scientists have proposed the use of nanotechnology in agriculture pest management [8, 9]. Some of the uses suggested by majority of researchers for agricultural pest management include liquid and solid synthesis of nanoparticles (NPs) [10], novaluron NPs [11], pesticides release by nanosuspensons [12], and nanomaterials produced from fungi [13]. As a substitute, botanicals are recommended for the protection and production of crop. Furthermore, botanicals have both benefits and drawbacks. Botanical insecticides, on the other hand are not commonly utilized in traditional production of crops; however, farmers in developed nations remember them.

The flower heads, leaves, stems, roots, and seeds of various species of plants are used in the synthesis of organic pesticides [14–22]. *Tanacetum cinerariifolium* (Pyrethrin) is extracted from pyrethrum flowers. Pyrethrin is an organic insecticide having a lengthy history of effective and safe utilization in pest control [23]. Pyrethrin is a bioactive substance made up of 6 bioactive constituents (pyrethrin I and pyrethrin II, cinerin I and cinerin II, as well as jasmolin I and jasmolin II). The ability of pyrethrin to destroy or control insects has been acknowledged for a very long time ago, and both ground and dried flower have conventionally been utilized globally for storage [24], management of agricultural pests [25, 26], and management of pests found in the household [27].

Rotenone (*Lonchocarpus utilis*) is viewed as a profoundly compelling administration specialist of numerous pest species, similar to aphids [28], moths [7], orthopterans [29], mites [30], and beetles and weevils [31]. Ryania (*Ryania speciosa* Vahl.) [32], bitter-wood (*Q. amara*) and feather-shank (*S. officinale*), end up being involved all over the planet as botanicals for pest control. As of late, progression of specific examination abilities, particularly connecting with the herbal-based nanoparticles has not been utilized hitherto in likely in the control of pest program. Subsequently, this appraisal is centered on botanical biopesticide techniques promptly accessible for the control of pests and the occupation of nanotechnology and nanomaterials as contemporary techniques on account of the control of pests in farming.

3.2 Nanotechnology and Its Application

Nanotechnology utilizes nanoparticles (NPs) fostering a component of hundred nanometer or a thousand nanometer. Nanoparticles produced from respectable metallic complexes include platinum, silver, and gold. Nanoparticles are also currently produced from metal oxidase materials (such as AgO, MgO, TiO_2, and ZnO), lipids, silicates, pottery, semiconductor quantum dots, ceramics as well as emulsions. Nanomaterials can be obtained from both inorganic and organic sources. These nanomaterials are combined to form products by organic, chemical and physical strategies. Nanobiopesticides, nanomicrobicides as well as nanopesticides are currently being utilized effectively in farming [33]. The field called nanotechnology is one of the fewest settled research field in the present material sciences. This nanotechnology field is evolving daily and it is making significant impact in every area of human existence. It also creates a developing sensation of satisfaction in the life sciences particularly biotechnology in addition to biomedical equipment [34]. The utilization of NPs is acquiring driving force in the current age since they have specific substance, mechanical properties and optical.

Nanotechnology is taken advantage of for most purposes inside the compound, organic, clinical, farming, and beauty care products ventures alongside much more. Moreover, the utilization of nanotechnology in farming still remains in its earliest stages. In the earliest age, biocides (imidaclaporid) are produced from mesoporous silica nanoparticles (MSNs), and water-soluble pesticides (validamycin) were produced from porous hollow silica nanoparticles (PHSNs) [35, 36]. The ability of nanostructured alumina to destroy insects was emptied on 2 main pests in stored food like beetles (*Sitophilus oryzae*) and grain borers (*Rhyzopertha dominica*). The outcome reports showed that nanostructured alumina led to increase in death rate after 72 hours of steady openness to cured wheat. Furthermore, a week and 2 days after the therapy, the lethal dose 50 increased (127–235 mg/kg) [37]. The diatomite was utilized to structure nebulous nanosized lipophilic, polar and nonpolar silica (Silica nanoparticle) at the range 15 to 30 nm size, and its toxicity against insects was tried on *Sitophilus oryzae* (rice weevil). The outcomes revealed that silica nanoparticle was found to be most efficient on rice weevil, which cause about 90% death rate, demonstrating the efficiency of silica nanoparticles to manage stored items pests [38]. Nanosilica produced out of silica is utilized as nanopesticides. The outer skeleton of pests comprises fats that can be utilized as a water screen to defend the pests and in this way avoid demise out of drying up. In any case, nanosilica which size range is between 3 and 5 nm becomes retained in the outer skeleton (cuticle) lipids by physisorption and in this way sets off demise of pests [39].

Chemical pesticides delivered using NPs enhance the effectiveness, improved consistency of protection for extremely bioactive constituents and reduced contact with representatives, similar to supplements solubilized in nonsynthetic solvents. For example, polymeric stabilizers bifenthrin NPs are suggested for the continuous delivery of bifenthrin NPs [40] and enhance pesticides stacking from 50% to 91%. Beforehand, Liu *et al.* [35] likewise discovered validacin, a pesticide utilizing permeable empty SiO_2 NPs. A few different substances have been suggested in control of pests [41–45] and also crop security [46]. Nanosilica can also be utilized on stored grains, worms, fungal organisms, parasites, household pests, and so forth [46]. Furthermore, various logical examinations confirmed that metal nanoparticles neutralize pests, insects, and plants microbes.

A researcher suggested the accompanying probably uses of nanotechnology in farming: release of nanocides-pesticides epitomized in nanoparrtcles for controlled discharge; adjustment of biological pesticides with nanoparticles; gradual delivery of nanomaterial-helped composts, biological fertilizers, in addition to trace elements for effective utility; and furthermore field utilizations of nanomaterial-helped conveyance as well as agrochemicals of hereditary material for the development of crops.

3.3 Nanotechnology and Biopesticides

Unpredictable utilization of pesticides in addition to manures prompts natural pollution raises the rate of pest and pathogenic microorganisms' resistance, and leads to reduction in the heterogeneity of animal and plant existence within the ecosystem. Diminished diversity of animal and plants existence in the soil and nitrogen fixation assist with the biological accumulation of pesticides, annihilation of the natural surroundings of animal, and well-being dangers to people and wild and domestic animals. Alongside synthetic substances, natural strategies, actual techniques, and mechanical methods end up being exclusively used or maybe in blend (integrated pest management [IPM]) for the destruction of illnesses and pests in order to shield the plants and income from the creation. A while later, a more extended idea, biointensive incorporated pest management (BIPM) was made and extent of natural material (or perhaps biological pesticides) has gotten to the next level [47]. Bionanomaterials by virtual of organic and nanomaterials related characteristics, has a prospective agricultural biotechnological utilization as an option of chemical pesticides, as well as simultaneously a part of BIPM. Remembering this theory, this specific section is deals delicately with both biopesticides and nanotechnology in all possible way. Fractions obtained from plants are used as plant-based pesticides and furthermore for bio reduced nanomaterial formulation. Biopesticides are distinct sorts of pesticides produced from biological substances as animals, microorganisms, plants, and a few minerals. These comprise of growths like *Beauveria* sp., microorganisms including *Bacillus* sp., neem fractions, alongside pheromones. Also, baking pop and canola oil possess pesticide utilizations and are seen as biopesticides. Among them, the two organisms and plants are used for the formulation of nanomaterials. For example, NPs of zinc, copper, cobalt, nickel, and bronze are likewise formulated inside the living plants.

3.4 Various Benefits of Biopesticides

In general, biopesticides utilize a limited objective choice alongside a truly specific method of activity; acts in a gradual (slow) process; possess sensibly vital times of utilization; subdue, as opposed to destroy, a pest populace; possess very little field assurance alongside a brief timeframe of realistic usability; are obviously better to individuals and furthermore the planet than ordinary pesticide; present no buildup issues; do not last in the surrounding; give to some degree little risk to beneficial predators and parasites (nontarget life forms); typically digest rapidly in the planet; are immediately processed by creatures getting sub lethal portions; partake in a wide range of movement; and thusly are not hard to approach and utilize. Observing these benefits, biopesticides could be used for the formulation of nanomaterials and used in agribusiness to improve people as well as also human nature [47]. Table 3.1 below showed the benefits and downsides of biopesticides.

Table 3.1 Benefits and downsides of biopesticides [48].

Benefits	Downsides
Highly efficient in reduced quantities and owing to biodegradability, degrades rapidly	Acts slowly in contrast to synthetic pesticides
Absence of residues	Target specific necessitating numerous methodologies
Has a low adverse effects to the environment and people	Low stability of some biopesticides
Can be efficiently utilized against plant pathogens	Less shelf life
Utilized together with the common pesticides	Numerous regulatory challenges and economic issues
Mechanism of action and biodegradable characteristics and host specificity are recognized	Possess little profit potential as a result of high selectivity
Low improvement expenses in contrast to traditional synthetic pesticides	

3.5 Different Kinds and Applications of Biopesticides

The consequences of the utilization of synthetic pesticides include the introduction of toxic compounds to nontarget organisms; pollution of the atmosphere, soil and aquatic environment modification in soil microflora; adverse effect on people's well-being; and also enhanced residues in food grains in addition to vegetative plants. However, creation of more secure options is required and has given stimulus to investigation and improvement of biological-based pesticides. Biological pesticides are based on natural control agents and do not have a global uniform definition. They include predators, parasites, semiochemicals, nematodes, microorganisms, and metabolites produced by plants. The benefits of biopesticides are well known. Generally, there is a rise in their utilization either by replacing or supplementing traditional pesticide [49]. There is a continuous rise in plant-incorporated products (PIPs) (which was accomplished by genes modifications) as essential biological control agents from 1990s. Presently, biopesticides' percentage in the global pesticide industry is highly notable. However, the application of biotic agents to manage illnesses and pests is not new. It was traced back to some decades ago when plants fractions had been successfully used as seed covers or even in areas of grain storage. The advancements of technology are useful in the improvement of aimed specific and also site-specific biological pesticides.

Plant-incorporated protectants, pesticides derived from microorganisms, and biochemical pesticides are the three major classes of biological pesticides [50]. The ability of numerous plants spp. to destroy insects was identified since the time ethnobotanical plants are used in farming; plants full of secondary metabolite constituents or maybe elements isolated out of the secondary metabolites are utilized as biochemical pesticides [51]. Essential oil as effective pesticide energy is taken advantage of from numerous species of plant [52, 53]. The most commonly utilized compounds to manage pests are pyrethrins and neem oils. Biochemical pesticide is classified into different categories. One of the classifications is the semiochemicals, synthetic indicators created by a microorganism, which cause a behavioral change in other organism. The well-known semiochemicals used in crop safety is the insect sex pheromones. Biopesticides obtained from microorganisms are mainly specific microscopic organisms, oomycetes, fungi, parasites, alongside protozoa that were appropriately used for regular control of insects (Figure 3.1). There are lots of differences in the creation, advancement and mechanism of action amidst various kinds of biopesticides derived from microorganisms. Biochemical pesticides are

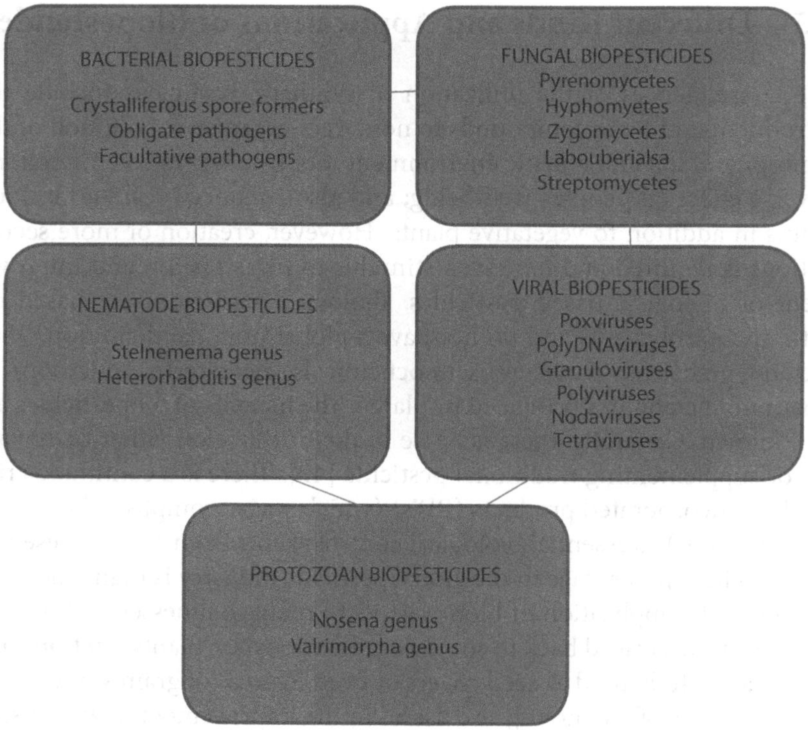

Figure 3.1 Representative types of microbial biopesticides.

compounds that destroy or perhaps deactivate the insect, obstruct reproduction, influence breath, and also act on the nervous system. A coordinated methodology for their utilization against a target pest is fundamental for the viable utilization of theirs in battling with pests in farming [54].

Microbial biopesticides end up being really applied to the type of splashes on extraordinary cultivating and agriculture regions and numerous business arrangements end up being notable [54]. Plant-incorporated protectants belong to a different group of biopesticides, where different vegetation is genetically engineered to create substances with pesticidal activity. One of the famous genetically engineered vegetation for pest resistance was developed with *Bacillus thuringiensis* (*Bt*), an insect poison which gives security against various bollworms and borers. *Bt*-based insect poisons are created by utilizing quality groupings, which make precious stone proteins. An assortment of plants is really worked with the *Bt* qualities. Such hereditarily adjusted crops are now made for safety against the corn borer, potato beetle, tobacco budworm, cotton bollworm, and so on.

The utilization of microscopic organisms and parasites as biological control specialists was confirmed in a choice of developed frameworks. Numerous biopesticides have additional advantages when used to develop techniques, which make them become incredible contender for management for pest. Some biopesticides synthesized from microorganisms upgrade the take-up of soil supplements by vegetation. They are efficient on plant microorganisms, which attack similar harvests [55].

Plant probiotic organisms as biocontrol specialists are as of now making progress. Plant probiotic microorganisms assist with plant security by solid threat with potential antagonisms through the arrival of substances inside the prompt climate. Biocidal metabolites, which incorporate unpredictable substances, are made of these microbes. One of the microbes, which deliver plant probiotics, includes *Pseudomonas chloraphis*. The main property of this specific species is the way that there is segregate fluctuation in the secretion of different metabolites, which could act as biological control specialists. The various metabolites it secretes include, phenazines, insecticidal proteins, certain hydrocarbons, butanediol, and HCN, Water-soluble mixtures which includes siderophores, pyrrolnitrins, and furthermore phenolics [56]. The worldwide biopesticides industry was calculated in 2013, to foster 5% of the whole market for plants safety products. A pesticide synthesized from microorganisms is the widest class under this. Pesticides synthesized from *Bacillus thuringiensis* constitute about 75% of the microbial biopesticides. The worldwide pesticides industries are ruled by North America, and there is a consistent development in Asia, Europe, and most of the Earth planet. The advancement of biopesticide industry or market have been cultivated by some factors which include product improvement, maintainability, effective utilization in integrated pest management, as well as demonstrable parts on natural farm applications [57].

In contrast with manufactured pesticides, biopesticides have a couple of benefits. The significant advantages of biological pesticides include the following: they have substantial low harmfulness than the traditional pesticides, influence simply the aimed pests in addition to other microbes, efficient in low quantities, and furthermore debase rapidly. They further enhance adequacy, improve yield, and are also cost efficient when utilized alone or maybe related to traditional products used for crop safety. Lessening the natural effect is nevertheless something significant about biopesticides, which licenses them to be a common sense answer for further develop proficiency and secure the climate. The development of biological pesticides is proposed to have been enacted by the quick reception of natural yield strategies in numerous nations [58]. However biological pesticides are presently utilized around the world, although there are several benefits and downsides attached to their use (Table 3.1). In any case, the advantages related with a safer ecological effect render bioproducts generally ideal.

3.6 Utilization of Nanotechnology in Agricultural Systems

Nanotechnology was utilized in a few fields like materials, biocomposities, biomass, food biofuel, sustenance paints, plant security, pesticides, manures, food, paper, geosensing innovation, alongside agrochemical enterprises to deliver novel medicines with a large scope of purposes. NPs (Nanoparticles) could happen in oligomeric, monomeric, or even polymeric particulate frameworks with a certain size ranges [59] with distinct physical characteristics like consistency and conductance, which make them engaging in materials science, as well as life science [60]. Silver nanoparticles, zinc oxide nanoparticles, copper nanoparticles, and silicon oxides nanoparticles all possess important characteristics of pest management alongside lowering the potential of soil, water, and ecological contamination, which then makes them ideal for the synthesis of biological pesticides [61]. Benefits of metal NPs include steadiness, slow energy, which could be increased for mass productions at ambient temperature, and creation of ecological derivatives for the syntheses of biological pesticides. Nanoparticles are utilized in the sector designated to agriculture to defend and secure vegetation, improve soil fruitfulness, manage unwanted plants, and also enhance soil supplements. Furthermore, they act as help and composts in plant improvement or to nurture host-specific pests' resistance. For instance, NPs and Zn are mixed together using biodegradable mixtures to produce Zn NPs (also known as zinc nanoparticles). Zinc nanoparticles are nontoxic to the environment, and also, they are utilized to enhance the fertility of soil. Soils treated with zinc nanoparticles fill three needs: enhance the zinc in the soil, the vegetable yield in such treated farmlands become a valuable source of zinc, and also control a wide range pests of agricultural significance. Also, herbicides, genes, synthetic compounds or fungicides might be mixed with NPs to work as magic bullet which aimed at specific parts of plants for pest management [62].

To be accurate, nanopesticides is more secure to be utilized and could possess enhanced potential to be utilized as pesticides, like through nanocoating of the items. The nanocoating includes nanosized particles of the dynamic pesticide substance, which are fixed using a tiny-walled sac. The nanocoating nematicides, fungicides, insecticides offer efficient pest management, stopping the buildups in the ground and its run off into aquatic environment. Furthermore, it possesses the benefit of protecting the bioactive pesticide constituent from debasement, developing its efficiency, and essentially reducing the pesticide contribution to at the very least 10 to 15 times under those utilized by with traditional formulations [63]. Other illustration of nanopesticides synthesized with reduced expenditure is the halloysite. Hallolysite has extensive delivery prospects, and furthermore improves correspondence with vegetation. Furthermore, it lowers pesticide utility by 70% to 80% in this way, which makes it a monetarily practical other option [64].

3.7 Nanobiopesticides

Nanobiopesticides are produced organically from synthetic compounds complexes together with NPs that have pesticidal properties. The effectiveness and efficiency of nanobiopesticides are improved through the use of bioactive molecules mixed with micelles, metal oxides, polymers, and so on. For a nanobiopesticides to be effective, it must have some of the intrinsic characteristics like superior dissolvability of deficiently solvent bioactive constituents, gradual delivery of bioactive constituents, and bioactive constituents which ought not to be debased prematurely [65]. There must be no compromise in the effectiveness of nanobiopesticides, even when used in little quantity. However, the aimed particularity and the result of it could be raised by formulating the nanobiopesticides in relation to utilizing the science of life cycle, as well as the activities of pathogenic microorganisms or pests. Nanoformulation gigantically affects the characteristics in addition to fate of its bioactive constituents. In this way, the shape, size range, or perhaps nature; the physical characteristics of nanoparticles; the adjuvants utilized, as well as their properties discharge with time in sensible circumstances required to begin before the utilization of theirs [65].

Nanobiopesticides comprise of NPs of either bioactive constituent with pesticidal activity or some other little modified structures that have important pesticidal activity [66]. A given illustration is the 4th instar hatchlings of the filariasis vector *C. quinquefasciatus* S., as well as the malaria vector *A. subpictus* G had been put through such assessments [67].

Silver nanoparticles of 60 nm showed top variability in a raw water solution on *A. subpictus* (LC50 = 27.85 and 5.14 mg/L; LC90 = 71.45 and

25.68 mg/L), and silver nanoparticles of 35 nm also showed top variability on *C. quinquefasciatus* (LC50 = 27.49 and 4.56 mg/L; LC90 = 70.38 and 13.14 mg/L) [68], which implies that the tiny mass of NPs have essential activity they carry out in the effectivity of the items. The plant extracts of the *F. benghalensis* (fig tree) and the *F. religiosa* (pipal tree) used for the production of Ag NPs is found to regulate the role of stomach proteolytic-enzyme movement in the cotton bollworm (*Helicorverpa armigera*) [69]. Other properties achieved through the change of the nanobiopesticides with biological fertilizers or perhaps polymer transporter development, which displays characteristics like biodegradability, crystallinity, solidness, solubility, vulnerability, and heat stability [70–72], include the ability to degrade slowly in vegetation with buildup levels underneath the normal conditions in food items, ability to degrade quickly in soils and increased ability to circulate [73]. For instance, nanoparticles (30–100 nm) synthesized from pyrethrin of novaluron is lethal to *S. littoralis* (Mediterranean brocade moth) hatchlings. By and by, such nanoproducts have ended up being hazardous to the climate, and their removal was an issue. This plainly represents an inquiry concerning harmless to the ecosystem security, that should be commonly looked to while producing nanoparticles based products [74].

3.7.1 Nanoformulations

One of the topics of interest in this present age is nanoformulations grounded on biological-based nanoparticles in addition to their synthesis [75]. For instance, biological formulation of silver nanoparticles incorporates the synthesis of plant fractions and organisms at ambient temperature using a technique that needs small or no input of energy [76]. Furthermore, the technique can be recycled and it is economically friendly [77]. The synthesis of stable nanoparticles with distinct composition, sizes, and morphologies can be done by picking the best microorganisms and improving the response conditions [78]. For instance, an entire of Ag nanoparticles formulation utilizing plant extracts of *O. sanctum* in certain circumstances and silver nitrate has been done by different researchers. This product has effectively managed pests like *H. armigera*. The monitored delivery of these nanobiopesticides is feasible by encapsulating the biological agent with a certain polymer like carbohydrate or protein which ties the bioactive constituent complex to NPs. This group of three structures of cross-linking a polymer with bioactive biopesticide-based NPs assists in the delivery of bioactive constituent nanoparticles in a controlled way and displays expanded pesticidal action [79].

3.7.2 Polymers

The normal polymers (regular and manufactured ones) utilized inside the controlled arrival of insecticide utilizations are summarized in Table 3.2 [79]. The biopolymer behaves as a transporter and furthermore assists in the conveyance of viable nanobiopesticides. The positive attributes of biopolymers include hydrophilic and lipophilic qualities, higher dissolution energy, in addition to negligible basic micelle concentration; they are formulated from polyoxyethylene and different aliphatic dicarboxylic acids. All these structures arrange themselves in nanomiceller totals when in the

Table 3.2 The list of polymers that have capacities for the synthesis of bionanoparticles for insecticides utilization [75].

Polymer	Active compound	Nanomaterials
Alginate-bentonite	Imidaclopeid	Clay
Carboxymethyl chitosan ricinoleic acid	Azadirachtin	Particles
Carboxymethylcellulose	1-naphthyl methylcarbamate	Capsules
Chitosan	Etofenprox	Capsules
Chitosan	Pyrethrins	Capsules
Glycol-ethylcellulose	Imidacloprid	Capsules
Lignin	Aldicarb	Gels
Lignin	Imidacloprid	Granule
Lignin-polyethylene	Imidacloprid	Capsules
Nanocellulose	Pesticide compound	Fibers
Polyamide	Pheromones	Fiber
Polyethylene glycol	Garlic essential oil	Capsules
Polyethylene glycol	Carbofuran	Suspension
Polyethyleneglycol-dimethyl ester	Carbofuran	Micelle
Sodium salt	Novaluron	Powder

fluid medium and also assist in controlled delivery syntheses, is demonstrated in the circumstance of an insect neurotoxin (imidacloprid), utilizing the nanocoating method [80]. Additionally, nanoitems based on biological products together with huge biologically based nanoitems with huge characteristics of being degraded by microbes can function as most prominent polymers for integrating nanosyntheses and consequently the aimless utilization of synthetic substance pesticides, which are naturally unsafe may be avoided. The polymers, like carbohydrates, gelatin, or chitosan, are utilized as fasteners, diluents, or decomposer alongside getting utilized as transporters for NPs. The thermal stability or temperature-delicate controlled delivery nanopesticides frameworks can be enhanced through the use of some polymers such as acrylamide, polyoxyethylene ether, poly(methoxyethylene), poly(N-isopropyl acrylamide), 2-hydroxypropyl ether, and polyvinylpyrrolidone. This can agreeably resolve issues related with farming creation [75].

Polymers could likewise help with building formulations of nanobiopesticide, which might be sensitive to pH, moisture delicate, light-touchy, temperature-delicate, and sensitive to enzyme items. The light-delicate polymers is categorized into vesicles and micelles (poly(ethylene oxide *b*-methacrylic acid), hydrogels (diazo compounds, organic halogen compounds, aromatic azide, chlorophyllin, aromatic nitro compounds, dichromate), nitrobenzenes (*O*-nitrobenzyl aldehyde, *O*-nitrobenzyl alcohol), liposomes (Bis-Azo-PC liposome and *trans*-liposomes), light-sensitive group like spiropyrans (Benzo-crown ether spiropyran, benzo thiopyran compound of monazo sulfide crown ether, benzo thiopyran compound of spiro monoaza crown ethers), coumarins (furocoumarins, psoralens, 5-methoxypsoralen, and 8-methoxypsoralen), azobenzenes (PAMAM, dendritic polyamide-amine).

Additionally, the heat-sensitive biopolymers can be likewise liposomes (daunorubicin liposomes, paclitaxel liposomes, cytarabine liposomes, 5 fluorouracil multiphase liposomes, adriamycin liposomes), polymers like (acrylamide, polyvinyl methyl ether, polyvinylpyrrolidone, hydroxypropylene cellulose, polyoxyethylene ether, poly-N-isopropyl), hydrogels (polyethylene glycol, poly (N isopropylacrylamide)), in addition to related NPs (inorganic and metal). In correlation, the power of hydrogen (pH)-responsive formulations can be categorized as polymer microspheres (polyacrylic corrosive, PAA), hydrogels (PEG-*b*-P4VP, polyethylene glycol-*b*-poly-4-vinylpyridine), micro-capsules (N,N dialkylaminoalkyl acrylate polymer, supramolecular graft polymer), or mesoporous materials (mesoporous silica, SBA-15), polymethacrylic corrosive (PMAA) [75].

Moisture-delicate materials include permeable metal oxides (permeable α-Fe_2O_3 nanospheres, $MeCr_2O_4$ Bi_2O_3 frameworks), hyperbranched quaternary ammonium salt (bactericide and sanitizer), or perhaps electrolytes (polystyrene sulfonate, polyacrylic corrosive), polyimides utilized in plastic. Among catalyst responsive items might be emamectin benzoate (acaricide and insect spray), chemical responsive nanoparticles (Fe NPs, fullerene subordinates, decreased graphene zinc ferrite attractive composite nanomaterials for example rGO-$ZnFe_2O_4$, Au NPs), enzyme-responsive polymers (kasugamycin-like cultivating fungicide), RADA16 I (oligopeptide material), polymethyl celluloseglycidyl methacrylate gel, polyethylene glycol diacrylate-methacrylic corrosive, hydrogels (glucose hydroxyethyl methacrylate dimethylaminoethyl methacrylate) [81]. In polymer based products, the polymer type, molecule size, subatomic weight, polymer breakdown, drug stacking, alongside collaborations between the nanoparticles in addition to polymer have an essential impact in picking the arrival of NPs from polymers for more effectiveness.

3.7.3 Nanoparticles Utilized in Biological Pesticides Controlled Released Formulations

The frequently utilized types of nanomaterials, which were utilizing in controlled released formulations (CRFs) for biocides release are:

1. Nanospheres: aggregate where the dynamic constituent is uniformly conveyed in the polymeric framework.
2. Nanocapsules: this is a mass where the dynamic constituent is concentrated close deeply, lined through the grid polymer.
3. Nanogels: hydrophilic (for the most part cross-connected) polymers that can take in higher quantity of H_2O.
4. Micelles: this is a mass laid out in fluid solutions by atoms with hydrophobic and hydrophilic segment of a molecule.

3.8 Strategies Used in the Production of Nanomaterials on the Basis of CRFs for the Application of Biocides

The procedures for CRFs preparation could be isolated from the physical or chemical ones (this can be seen in Figures 3.2 and 3.3) [82]. The synthetic compounds strategies are obtained from covalent synthetic substances

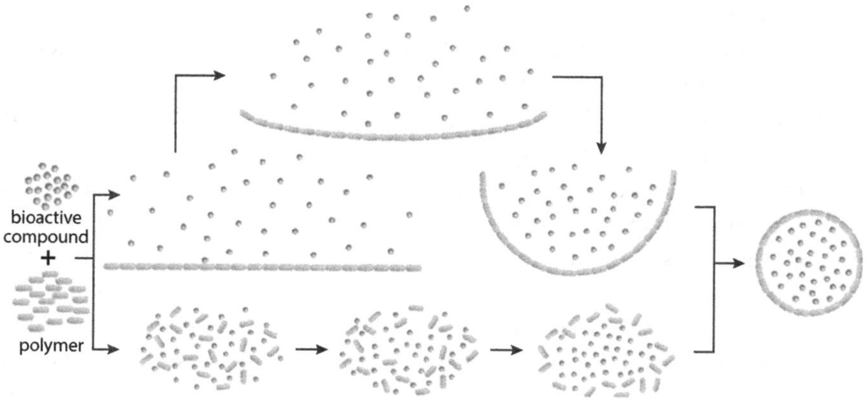

Figure 3.2 Physical methods for CRF preparation.

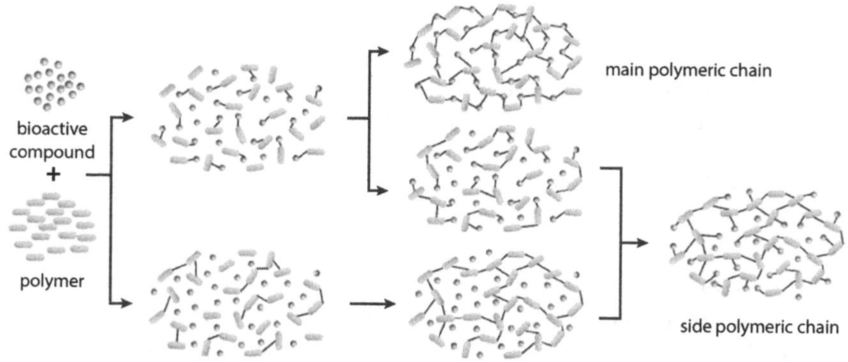

Figure 3.3 Physical methods for CRF preparation.

made between the bioactive constituents and the covering lattice, like a polymer. This bond might be set in two locales: in the side chain or even in essential polymeric chain. The novel macromolecule is named probiocide in the first one, since the mixture would obtain the characteristics after it has been distributed. In the following one, the insecticide particle can bind at first towards the side chain of only 1 monomer after which the polymerization reaction occurs or perhaps the polymerization happens first and then the biocide is bonded to the side chain. The last method is reliant upon the intermolecular interaction. In this specific circumstance, the immobilization of the biocide occurs in the web made by the cross-linkages in the polymer. The actual strategies can moreover be divided in 2 particular classifications. In the absolute initial, a blend of biocide and polymer is made. In another, the polymeric chain frames a “membrane” disconnecting the active constituents from the ecosystem. This interaction creates the nanocapsules. Despite the fact that there are a few different types of nanomaterials which might be utilized in controlled release formulations, the microcapsules and nanocapsules are without a doubt generally utilized for controlled arrival of biocides [79].

3.9 Impacts of Nanobiopesticides

Nanoparticles can be tracked down in nature, yet could be additionally produced through man-made actions. Modified NPs have been produced in different disciplines of science such as biotechnology, mechanical engineering, chemistry, and electronics. Silver nanoparticles are utilized for their ability to conduct electricity, low sintering temperature and stability. Silver nanoparticles can be applied in biomedical equipment, antibacterial encapsulation, consoles, clothing materials, and wound dressings. NPs can be formulated from different compounds. Some of these compounds are dendrimers, organic particles, polymers, metallic oxides, silicates, semiconductor quantum dots, and ceramics, and so on. Natural origins sources like microorganisms (bacteria, fungi, viruses, and fungi), and plants are also used to produce nanoparticles. In a few organic systems, the significant uses of NPs could be above hundred nanometers also. Use of NPs was also tracked down to crop development and protection.

Nanoparticles have been utilized for composts, insecticides, pesticides and organic material conveyance in sensors and pesticide debasement. Few of the promising improvements in the utilization of nanotechnology in farming include the synthesis of nanocides, controlled delivery of nanofertilizers, nanosensors for the recognition of pathogenic microorganisms and pesticides for the elimination or degradation of agrochemical to enhance soil fertility [83, 84]. Results obtained from several researchers show that NPs has the ability to enhance the steadiness or stability of pesticides produced from synthetic substances in the environment, i.e., from rain, heat and sun. Pesticides encapsulation is an appealing decision for protection on the field. Nanospheres are used to assess the enhanced conveyance of pesticides through the phenylpyrazole group [85]. Permeable SiO_2 NPs are demonstrated to defend avermectin against Ultra Violet inactivation [86]. Additionally, nanoclays are demonstrated to serve as a source of protection to agrochemicals [87]. The utilization of polymeric NPs may enhance the release of chlorothalonil and tebuconazole, fungicides utilized for strong wood conservation [88]. Pesticides release might be upgraded by utilizing nanocapsules. Polymeric nanocapsules are utilized for successful release of the pesticides ivermectin and acetamiprid [89]. Nanosized Al_2O_3, Fe_2O_3, SiO_2, or TiO_2 are proposed to be utilized as pesticide transporters for enhanced biological activity and furthermore residue reduction [90]. Apart from their actions on large scale of pests and insects, Ag NPs, ZnO NPs and SiO_2 NPs are also said to be effective in the reduction of environmental pollution.

ZnO nanoparticles coated with *B. thuringiensis* (19–20 nm) have an effect in the fertility larval improvement phase, and also death rate when tried on *C. maculatus*. Observation for the total death rate was made at 25 μg/ml. Moreover, a defer occurred in the whole growth phase of the insects. As much as 84 percent decline in decay illness was found with the guide of glucan NPs produced from *P. aphanidermatum* [91]. The nanoformulation utilized delayed the larval stage by 92 hours.

A synergistic impact of SiO_2 NPs with *Bacillus thuringiensis* has been tracked down to the circumstance of *C. elegans*, in which *Bacillus thuringiensis* chitinase immobilized onto SiO_2 NPs was assessed for their ability to destroy nematodes [92].

The utilization of nanocapsules in addition to nanospheres delivered from need seed remove on the prewinter armed force worm, *S. frugiperda*, brought about larval mortality. As much as 48.75% alongside 45% death rate was seen with nanosphere-based and nanocolloidal-applications, individually [93]. The best usage of *M. verrucuria*, a nonlignolytic parasite, as a biological control delegate is shown. It produces extracellular hydrolytic catalysts which might possibly hydrolyze the bug fingernail skin and the parasitic cell divider. The dissolvability and adsorption of the follicle debasing/mycolytic compound of *M. verrucaria* was tried. The protein complex was found to attempt to oversee *H. armigera* pervasion [94]. Undefined nanosilica from different biological origin displayed the capability that is perfect just like a biological pesticide by upsetting the cuticular layer and destroying the insects [39]. Nanosilica particles initiated hundred percent death rate when utilized on the rice weevil, *S. oryzae* [95].

3.10 Conclusion

The utilization of biopesticides have been accepted in the global world, whether used with traditional pesticides or used singly. However, one of the incessant issues farmers encounter in agriculture is the issue of pesticide pollution. The overuse of synthetic pesticides has been said to have a critical function in the lessening of the resources found in the soil, enhancement of pest resistance and also affects food quality. It also has negative impacts on the food chain. The production of nanobiopesticides is quite related to products synthesized from plants and microorganisms. This review shows the effect of nanobiopesticides on delaying insect growth or even death of the insects. To understand the advantages of nanobiopesticides at different stages, a lot of research needs to be done.

References

1. Casida, J.E. and Quistad, G.B., Insecticide targets: Learning to keep up with resistance and changing concepts of safety. *Agric. Chem. Biotechnol.*, 43, 185–191, 2000.
2. Misra, A.N., Misral, M., Singh, R., Nanotechnology in agriculture and food industry. *Int. J. Pure Appl. Sci. Technol.*, 16, 2, 1–9, 2013.
3. Prasann, B.M., *Nanotechnology in agriculture*, ICAR National Fellow, Division of Genetics, IARI, New Delhi, 2007.
4. Mousavi, S.R. and Rezaei, M., Nanotechnology in agriculture and food production. *J. Appl. Environ. Biol. Sci.*, 1, 414–419, 2011.
5. Brecht, M., Datnoff, L., Nagata, R., Kucharek, T., *The role of silicon in suppressing tray leaf spot development in St. Augustine grass*, pp. 1–4, University of Florida, USA, 2003.
6. Huang, B., Chen, F., Shen, Y., Qian, K., Wang, Y., Sun, C., Zhao, X., Cui, B., Gao, F., Zeng, Z., Cui, H., Advances in targeted pesticides with environmentally responsive controlled release by nanotechnology. *Nanomaterials (Basel)*, 8, 102, 2018.
7. Wang, Y.A., Li, J.J., Chen, H.Y., Peng, X.G., Stabilization of inorganic nanocrystals by organic dendrons. *J. Am. Chem. Soc.*, 124, 2293–2298, 2002.
8. Rai, M. and Ingle, A., Role of nanotechnology in agriculture with special reference to management of insect pests. *Appl. Microbiol. Biotechnol.*, 94, 287–293, 2012.
9. Margulis-Goshen, K. and Magdassi, S., Nanotechnology: An advanced approach to the development of potent insecticides, in: *Advanced Technologies for Managing Insect Pests*, pp. 295–314, Springer, Dordrecht, 2013.
10. Goswami, A., Roy, I., Sengupta, S., Debnath, N., Novel applications of solid and liquid formulations of nanoparticles against insect pests and pathogens. *Thin Solid Films*, 519, 3, 1252–1257, 2010.
11. Elek, N., Hoffman, R., Ravi, U., Resh, R., Ishaaya, I., Magdassia, S., Novaluron nanoparticles: Formation and potential use in controlling agricultural insect pests. *Colloids Surf. A Physicochem. Eng. Asp.*, 372, 66–72, 2010.

12. Chin, C.P., Wu, H.S., Wang, S.S., New approach to pesticide delivery using nanosuspensions: Research and applications. *Ind. Eng. Chem. Res.*, 50, 7637–7643, 2011.
13. Kashyap, P.L., Kumar, S., Srivastava, A.K., Sharma, A.K., Myconanotechnology in agriculture: A perspective. *World J. Microbiol. Biotechnol.*, 29, 191–207, 2013.
14. Adetunji, C.O., Akram, M., Samuel, M., Shahzad, K., Ayeni, A.E., Hasan, S., Adetunji, J.B., Hasan, S.M., Olaniyan, M., Muhibi, M.A., Polysaccharides derived from natural sources: A panacea to health and nutritional challenges, in: *Polysaccharides: Properties and Applications*, pp. 701–738, John Wiley & Sons, Inc, USA, 2021.
15. Adetunji, C.O., Palnam, W.D., Adetunji, J.B., Ubi, B.E., Ayeni, A.E., Thangadurai, D., Akram, M., Kham, F.S., Michael, O.S., Olaniyan, O.T., Temidayo, A.K., Islam, S., Adebisi, A.K., Hefft, D.I., Single cell sequencing of microbial species, in: *Sequencing Technologies in Microbial Food Safety and Quality*, USA, pp. 349–358, CRC Press, 2021.
16. Adetunji, C.O., Michael, O.S., Anani., O.A., Pharmafoods for body cleansing of toxic exposure to chemical and biological warfare agents, in: *Phytochemistry, the Military and Health*, pp. 239–255, Amsterdam, Elsevier, 2021.
17. Adetunji, C.O., Anani, O.A., Michael., O.S., Application of nanoceutical technology for fast and efficient control of illness, in: *Phytochemistry, the Military and Health*, pp. 497–508, Elsevier, Amsterdam, 2021.
18. Michael, O.S., Adetunji, C.O., Ayeni, A.E., Akram, M., Adetunji, J.B., Olaniyan, M., Muhibi., M.A., Marine polysaccharides: Properties and applications, in: *Polysaccharides: Properties and Applications*, pp. 423–439, John Wiley & Sons, Inc, USA, 2021.
19. Michael, O.S., Bamidele, O., Ogheneovo, P., Ariyo, T.A., Adedayo, L.D., Oluranti, O.I., Soladoye, E.O., Adetunji, C.O., Awobajo, F.O., Watermelon rind ethanol extract exhibits hepato-renal protection against lead induced-impaired antioxidant defenses in male Wistar rats. *Curr. Res. Physiol.*, 4, 252–259, 2021.
20. Adetunji, J.B., Adetunji, C.O., Olaniyan, O.T., African walnuts: A natural depository of nutritional and bioactive compounds essential for food and nutritional security in africa, in: *Food Security and Safety*, O.O. Babalola (Ed.), Springer, Cham, 2021.
21. Adetunji, C.O. and Fawole, O.B., *In-vitro* effects of crude extracts of *Jatropha curcas on Aspergillus niger* v. tieghem associated with post harvest rot of onion bulb. *Int. J. Phy. All. Sci.*, 1, 1, 1–5, 2012.
23. Adetunji, C.O., Fawole, O.B., Olaleye, O.O., Williams, J.O., Arowora, K.A., Adetunji, J.B., Abiodun, A.A., Kazeem, M.O., Antibacterial activity and phytochemical screening of extracts from the stem of Aspila Africana, in: *Science Focus*, vol. 17, pp. 116–121, Faculty of Pure and Applied Sciences, LAUTECH, Nigeria, 2012.

24. Akram, M., Michael, O.S., Saeed, M., Adetunji, C.O., Zahid, R., Adetunji, J.B., Laila, U., Mtewa, A.G., Ozolua, P., Egbuna, C., Ethnopharmacological properties of asian medicinal plants during conflict-related blockades, in: *Phytochemistry, the Military and Health*, pp. 53–68, Elsevier, Amsterdam, 2021a.
25. Akram, M., Adetunji, C.O., Laila, U., Michael, O.S., Samson, E.O., Kadiri, O., Ansari, R., Juliana Bunmi Adetunji, J.B., Ozolua, P., Mtewa, A.G., Egbuna, C., Overview of the traditional systems of medicine in different continents during postwar recovery, in: *Phytochemistry, the Military and Health*, pp. 37–52, Elsevier, Amsterdam, 2021b.
26. Grdiša, M., Babić, S., Periša, M., Carović-Stanko, K., Kolak, I., Liber, Z., Satovic, Z., Chemical diversity of the natural populations of dalmatian pyrethrum (*Tanacetum cinerariifolium* (Trevir.) Sch. Bip.) in croatia. *Chem. Biodivers*, 10, 3, 460–472, 2013.
27. Campos, M. and Phillips, T.W., Laboratory evaluation of attract-and-kill formulations against the Indianmeal moth, *Plodia interpunctella* (Hübner) (*Lepidoptera: Pyralidae*). *J. Stored Prod. Res.*, 2013, http://krex.ksu.edu.
28. Dubey, N.K., Shukla, R., Kumar, A., Singh, P., Prakash, B., Prospects of botanical pesticides in sustainable agriculture. *Curr. Sci.*, 98, 4, 479–480, 2010.
29. Prota, N., Bouwmeester, H.J., Jongsma, M.A., Comparative antifeedant activities of polygodial and pyrethrins against white flies (*Bemisia tabaci*) and aphids (*Myzus persicae*). *Pest Manage. Sci.*, 70, 4, 682–688, 2013.
30. Fei, Y., Chunhua, Z., Qiongbo, H., Meiying, H., The joint action of destruxins and botanical insecticides (Rotenone, Azadirachtin and Paeonolum) against the cotton aphid, aphis gossypii glover. *Mol.*, 17, 7533–7542, 2012.
31. Sandoval-Mojica, A.F. and Capinera, J.L., Antifeedant effect of commercial chemicals and plant extracts against *Schistocerca americana* (Orthoptera: Acrididae) and *Diaprepes abbreviatus* (Coleoptera: Curculionidae). *Pest Manage. Sci.*, 67, 7, 860–868, 2011.
32. Satta, A., Floris, I., Caboni, P., Cabras, P., Eguaras, M., Velis, G., New experimental data on use of rotenone as an acaricide for control of Varroa destructor in honey bee colonies. *J. Econ. Entomol.*, 101, 4, 1075–1080, 2008.
33. Regnault-Roger, C., Vincent, C., Arnason, J.T., Essential oils in insect control: Low-risk products in a high-stakes world. *Annu. Rev. Entomol.*, 57, 405–424, 2012.
34. Owolade, O.F., Ogunleti, D.O., Adenekan, M.O., Titanium dioxide affects disease development and yield of edible cowpea. *Agric. Food Chem.*, 7, 50, 2942–2947, 2008.
35. De Villiers, S.M. and Hoisington, D.A., The trends and future of biotechnology crops for insect pest control. *Afr. J. Biotechnol.*, 10, 23, 4677–4681, 2011.
36. Liu, F., Wen, L.X., Li, Z.Z., Yu, W., Sun, H.Y., Chen, J.F., Porous hollow silica nanoparticles as controlled delivery system for water-soluble pesticide. *Mater. Res. Bull.*, 41, 2268–2275, 2006.

37. Popat, A., Liu, J., Hu, Q., Kennedy, M., Peters, B., Lu, G.Q.M., Qiao, S.Z., Adsorption and release of biocides with mesoporous silica nanoparticles. *Nanoscale*, 4, 3, 970–975, 2012.
38. Stadler, T., Buteler, M., Weaver, D.K., Novel use of nanostructured alumina as an insecticide. *Pest Manage. Sci.*, 66, 577–579, 2010.
39. Debnath, N., Das, S., Seth, D., Chandra, R., Bhattacharya, S.C., Goswami, A., Entomotoxic effect of silica nanoparticles against Sitophilus oryzae (L.). *Pest Sci.*, 84, 99–105, 2011.
40. Barik, T.K., Sahu, B., Swain, V., Nano-silica from medicine to pest control. *J. Parlog Res.*, 103, 1, 253–258, 2008.
41. Liu, Y., Tong, Z., Prudhomme, R.K., Stabilized polymeric nanoparticles for controlled and efficient release of bifenthrin. *Pest Manage. Sci.*, 64, 808–812, 2008.
42. Joseph, T. and Morrison, M., *Nanotechnology in agriculture and food a nanoforum report*, 2006, www.nanoforum.org.
43. Scrinis, G. and Lyons, K., The emerging nano-corporate paradigm: Nanotechnology and the transformation of nature, food and agri food systems. *Int. J. Sociol. Food Agric.*, 15, 2, 1–23, 2007.
44. Bhattacharyya, A., Nanoparticles from drug delivery to insect pest control. *Akshar*, 1, 1, 1–7, 2009.
45. Harper, S., New approaches needed to gauge safety of nanotech-based pesticides, researchers urge. *Phys. Chem.*, 4, 33, 2010–2012, 2010.
46. Bhattacharyya, A., Bhaumik, A., Rani, P.U., Mandal, S., Epidi, T.T., Nanoparticles-a recent approach to insect pest control. *Afr. J. Biotech.*, 9, 3489–3493, 2010.
47. Ghormade, V., Deshpande, M.V., Paknikar, K.M., Perspectives for nanobiotechnology enabled protection and nutrition of plants. *Biotechnol. Adv.*, 29, 6, 792–803, 2011.
48. Sathiyabama, M. and Charles, R.E., Fungal cell wall polymer based nanoparticles in protection of tomato plants from wilt disease caused by *F. oxysporum f.sp.* lycopersici. *Carbohydr. Polym.*, 133, 400–407, 2015.
49. Manjunatha, S.B., Biradar, D.P., Aladakatti, Y.R., Nanotechnology and its applications in agriculture: A review. *J. Agric. Sci.*, 29, 1–13, 2016.
50. Seiber, J.N., Coars, J., Duke, S.O., Gross, A.D., Biopesticides: State of the art and future opportunities. *J. Agric. Food Chem.*, 62, 11613–11619, 2014.
51. Chandler, D., Bailey, A.S., Tatchell, G.M., Davidson, G., Greaves, J., Grant, W.P., The development, regulation and use of biopesticides for integrated pest management. *Philos. Trans. R. Soc. B.*, 366, 1987–1998, 2011.
52. Koul, O., *Naturally occurring insecticidal toxins*, p. 850, CAB International, Wallingford, 2016.
53. Koul, O., Walia, S., Dhaliwal, G.S., Essential oils as green pesticides: Potential and constraints. *Biopestic. Int.*, 4, 63–84, 2008.
54. Chowdhury, K., Kumar, A., Sharma, S., Pathak, R., Jangir, M., Ocimum sp.: Souce of biorational pesticides. *Ind. Crop Prod.*, 122, 686–701, 2018.

55. Koul, O., Microbial biopesticides: Opportunities and challenges. *CAB Rev.*, 6, 056, 1–26, 2011.
56. Glare, T., Caradus, J., Gelernter, W., Jackson, T., Keyhani, N., Kohl, J., Marrone, P., Morin, L., Stewart, A., Have biopesticides come of age? *Trends Biotechnol.*, 30, 250–258, 2018.
57. Anderson, A.J. and Kim, Y.C., Biopesticides produced by plant-probiotic pseudomonas chloraphis isolates. *Crop Prot.*, 105, 62–69, 2018.
58. Olson, S., An analysis of the biopesticide market now and where it is going. *Outlook Pest Manage.*, 26, 203–206, 2015.
59. Eze, S.C., Mba, C.L., Ezeaku, P.I., Analytical review of pesticide formulation trends and application: The effects on the target organisms and environment. *Int. J. Sci. Environ. Technol.*, 5, 253–266, 2016.
60. Chinnamuthu, C.R. and Boopathi, P.M., Nanotechnology and agroecosystem. *Madras Agric. J.*, 96, 17–31, 2009.
61. *Nature,* 2018, https://www.nature.com/subjects/nanoparticles (Accessed April 18, 2018).
62. Hameed, A. and Al-Samarrai, M., Nanoparticles as alternative to pesticides in management plant diseases—A review. *Int. J. Sci. Res. Pub.*, 2, 261–264, 2012.
63. Perea-de-Lugue, A. and Rubiales, D., Nanotechnology for parasitic plant control. *Pest Manag. Sci.*, 65, 540–545, 2009.
64. OECD Allianz, *Opportunities and risks of nanotechnologies*, pp. 1–44, OECD International Futures Programme, 2008, https://www.oecd.org/science/nanosafety/44108334.pdf.
65. Ragaei, M. and Sabry, A.H., Nanotechnology for insect pest control. *Int. J. Sci., Env. Tech.*, 3, 2, 528–545, 2014.
66. Bergeson, L.L., Nanosilver pesticide products: What does the future hold? *Environ. Qual. Manag.*, 19, 73–82, 2010.
67. Lade, B.D., Gogle, D.P., Nandeshwar, S.B., Nano-biopesticide to constraint plant destructive pests. *J. Nanomed. Res.*, 6, 1–9, 2017.
68. Rajakumar, G. and Rahuman, A.A., Larvicidal activity of synthesized silver nanoparticles using Eclipta prostrata leaf extract against filariasis and malaria vectors. *Acta Trop.*, 118, 196–203, 2011.
69. Kantrao, S., Ravindra, M.A., Akbar, S.M.D., Jayanthi, P.D.K., Venkataraman, A., Effect of biosynthesized silver nanoparticles on growth and development of helicoverpa armigera (Lepidoptera: Noctuidae): Interaction with midgut protease. *J. Asia Pac. Entomol.*, 20, 583–589, 2017.
70. Jianhui, Y., Kelong, H., Yuelong, W., Suqin, L., Study on anti-pollution nano-preparation of dimethomorph and its performance. *Chin. Sci. Bull.*, 50, 108–112, 2005.
71. Bouwmeester, H., Dekkers, S., Noordam, M.Y., Hagens, W.I., Bulder, A.S., de Heer, C., ten Voorde, S.E.C.G.S., Wijnhoven, W.P., Marvin, H.J.P., Sips, A.J.A.M., Review of health safety aspects of nanotechnologies in food production. *Regul. Toxicol. Pharmacol.*, 53, 52–62, 2009.

72. Bordes, P., Pollet, E., Avérous, L., Nano-biocomposites: Biodegradable polyester/nanoclay systems. *Prog. Polym. Sci.*, 34, 125–155, 2009.
73. Bergeson, L.L., Nanosilver: US EPA's pesticide office considers how best to proceed. *Environ. Qual. Manag.*, 19, 79–85, 2010a.
74. Elek, N., Hoffman, R., Raviv, U., Novaluron nanoparticles: Formation and potential use in controlling agricultural insect pests. *Colloid Surf A*, 372, 1–3, 66–72, 2010.
75. Lade, B.D., Gogle, D.P., Nandeshwar, S.B., Nano-biopesticide to constraint plant destructive pests. *J. Nanomed. Res.*, 6, 1–9, 2017.
76. Benelli, G., Canale, A., Toniolo, C., Higuchi, A., Murugan, K., Pavela, R., Nicoletti, M., Neem (Azadirachta indica): Towards the ideal insecticide? *Nat. Prod. Res.*, 31, 369–386, 2016.
77. Ingale, A.G. and Chaudhari, A.N., Biogenic synthesis of nanoparticles and potential applications: An eco-friendly approach. *J. Nanomed. Nanotechol.*, 4, 165, 2013.
78. Iravani, S., Bacteria in nanoparticle synthesis: Current status and future prospects. *Int. Sch. Res. Notices*, 2014, 1–18, Article ID: 359316, 2014.
79. Ragaei, M. and Sabry, A.H., Nanotechnology for insect pest control. *Int. J. Sci. Environ. Technol.*, 3, 528–545, 2014.
80. Nuruzzaman, M., Rahman, M.M., Liu, Y., Naidu, R., Nanoencapsulation, nano-guard for pesticides: A new window for safe application. *J. Agric. Food Chem.*, 64, 1447–1483, 2016.
81. Huang, B., Chen, F., Shen, Y., Qian, K., Wang, Y., Sun, C., Zhao, X., Cui, B., Gao, F., Zeng, Z., Cui, H., Advances in targeted pesticides with environmentally responsive controlled release by nanotechnology. *Nanomaterials (Basel)*, 8, 102, 2018.
82. Wilkins, R.M., Controlled release technology, agricultural, in: *Kirk Othmer Encyclopedia of Chemical Technology*, 5th Ed, A. Seidel (Ed.), John Wiley & Sons, New Jersey, 2004.
83. Chormade, V., Deshpande, M.V., Paknikar, K.M., Perspectives for nanobiotechnology enabled protection and nutrition in plants. *Biotechnol. Adv.*, 29, 792–803, 2011.
84. Bipin, D.L., Dayanand, P.G., Sukhdev, B.N., Nano biopesticide to constraint plant destructive pests. *J. Nanomed. Res.*, 6, 1–9, 2017.
85. Boehm, A.L., Martinon, I., Zerrouk, R., Rump, E., Fessi, H., Nanoprecipitation technique for the encapsulation of agrochemical active ingredients. *J. Microencapsul.*, 20, 433–441, 2003.
86. Li, Z.Z., Chen, J.F., Liu, F., Liu, A.Q., Wang, Q., Sun, H.Y., Study of UV shielding properties of novel porous hollow silica nanoparticle carriers of Avermectin. *Pest Manag. Sci.*, 63, 241–246, 2007.
87. El-Nahhal, Y., Nir, S., Marguiles, L., Rubin, B., Reduction of photodegradation and volatilization of herbicides in organoclay formulations. *Appl. Clay Sci.*, 14, 105–119, 1999.

88. Liu, Y., Yan, I., Heiden, P., Laks, P., Use of nanoparticles for controlled release of biocides in solid wood. *J. Appl. Polym. Sci.*, 79, 458–465, 2001.
89. Zheng, H. and Shang, Q., Water suspension acetamiprid nano capsule preparation and its repairing method. CN1491558. *Chem. Abstr.*, 143, 73729, 2005.
90. Wang, X., Wang, Y., Xiong, X., Li, T., Liang, J., Chen, J., Aqueous nano insecticide suspension and its preparation process. CN1486606. *Chem. Abstr.*, 142, 213751, 2004.
91. Sathiyanarayanan, A. and Muthukrishnan, S., Protection of turmeric plants from rhizome rot disease under field conditions by -d-glucan nanoparticle. *Int. J. Biol. Macromol.*, 77, 9–14, 2015.
92. Qina, X., Xianga, X., Suna, X., Ni, H., Li, L., Preparation of nanoscale Bacillus thuringiensis chitinases using silica nanoparticles for nematicide delivery. *Int. J. Biol. Macromol.*, 82, 13–21, 2016.
93. Giongo, A.M.M., Vendramim, J.D., Forim, M.R., Evaluation of neem-based nanoformulations as alternative to control fall armyworm. *Cienc. Agrotechnol.*, 40, 26–36, 2016.
94. Chavan, S.B., Vidhate, R.P., Kallure, G.S., Dandawate, N.L., Khire, J.M., Deshpande, M.V., Stability studies of cuticle degrading and mycolytic enzymes of myrothecium verrucarian for control of insect pests and fungal phytopathogens. *Indian J. Biotechnol.*, 16, 404–412, 2017.
95. Debnath, N., Das, S., Seth, D., Chandra, R., Bhattacharya, S.C., Goswami, A., Entomotoxic effect of silica nanoparticles against sitophilus oryzae (L.). *Pest Sci.*, 84, 99–105, 2011.

4

Application of Plant-Based Nanobiopesticides as Disinfectant

Babatunde Oluwafemi Adetuyi[1]*, Peace Abiodun Olajide[1], Oluwakemi Semiloore Omowumi[1] and Charles Oluwaseun Adetunji[2]

[1]Department of Natural Sciences, Faculty of Pure and Applied Sciences, Precious Cornerstone University, Ibadan, Oyo State, Nigeria
[2]Applied Microbiology, Biotechnology and Nanotechnology Laboratory, Department of Microbiology, Edo University Iyamho, Auchi, Edo State, Nigeria

Abstract
Most plants are affected by pest globally, and this has led to the continuous use of manures, fertilizers and some other chemicals, which are in the form of pesticides, which affects plants, animals, and the immediate environment, by causing serious health challenge to both citizens and farmers. Due to the misuse of some of this biopesticides, negative effects like the pest building resistance to it, reduction in the fertility of soil. Hence, this led to the emergence of new products to mitigate this situation. Nanotechnology and nanoscience, which are new disciplines, have been linked to a lot of other fields, most especially in the agricultural, chemical, protection of plants, and crops in order to control pest to an extent. Recent developments include nanoparticle-mediated gene or DNA transfer in plants, biosensors for distant sensing, and nanobiopesticides and insecticides in various formulations, including those that are encapsulated. In this chapter, the application of plant based nanobiopesticide, basically used for plants as disinfectants are being reviewed extensively and a deep analysis is also being done in relations to pest in field and the use of various known solutions and nanobiopesticides for the protection of crops and plants in general.

***Keywords*:** Nanobiopesticides, disinfectant, plant, pest, crop protection

**Corresponding author*: badetuyi@pcu.edu.ng

Charles Oluwaseun Adetunji and Julius Kola Oloke (eds.) Handbook of Agricultural Biotechnology: Volume I Nanopesticides, (63–130) © 2024 Scrivener Publishing LLC

4.1 Introduction

The role of agriculture to ensure a good life has been deduced to serve as a strong piller to provide food and develop the economy of a country [1, 2, 4]. But of recent, the agricultural sector is faced with a lot of challenges and situations like change in climate, soil contamination by a lot of pollutants that care known to be dangerous to the environment like fertilizers and pesticides, thereby increasing or making outrageous, the demand for food globally, viz-a-viz the rapid increase in the world population [7–11]. In the just published report of the United Nation about the increase in world population to over 8.5 billion and 9 billion by 2030 and 2050 respectively, then there must been an urgent set down goal to meet over 50 % of the demands of the populace in the provision and production of adequate food by the agricultural sector [12–16]. Although, the thought of industrialization, also helping in the provision of this demand has been the focus of the government but this has drastically led to the drastic reduction of some natural resources like oceans, forest and ecological bioderversity, that most countries and populace depend on, for their livelihood [17–20, 167]. Hence, seeking the need for continuous production of crops and securing great security for foods. So as to maintain the true balance between the production of food and the environment, ecological biodiversity plays a vital role, because it helps to build the resilience of agriculture and the stress induced by the environment which leads to crop failure [153, 154, 156]. Livestock, crops, forestry and acquaculture systems are the type of specific food and agriculture biodiversity that sustain human being [23–26, 41], while most animals and plant species that depend on the global population for fuel, fibers, and food uses ocean and forest for their home. Therefore, food security is affected by issues that cause disequilibrium in ecology [35, 38, 285]. Hence, there are novel methods being invented by several researchers to tackle this important issue, and among these methods used in revolutionizing some of the new agriculture practice is the use of nanomaterial (NM) based product [102–104]. These materials have some unique antioxidant properties that makes them very important and useful for their respective function. Chemistry, pharmaceutical research, diagnosis, and therapies are a few of the industries using nanotechnology [96, 98, 186, 320]. The leading nanotechnological technique is the nanoparticles (NPs), that are inserted or incorporated into a lot of products consumed by humans because of their respective ability of exhibiting some phytochemical properties when compared to their other products [91, 93, 94, 159], hence making this novel properties of nanoparticles important in the food and agricultural sector [107, 110]. Agriculture of this century is changing into another form of precision via this nanoformulations that functions to acquire large output from the resources that are available. Examples of nanomaterials that has a novel surface chemistry, optical properties and electrical properties include nanoclays, nanotubes, and nanowires [365] and consequently, they have improved detection limits, functional sensitivity, and fast response times [85, 86, 88, 177]. In order to increase the production of crop, fertilizer is needed but studies have shown that it also lowers the fertility of soil by disaggregating the mineral balance in the soil [117, 120, 312]. These pesticides are typically sprayed on crops, which results in a surfeit impact. It should also be noted that the cost of producing these pesticides are extremely high, hence it needs to be controlled [290–293]. The agricultural use of nanomaterial focus on the reduction in the loss of nutrient in order to increase the end yield, and the amount of product for the protection of plants [28–30, 337], and maximize output while reducing production costs. We establish and provide evidence for the assertion that nanomaterials are a superior means of improving agriculture and food systems [44, 49, 50]. We looked at the importance of assessing the level of toxicity of products made with NM before releasing them into the market [52–54]. Nanotechnology-based products are being developed all the time, such as intelligent agrochemical delivery systems that use nanomaterials to transport active chemicals [58–60]. Innovative bionanocomposites with enhanced mechanical and physical properties can be created from agrowaste materials for application in industry [60–62]. Still needed are improvements in process design, risk evaluation of nanopesticides and nanofertilizers, and regulations for the commercialization of nanoagroproducts in order to fully reap the benefits of nanotechnology [32, 67, 83]. We are adamant that a broad perspective Pest animal populations are insects, which can be found in a wide range of environments and species. The many insects harm crop plants and act as carriers of various diseases [66, 69, 71]. These have an impact on the economy, crop plant yields, and domestic and foreign markets. Several pesticides have been tried to control insects in agriculture to reduce agricultural output losses caused by them [72, 73, 270]. One of the main factors affecting agricultural output is insect pests, which results in an annual loss of billions of dollars [80, 82, 280]. For the majority of crop plants, the two primary types, larva and adult, have proved the most deadly [122–125]. For instance, the hazardous pest *Helicoverpa armigera* damages horticultural and field crops all around the world. It has been severely prevalent in South America, Asia, Europe, Africa, Oceania, and Asia [77, 79, 325]. Due to 150 different insect attacks at different life phases, the cotton yield is low [12, 130–133]. The need for researcher to gather both conventional and cutting-edge methods to deal with distinct pest concerns globally has arisen as a result of this [133–137]. The contact between the insect and the host plant causes the plant's immune system to become active, and different production pathways for both primary and secondary metabolites interact with one another [138–140]. The cell wall serves as a barrier to insects, but when it becomes permeable to deadly insects, the plant is ultimately destroyed [143–145]. When an insect interacts with the plant, certain defence mechanisms are engaged by the secondary metabolites produced [152, 160, 163]. The metabolites of the phenylpropanoid pathway play a significant role in the synthesis of several metabolites with distinct properties [164, 170, 171]. The plant's secondary metabolites aid it in a number of ways, including I defense against herbivore and pathogen attack, enticement of pollinators and symbionts, and communication between plants [111, 178, 180, 182]. Consequently, biopecticides based on secondary metabolites are a natural source that might be used to target pests and are safe for the environment [183, 187, 191, 192]. Unlike synthetic pesticides, these biopecticides are natural, low-risk, and environmentally beneficial. Therefore, the main strategy for reducing their population and preventing agricultural losses is the study of plants that are potentially resistant to insect pests [195–198].

It is possible to dissect weed plants and other nonagricultural plants to find numerous secondary metabolites that are beneficial for pest control. Some plants have the innate ability to combat pests; their extracts display several insect-resistant chemicals, like b-caryophyllene, limonene, and camphor from *Ocimum kilimandscharicum* [87, 206–209]. Neem extract was developed for its larvicidal and pesticidal properties [210–213, 247]. The weed plants *Calotropis procera, Dathura, Catharanthus roseus,* and *Argimone maxicana* contain phytochemicals and secondary metabolites that are crucial to reducing insect populations [222–225, 248]. By combining diverse secondary metabolites in varied ratios, it is attempted to create a biopesticide with effective insect killing properties [226–228]. However, the best-selling biopesticide is now at risk of insect resistance due to its frequent use [229–231]. It has made it possible for researchers and scientists around the globe to put together secondary metabolites that are effective against insect pests [218–221]. With the aid of current technologies, the other pest-repelling option concealed in nature must become available. The ideal alternative is to utilize a greener method that causes less pollution and is less hazardous to plants and animals [338, 344, 388]. The Green Chemistry topic, which starts with green, all-natural materials and creates nanosized particles capped with pest-resistant secondary metabolites, may offer a remedy for faulty pest in farms [214–217].

Table 4.1 Pesticide's classification based on their target.

Types of pesticides	Targetted pest
Nematicides	Nematodes (roundworms)
Rodenticides	Rodents
Piscicides	Fish
Insecticides	Insect
Avicides	Birds
Molluscicides	Snail and slugs
Bactericides	Bacteria
Fungicide	Fungi
Herbicide	Plants
Algicides	Algae

Other pesticides are effective against a number of pests and available in a wide range of varieties. The Latin verb "to kill" is where the term "cide" comes from [116, 234–237]. In various nations, there are tens of thousands of independent producers, formulators, and manufacturers of pesticides. The chemicals of pesticide products typically include both active agents and inactive substances. The other substances contribute in a variety of ways, such as by luring the insect, distributing the chemically active ingredients, and/or reducing drift. The active ingredients are the primary component that kills pests [116, 244, 245]. As stated in Table 4.1, pesticides are categorized based on their goals.

4.2 The Need for Biopesticides Worldwide

In today's society, developing ecofriendly pesticides that can be manufactured with a high level of efficiency for a little investment is of utmost importance [251, 254, 255]. The formulation of the ecofriendly bio pesticides and farmers' demand for them as a product both increased [257, 261–263]. The world's governments continuously work to produce effective biopesticides [266]. The market for synthetic and biological pesticides was noted to have increased from \$54.8 to \$61.8 billion in 2013 to 2014 and is anticipated to reach \$83.7 billion by 2019 [185, 271, 274–277]. Additionally, it is anticipated that by 2019, the global market for biopesticides will reach USD 6.9 billion. The Asia-Pacific area is predicted to have a 20% demand for biopesticides, but North America is predicted to have a 40% market share [282, 283, 310, 314]. Many nations rely on biological insecticides for agricultural biotechnology pest management to ensure food security [166, 304, 309]. As a result, sales of bio pesticides have surged in comparison to pesticides made of chemicals [299]. Chemical pest control (traditional trend): Local pest stores, internet retailers like Amazon, as well as a number of manufacturers, distributors, and exporters all carry the most popular chemical pesticides [315, 322–324]. Early sulfur-containing chemicals included phosphate fertilizers, sulfuric acid, fungicides, and insecticides. In addition to sulfur, some of the typical active chemical components are boricacid, abamectin, cyfluthrin, fipronil, permethrin, and bifenthrin [328, 329]. Pests with chemical bases, such as organophosphates, carbamates, organochlorines, pyrethroids, and neonicotinoids, are categorized based on their chemical makeup [334, 336]. A number of major compounds that are used as pesticides are listed in Table 4.2 along with their effects on humans, animals, plants, and other insects [339, 347, 349]. Only a handful of the numerous insecticides that are offered for sale in the U.S. The usage of various chemical insecticides, pesticides, and synthetic insecticides, such as pyrethroids, has increased as a result of pest management. Due to constant exposure to synthetic pyrethroids, the insect has developed a tolerance that allows it to survive [101, 350, 351]. Therefore, a pest on crop plants cannot be managed by using conventional chemical-based insecticides. Phytochemicals, secondary metabolites, and biologically generated natural molecules from plants are being used as an alternative to pesticides to protect plants [354, 356, 366]. The effectiveness of the chemically based nanoparticles against larvae is tested after they are manufactured and used. In trials with Egyptian cotton leafworm *Spodoptera littoralis* larvae, the water-insoluble insecticide novaluron's nanoparticles, which have a size range of 30 to 100 nm, exhibit toxicity [367–370]. However, because the results of the experiments were dangerous substances, the chemicals were not chosen for the development of nanoparticles in the agriculture chemicals sector [118].

Biologically based pest control (old trend): Plant-derived phytochemicals have been used as biopesticides due to their nontoxicity to both humans and animals. Alkaloids, terpenoids, flavonoids, phenol, polyphenons, glycosides, and tannins are examples of secondary plant metabolites having a variety of bioactivities against dangerous insects. The pest Helicoverpa

Table 4.2 The chemical composition and impact on humans, other animals, and insects.

Chemicals	Effect	Activity
Organophosphate	Humans, other animals, and insects	By disrupting the nerve and brain systems, It can kill insects.
Carbamates	They decompose faster and provide less of a threat to people.	By disrupting the nerve and brain systems, It can kill insects.
Organochlorines Dichlorodiphenyl-trichloroethane	Discovered in soil, animal, and human bodies.	Persistent pesticide harms human endocrine systems, damages genes, and increases the risk of cancer.
Pyrethroid	Less impact on people	Numbness, itchiness, burning, stinging, and tingling sensations.
Neonicotinoid	Nicotine is a plant chemical that affects numerous animals, including mammals, and is the source of neonicotinoids.	Obstruct an insect neural system pathway and play a part in honeybee colony collapse disorder

armigera has been somewhat controlled by these biopesticides based on secondary metabolites [233]. Numerous research studies have shown that powdered medicinal plant extracts are effective against pests including mosquitoes and Helicoverpa armigera, among others. Numerous medicinal herbs, including neem extract [5], *Annona squamosa, Vitex negundo,* and *Acorus calamus* [246], *Gnidia glauca, Toddalia asiatica* [6], *Argimone maxicana*, and *Calotropis procera*, have been [43]. The study found that neem seed kernel extract (30.08%) and tobacco leaf extract (28.68%) had the highest percent infection decrease. The performance of neem seed kernel extract (NKAE) as an antifeedant, ovicidal, nymphal duration, and hatching agent on the tea mosquito bug was evaluated in a lab setting (*Helopeltis theivora*). The effectiveness of the extracts made from plant materials such fruit stems, roots, fruits, and leaves has been tested against insect pests. The extract contains a number of bioactive metabolites with pesticide properties. However, the uncontrolled use of biological pesticides made from plant components would result in the development of resistance, and new products to combat the pest problem would start to appear. Therefore, it is necessary to investigate another scientific option to reduce the impact of pests on crop plants [372–375].

Nanobased pest control is a recent development. An extensive amount of nanotechnology has been proposed, and it has been used in numerous formulations to make a variety of novel goods with several applications in the agrochemical, textile, paper, food, fertilizer, pesticide, plant protection, nutrition paint, and paint sectors [380]. Thus, as compared to conventional pesticides, nanobased innovative nanopesticides like Ag, Cu, SiO_2, ZnO, and nanoformulations are more effective at protecting against a wide range of pests while also lowering pollution levels in water, soils, and the environment. Zn metal is a significant target for the production of Zn nanoparticles since it is a crucial nutrient for plants but is less readily available in soil. This makes Zn nanoparticles useful for controlling pests and good for plants. ZnO NP are affordable and secure. They can be applied to a wide range of industries, such as those that deal with agrochemicals, anticancer, antidiabetic, antibacterial, and antifungal qualities. Silver could be a different option. Silver (Ag), which has greater effectiveness and activity, is the green method's preferred target for antibacterial, microbial, fungal, larvicidal, pesticidal, anti-inflammatory, antiplatelet activity, anti-angiogenesis, and antiviral activity in biological systems, living organisms, medicine, plant management, pest control, and agricultural aspects. ZnO nanoparticles exhibit antifungal action against the plant disease *Fusarium graminearum* [249]. The advantages of using metal nanoparticles are their stability, sluggish kinetics, capacity to operate at room temperature, and formation of ecofriendly byproducts. Therefore, the greatest alternative for creating pest-control tools for hazardous insects could be a nanoformulation of biopesticide. The created bio nanoparticles (shaped like nanocorns) were put in a spray bottle with a predetermined amount of water before being applied to the pest insect *H. armegera*. In order to control the mosquito, aqueous extract from *Eclipta prostrata* was used to create silver nanoparticles (AgNPs) with larvicidal action [247].

4.3 Structures for Possible Nanobiopesticides

Researchers are working hardest in the field of plant protection to develop practical alternatives to synthetic commercial pesticides. Traditional methods are more effective when combined with nanoscience since they are more soluble, release energy more slowly, and don't degrade as quickly [239]. The biological substances serve as capping and reducing agents for the silver salt, which results in the production of stable nanoparticles. These nanobiopesticides or nanomaterials can be used directly or indirectly as pesticides due to their fundamentally intrinsic features like stiffness, permeability, crystallinity, thermal stability, and biodegradability. These qualities make them preferable to chemical or conventional pesticides. Insects may experience dyspepsia when exposed to nanocarrier materials that contain certain pesticide secondary metabolites [391].

4.4 Plant Interactions Between NPs

Science and technological advances help the agricultural industry by supplying us with fresh approaches and answers to challenging issues. With the development of nanotechnology, nanoformulations for sustainable agriculture are regularly generated that are more effective and contain less contaminants [114]. After entering the intricate plant-soil system, these modern materials have the capacity to change the physiology of plants, which can be used to comprehend the impacts. Additionally, crucial understanding of these NMs' interactions, whether favorable or unfavorable, with plants is necessary for the regulated delivery of active chemicals. They might therefore open up fresh paths for creating superior nanomaterial-based goods. Additionally, it is thought that NM concentrations in the natural environment are far lower than those that are deemed dangerous. However, there are still some loopholes that require thorough safety analyses to close [51].

4.5 Systemic Method for NMs Selectivity: Uptake and Interaction Based on Physicochemical Properties

The uptake and transformation of NPs are influenced by a number of parameters, including plant physiology, the NP's own physiochemical characteristics, size, surface charge, shape, and potential interactions with plants [260].

4.6 NP Uptake Dependent on Size

The size of the NPs must be taken into account as a crucial parameter to study uptake in plants because many barriers found inside the plants are in the micrometer (mm) to nanoscale (nm) range. Like, the foliar epidermis contains the cells that make up the cuticle membrane. When the stomata in the epidermis open to allow for gaseous exchange, they produce a pore that is 3–12 mm wide and 10–30 mm long. As a result, NPs can move through the plants through these stomatal openings. The stomata's trichrome and the cuticle layer of the epidermis also exhibit extremely different permeability characteristics. However, the cuticle layer is more noticeable and has a size exclusion limit in the nm range on the leaf epidermis [346]. It has been documented that fluorescently tagged NPs larger than 50 nm can collect in the epidermis behind the cuticle, where stomata are absent. According to one study, NPs with a size range of 4 to 100 nm can disrupt the waxy layer to pass through the cuticle [235]. When compared to particles of 1 m, only the stomata of *Vicia faba* leaves can accommodate polymeric NPs with a diameter of 43 nm [100]. NPs that enter through the stomata frequently end up being deposited on the substomatal cavity's cell wall. It was feasible to identify the potential penetration of tiny NPs, such as 20 nm Fe_3O_4 NPs, in the *Nicotiana benthamiana* plant by TEM tests [60]. These plant cell walls, which are normally in the size range of 3.5 to 20 nm and most frequently around 5 nm, were the main source of the major size exclusion limit [65, 256, 333]. Because of this, undamaged cells' cell walls are easily penetrated by NPs smaller than 5 nm. SWCNT was able to develop an endocytosis-like structure in the entire Arabidopsis thaliana plant cell wall in one experiment [307]. The cell wall states that NPs with sizes up to a particular threshold can pass through it. Because quantum dots (QDs) fall within the range of plant cell wall pores (sub-10 nm), they can easily enter plants [359]. Lettuce leaves contained AgNPs with diameters ranging from 24.8 to 38.6 nm when administered by the foliar technique. Additionally, they demonstrated how $AgNO_3$ changed in plants when exposed to AgNPs, demonstrating another aspect of the biotransformation phenomena [190].

Large-size pores are created when NPs interact with the cell wall, which makes it easier for NPs to enter the body [63]. Despite findings on the accumulation of larger sized NPs in the cell wall and subsequently in the cytoplasm, a discrepancy between the reported pore size of the cell wall (depends on several factors) and the accumulation of larger sized NPs is still maintained. As a model system, plant protoplasts have been used in studies on NP uptake via the plant cell membrane. Additionally, this avoids the technical challenges associated with measuring the uptake of NPs inside plants, which are primarily brought on by internal plant cell fluorescence and the impermeability that plants impose to some staining dyes [335]. It is crucial to understand the path that NPs take after entering plants since it provides insight into potential accumulation sites. Two important channels for NPs to ascend and descend within a plant are the apoplast and the symplast. The symplastic pathway facilitates movement into the cytoplasm of nearby cells, whereas the apoplastic pathway allows flow through extracellular gaps involving the cell walls of surrounding cells and xylem arteries [288]. Cell to cell mobility is made possible by the plasmodesmata, which serve as a cytoplasmic bridge for particle transport [296]. They are surrounded by 3-nm particles and range in diameter from 20 to 50 nm [89].

4.7 Surface Charge-Related NP Uptake

Because they are ingested by, absorbed by, and traded among plants, NPs enjoy a high degree of freedom. The NMs' physical–chemical interactions with plants, the surface receptors, transporters, and specific membrane proteins are affected by their energy and surface charge [165]. In contrast to their pure counterparts, NPs with varied surface charges have different capacities for aggregation and surface characteristics [147].

Milewska-Hendel *et al.* convincingly show how AuNPs are taken up by roots and transported from roots to shoots in a surface charge-dependent manner [219]. Negatively charged NPs are expected to facilitate symplastic and apoplastic trafficking into the vascular system via roots and shoots. As a result, surface charge alteration greatly alters the NPs' ability to move across plants. Based on their physical–chemical characteristics, NPs and agrochemicals have an impact on plant health that can be both favorable and harmful [157].

4.8 Differences in Anatomy and Application-Related NP Uptake

The surface charge of the NPs and the variations in a plant's anatomy are strongly connected. Dicot plants, such as lettuce and tomato, have a better association of positively charged cerium oxide NPs due to their larger root surface area. However, neutral and negatively charged NPs show improved root-to-shoot translocation [316]. Additional evidence that anatomical and physiological variations in plants result in different NM uptake patterns comes from the observations that different crops belonging to different species exhibit a varying pattern of accumulation and absorption of distinct classes (such as TiO_2, carbon-coated, and Au) of NMs [74, 181].

Due to the cuticles' increased permeability, dicots exhibit higher NP uptake than monocot plants. This is demonstrated through the use of lipid-based liquid crystalline nanostructures as a delivery mechanism [235]. The differing cell wall architecture between monocots and dicots is thought to be the cause of this variation. Pectin polymer is loosely linked to cellulose in monocots [64]. Another notable anatomical variation is the shape of the leaf, such as the presence of dumbbell-shaped stomata in monocot plants and kidney-shaped stomata in dicot plants. As a result, the efficiency of NP absorption and translocation varies.

Additionally, monocots' leaves feature parallel veins, which are the most tear-resistant and consequently have a high strength [40, 148]. Because leaves are used for gas exchange, cuticles prevent the free penetration of any chemical, and roots and leaves are used to absorb nutrients from the soil, how these NMs are applied also affects how well plants can absorb them [298]. However, plants that can withstand droughts grow tougher cell walls, cuticles, Casparian strips, and other barriers that are useful for reducing water loss when water is in short supply [287]. The stability of several key soil components, such as humic acid and organic matter, increases the bioavailability of NMs [238]. Mycorrhizal fungi and bacteria that coexist with plants have an impact on how well plants absorb NPs [112]. CuO NPs are transported by xylem and phloem, respectively, from the roots to the shoots and back from the shoots to the roots [352].

4.9 The Plants' Physiochemical Reaction to NPs and the Effects on Plant Growth and Seed Germination

The size, chemical make-up, functional groups found on their surface, and kind of coating all have a significant role in the uptake of NMs inside plants. The interaction and uptake of NMs result in molecular alterations that have an impact on the general physiology of plants [161]. A novel molecular mechanism underlies the potential impact of multiwalled carbon nanotubes (MWCNTs) on the expansion (55–64% increase) of tobacco cells. The majority of the active genes are CycB, NtLRX1, and NtPIP1, which are involved in cell division, cell wall elongation, and tobacco aquaporin, respectively [172]. Increased development and vigor depend on NMs' capacity to pierce the tough seed coating and permit water importation.

Additionally, a promising tactic that further validates the promise of high-yield value crops before sowing is the seed priming method using nanotechnology [3, 34].

Using MWCNTs has an impact on the growth of various significant crops, including maize, soybeans, and barley. Raman spectroscopy and TEM imaging have both demonstrated that MWCNTs congregate within the endosperm of exposed seeds [179]. The movement of NPs throughout the plant and their interactions with cellular machinery helped the plant flourish. Mesoporous silica nanoparticles (MSNs) can interact with chloroplasts to stimulate photosynthesis, enhancing seed germination as well as total protein and chlorophyll content. The absence of stress conditions in any of the plants at the maximum concentration (2,000 mg/l) of MSNs suggests that its use as a smart delivery method is safe [326]. Silica plays a crucial role in supplying nutrients since a lack of it renders plants weaker and more vulnerable to biotic and abiotic stress [269]. Additionally, silica application reduces the impact of drought stress and boosts biomass in stressed-out plants [22]. Compared to its bulk counterpart and control, maize plants respond better to the application of silica nanoparticles (SiO_2 NPs) made from rice husk and comprehensively described. The maize plants' roots grew longer and their silica intake was improved; this makes them more drought-resistant. Additionally, the direct uptake and buildup of SiO_2 NPs in the maize epidermal cells was revealed by root sectioning [330]. Rastog [281] have evaluated the specific effects of SiO_2 NPs on agriculture in great depth.

TiO_2, another possible NP, also encourages plant growth, as seen in canola (*Brassica napus*), which led to enhanced seed vigor and seed germination. When compared to the control, the maximum effect was observed at concentrations between 1,200 and 1,500 mg/l, with increased plumule growth and a big radicle [210]. Depending on its additive concentration and size, the effect of each NP might be either positive or negative. In comparison to larger particles, plants acquire more of the tiniest TiO_2 NPs [181]. Additionally, these NPs can influence miRNA levels to start plant growth-promoting pathways [57].

Fe NPs had a beneficial effect on the Capsicum annum plant at lower concentrations, promoting plant development by increasing the number of chloroplasts and grana stacking. However, when these NPs were utilized at larger concentrations, the plant was harmed [371]. Other metal oxide NPs, such as ZnO and Cu, have been demonstrated to have an impact on the germination of seeds in the black gram plant *Vigna mungo* [273]. Less well known polyvinylpyrrolidone (PVP) coated platinum nanoparticles in

Pisum sativum have been shown to improve seed germination and plant growth [272]. Even at a low concentration, the biocompatible nanoparticle chitosan promoted the levels of indole-3-acetic acid (IAA) in wheat or *Triticum aestivum* L., which in turn influenced seed germination and growth [189].

4.10 Modern NPs for Plant Protection Advances

Photosynthesis, seedling vigor, root and shoot growth, and other significant aspects of the plant's growth and development are all influenced by the application of NPs (detailed in the preceding sections). Being ubiquitous, plants must endure harsh climatic conditions like drought, salt stress, high temperatures, UV radiation, and salt stress. Plants exhibit a variety of distinct stress responses, including changes in molecular processes, the expression of genes that are responsive to stress, and the synthesis of antioxidative enzymes that are essential for scavenging the plants in challenging environmental conditions [284]. Plants produce a variety of organic osmolytes, including polyols and trehalose, as well as a variety of amino acids, including proline and glycine, to protect themselves against osmotic stress. NPs offer an effective solution, or, to put it another way, help to plants in reducing this defense mechanism.

4.11 NPs Reduce Abiotic Stress Reaction

NPs not only promote plant growth but also shield it from abiotic stress. Due to the NP's huge surface area and small size, toxic metal binds to it, decreasing its availability. Drought, salinity, alkalinity, temperature swings, and metal and mineral toxicity are examples of abiotic stress. NPs can operate as nanoenzymes that can scavenge from oxidative stress and mimic the activities of antioxidant enzymes [305]. Since photosynthesis is a very vulnerable process and a crucial component of plant metabolism, its proper operation can be preserved by reducing oxidative and osmotic stress. Under stressful circumstances, photosystem II (PS II), rubisco, and ATP are the main targets in the photosynthetic apparatus.

It has been researched how plants respond to abiotic stress. For instance, the *Cucurbita pepo* plant's response to a defensive mechanism against salt stress was improved by SiO_2 NPs, which also increased plant transpiration rate, water use efficiency, total chlorophyll, and carbonic anhydrase activity [308]. The electron transport chain's linolenic acid is inhibited by

TiO_2 (anatase), which was also discovered to alter photoreduction activity. Additionally, it slowed down chloroplasts' rate of oxygen evolution [220]. The hallmark of abiotic stress is the production of ROS by cell organelles under stressful conditions. Additionally, plants have the enzyme tools necessary to combat the oxidative stress their environment places on them. However, when stress overwhelms the defensive system, plants must deal with the repercussions. By turning on particular genes, storing osmolytes, supplying free nutrients, and delivering amino acids, NMs aid in reducing such stress.

4.12 NPs of Cerium (CeO NPs)

Depending on the exposure dose, coating, surface charge, plant species, and growing conditions, nanoceria has extensive effects on plant health that can be both favorable and negative [218]. A class of CeO NPs known as "nanoceria" finds extensive use in the biomedical sector as an antioxidant [200]. NPs can have certain harmful effects on plants, but when we consider their beneficial characteristics, they outweigh these effects and can be used to boost plants' health. These NPs could increase the endurance of plants to diverse abiotic stresses by modulating key pathways, as we discussed earlier in relation to the potential interaction at the nano-biointerface [297]. For instance, abiotic stress raises excessive ROS levels, which lowers plant photosynthetic efficiency and may result in biomolecule oxidation [343]. Due to its special redox potential, which is based on the simple transition between Ce3+ and Ce4+ oxidation states [76], nano-Ce is well suited to minimize this effect. As a result, it functions as a ROS scavenger.

Additionally, a CeO NP mimics catalase activity at low Ce3+/Ce4+ ratios and exhibits the scavenging effect while mimicking superoxide dismutase activity at a high Ce3+/Ce4+ ratio [265]. The oxidative scavenging effect is also applied to other stimuli, such as too much light, heat, and darkness. Additionally, it results in a decline in the quantity of photosystem II, photochemical efficiency, chlorophyll concentration, and altered plant shape [241]. The ameliorative action of anionic CeO NPs in the A. thaliana plant against abiotic stress, such as excessive heat, light, dark, and chilling [361]. The physiology of plants is similarly threatened by the salinity stress through a similar mechanism to oxidative stress. CeO NPs were discovered to be able to reduce the oxidative stress in Brassica napus. At concentrations of 200 and 1,000 mg/kg, NaCl (100 mM) induced the impact via altering physiobiochemical characteristics, such as increased plant biomass, chlorophyll content (which, in turn, enhances the Mg_{2+} absorption), and effective photosystem [289].

One of the characteristics of salinity stress is the maintenance of the cytosolic Na+/K+ ratio, and NPs can effectively impact this transport [141]. Nanoceria has been shown to increase plant resistance to salinity stress. Improves plant photosynthetic efficiency and biomass by directly influencing hydroxyl radical (OH) production and potassium fluxes (reducing K+ efflux and improving K+ retention) through the plasma membrane of mesophyll [360]. CeO NPs' catalytic scavenging ability has also been used to lessen drought stress, which has serious effects on plant development and yield globally. At a low concentration of 10 mg/L, the extremely effective, drought-resistance activity of CeO NPs was seen in foliar-sprayed sorghum plants.

Additionally, it reduced the amount of ROS and lipid peroxidation, which led to increased rates of carbon absorption and pollen germination [95]. CeO NPs may be able to reduce the effects of salt stress on seedlings. When seeds primed with poly (acrylic acid)-coated CeO NPs (500 mg/L in water for 24 h) were germinated under salinity stress (200 mM NaCl), the root length, weight, and viability of the seedlings improved dramatically when compared to the control. As a result of the ensuing disruption of the antioxidative enzyme system, ion binding and Ca_{2+} signaling, and terpene production pathways, oxidative stress is reduced and salinity stress tolerance is increased [33].

4.13 NPs of Silicon (Si NPs)

After oxygen, silicon (Si) is recognized to be the second most prevalent element on Earth, and it has grown to be quite important in agriculture. Si is regarded as a transitional element for plants because, in addition to ensuring their existence, it also provides them with sufficient benefits when present [205]. These NPs can interact with plants either directly or indirectly, causing morphological and physiological changes that provide a tolerance to stress. It promotes plant growth, boosts biomass, changes architecture and physiology, modulates tissue differentiation, activates defense mechanisms, and aids in acclimating to stressful situations in plants [45].

When administered at varying concentrations, SiNPs displayed antistress benefits against moderate to severe drought stress in Hawthorns (*Crataegus* sp.). The responses varied in the seedlings depending on the concentration. Increased membrane electrolyte leakage, relative water content, improved photosynthetic capacity, and higher concentrations of chlorophylls, carotenoids, and proline are some of these consequences [37]. By accumulating proline, the application of Nano–SiO_2 increased tolerance in saline environments by maintaining ionic balance, enhancing the antioxidant system, and increasing levels of different phytopropanoids, which resulted in osmotic adjustments [313]. SiO_2 NPs have also demonstrated to improve water use effectiveness, stomatal conductance, transpiration rate, and minimize chlorophyll degradation during salt stress, resulting in tolerance to external insult. The epicuticular wax layer has undergone major alterations, and the main cause is salt stress.

Interestingly, strawberry plants treated with nano-Si had thicker and better-structured epicuticular wax than salt-stressed plants did [42]. Another study that looked at the effects of nano-Si to reduce salt stress on sweet pepper plants (Capsicum annum L.) found substantial differences in comparison to their bulk equivalent [331]. The plants received low concentrations of nano-Si (1.0 and 2.0 cm^3/L) through an irrigation system for weeks, which led to improved leaves and fresh weight. Potato plants subjected to various salt stresses similarly showed the concentration-dependent effect of SiO_2 NPs (NaCl; 50 and 100 mM). At both a lower dosage (50 mg/L) and a greater concentration (100 mg/L), the NPs were able to demonstrate improved stress reduction [131]. Lower concentration, however, was discovered to be more efficient. Contrary to the presumptions that NMs solely have poisonous properties, these researches particularly offer a better perspective on the beneficial and negative influence of NPs. Therefore, before using NPs in any discipline, it is necessary to first understand their features.

Plant hormones are crucial in times of external stress and help plants better adapt to a variety of shifting environmental situations [340]. Gibberellin, a plant growth hormone, plays a significant part in how plants react to a variety of abiotic stimuli, including as drought, shade, flooding, and low temperatures, mostly by slowing down plant growth and shifting the emphasis to stress resistance [75]. NMs can also be used to deliver plant hormones to the plant; for example, mesoporous silica NPs (MSN) that contained abscisic acid were tested on the plant. It considerably demonstrated the delayed hormone release and enhanced the seedlings' capacity to withstand drought [327].

4.14 NPs of Titanium Dioxide (TiO_2 NPs)

There are other NPs that exert their effects through alternative pathways involving gene regulation, in addition to the antioxidative actions of certain

NPs that scavenge ROS produced in response to threat or stress in plants that we have already described. By activating a mechanism of antioxidant enzyme activities, for instance, TiO_2 NP exposure (0.01%) raised the chlorophyll content and biomass and lowered malondialdehyde and hydrogen peroxide while increasing the formation of proline and soluble carbohydrates, maintaining the osmotic equilibrium [1]. Similar to this, nano-TiO_2 was able to promote the expression of a certain important noncoding RNA thought to be crucial to the mechanism of resistance to abiotic stressFrazier *et al.* noticed the expression of 11 conserved miRNA in tobacco seedlings in response to TiO_2 NPs (0.1%, 1%, 2.5%, and 5%), which protect plants from heavy metal stress [115].

Lack of water poses a serious problem for plant production because it reduces plant vigor and severely harms crops. By enhancing the activity of the nitrate reductase (NR) enzyme, which in turn improves the accumulation of osmolytes, nano-TiO_2 can improve the plant's state of hydration. Nitric oxide (NO) is produced as a result of the enhanced NR enzyme activity, and this NO eventually triggers the synthesis of proline and glycine betaine [168]. TiO_2 NPs frequently exhibit a defensive mechanism against stress in plants that is both enzymatic and nonenzymatic. It is interesting to note that TiO_2 NPs are recognized for controlling other enzymes like glutamate hydrogenase and glutamine synthase, which results in the accumulation of additional nutrients and the creation of essential oils [21]. In order to assess the impact of nano-TiO_2 (0, 50, 100, and 200 mg L1) on Moldavian balm (Dracocephalum moldavica L.) plants cultivated under aggravated salinity stress, Gohari *et al.* conducted a greenhouse experiment (0, 50, and 100 mM NaCl). Under typical conditions, the TiO_2-treated plants (100 mg/L) significantly reduced the production of the following essential oils: geranial, z-citral, geranyl acetate, and geraniol (1.19%). This suggested that aromatic plants could be protected from the stress state by altering their essential oil production profile and composition directly [129].

4.15 Nanopesticides

Since more than half of the world's population depends on plants and their byproducts for sustenance, increasing plant output and preserving their health are important issues [382–392]. Pathogens, nutritional deficiencies, and soil/air pollution are a few biotic and abiotic elements that can all have an impact on a plant's health by causing harm that, in turn, reduces crop and fruit yield. A variety of disease-causing microbes frequently cause plant disease (or pests). Therefore, stopping them from making more plants wilt is an urgent necessity. In agriculture, insects are important because they destroy crops and overrun food storage, resulting in deteriorating food quality and the spread of plant diseases. Pesticides are defined by the U.S. Environmental Protection Agency as compounds that intentionally ward off, repel, and eradicate pests.

By defending plants against harmful elements like plant diseases and insects, pesticides are utilized to increase and improve crop productivity and efficiency [158]. The usage of pesticides, on the other hand, has been found to be detrimental to human and animal health and poisonous to the environment. Consequently, numerous pesticides are prohibited by national or international agencies. This entails a number of important and crucial issues, such as the widespread use of pesticides at high concentrations, which degrade the environment, increase bioaccumulation, render the soil infertile, and alter its microbiota [216]. To boost agronomic output and efficiency and lessen the impact on the environment, some factors must be taken into account.

Nanotechnology offers a unique and enhanced approach to the challenging and expensive task of creating efficient and secure pesticides [295]. Pharmacy has already been significantly impacted by nanotechnology (drug delivery). It is encouraging given the most recent advancements in the food and agriculture industries. Nanotechnology is not only used to protect plants from pests; it is also used to reduce waste, track plant growth, ensure better food quality, and increase global food supply [158]. Nanoherbicides, nanoinsecticides, nanonematocides, and nanofungicides are the nanopesticides that are most frequently tested. Nanostructured alumina (NSA), a substance that was nanoengineered and acts as an insecticide, is the best illustration of a nanopesticide. It is produced by oxidizing metal, which has a constant electric charge. NSA was created using advanced nanotechnology so that it may use its system to fight insects. Triboelectrification produces the electric charges that insects display. Tribo-charge is a form of contact electrification created when two nonconductive objects or materials come into touch with one another as a result of friction. Additionally, NSA has a fascinating method of action that involves two successive mechanical stages and insect tribo-charging. Positively charged bug charges interact with negatively charged NSA particles mostly through powerful electrical bonding. Through the powerful sorptive ability of the NSA particle, which detaches the insect's cuticle, a fibrous composite of chitin, this interaction leads the insect to become dehydrated, leading to the bug's eventual death from dehydration. According to the theory, the insecticidal effect of NSA powder or particles is based on the particle's attachment to the insect's cuticle, which furthers upsets the water balance and increases ambient humidity. Because of its small size, inherent electric charge, sorbative abundance, and vast surface area, NSA has a high insecticidal effect [318].

Even when used in lesser quantities, nanofertilizers deliver nutrients to plants or improve the effects of fertilizers. The fertilizers can be encapsulated in nanoform to improve nutrient uptake. Finally, it lessens the danger of environmental degradation, reduces nutrient loss, and enhances crop quality and production.

Additionally, insect-resistant transgenic (Bt cotton) crops and conventional cotton are being helped by engineered nanopesticides, which were created to combat the bollworm. Cu NPs' insecticidal activity has the ability to increase the production of an exogenous microbial protein from Bacillus thuringiensis programmed through the Bt toxin in plant tissue to increase resistance to the bollworm at a low dose (10 mg/L) [378]. TiO_2 and ZnO NPs, which are photochemically active, function as antibacterial agents when spot disease affects tomato plants. TiO_2 and ZnO NPs, which are photochemically active, produce excited electrons when they are exposed to light. These excited electrons then directly transfer to superoxide radicals in the presence of oxygen. NPs are designed to have antibacterial qualities and have a significant impact on agriculture as nanoinsecticides using photochemically active TiO_2. With regard to X. perforans, which causes spot disease in tomato plants, it exhibits significant photocatalytic activity as well as greater antibacterial capability [258].

Additionally, the pesticide industry places a high demand on the packaging of conventional pesticides with NPs or polymers because they are expected to address these issues by improving pesticide efficiency, raising production, reducing excess runoff by slow release of active ingredients over a prolonged period of time, and ultimately protecting the environment. In order to enhance current pest control methods and develop advanced nanoformulations that are delivered at the desired site, are stable, quickly penetrate organisms, are cost-effective, and are more active during field applications, this new field of research examines how NPs interact with insects. The active ingredient is included in the core of the nanoformulation or nanocarrier, which serves as the vehicle to transport and regulate distribution or release of the compound. This allows for the preservation of an acceptable amount of chemical throughout the entire period of insect growth (from the transition from larva to the maturity stage). Etofenprox or alpha-cypermethrin, for example, are used in the sustained nanocarrier assembly of chitosan liposomes, which coat the inner core. The development of various NMs and nanoformulations of conventional insecticides has also had a substantial impact on pest management.

4.16 Nanoemulsions

The term "nanoemulsions" describes an oil-in-water emulsion in which a pesticide is dissolved in water as nanoscale droplets. Increasing the solubility of pesticide active components is the main benefit of creating nanoemulsions. Neem oil nanoemulsions demonstrate that smaller droplets can be efficiently absorbed because the LC50 is seen to decrease as droplet size [36]. However, the permethrin nanoemulsion's superior effectiveness against the yellow fever mosquito (Aedes aegypti) compared to permethrin macroparticles is a sign of its improved absorption [177]. According to one study, Emamectin-benzoate nanoemulsion is superior to coarse emulsions at repelling the Asian rice borer (Chilo suppressalis) [108]. Some studies reported no significant differences in the LC50 values of a commercially available coarse emulsion of chlorpyrifos and a nanoemulsion of the pesticide against the cotton bollworm (Helicoverpa armigera) when using pesticide colloids [267]. Their regulated or gradual release from the colloidal formulations, which leads to more sustained and protracted exposure, is another crucial factor.

4.17 Polymer Nanopesticides

These polymer-based nanocarriers operate as a protective reservoir containing polyesters and polysaccharides and are used to encapsulate insecticides. Additionally, they often lead to an increase in the dispersion of pesticides in aqueous environments, enabling their slow and controlled release. Because of their adaptability in design, biodegradability (corn oil, beeswax, and cashew gum), and biocompatibility, polymer nanopesticides have drawn a lot of interest. A better alternative would be polymer-based nanoformulations like nanogels, nanospheres, or nanofibers [242].

4.18 Nanopesticides as Solid NPs

Some NPs that are thought to work as antibacterial or pesticides instead merely serve as nanocarriers, such as inert dust made of alumina, silica, and clays that harm an insect's cuticle's inner wax layer, causing the insect to lose water and perish from desiccation. Since it is unlikely that insects may acquire resistance, this method of treatment is regarded as being safer [301]. SiNPs physiosorbed on lipids destroy the protective epidermal layer, resulting in the death of the insects [48]. Stadler *et al.* [317] identified nanostructured alumina as an effective insecticide against the two grain pests R. dominica and S. oryzae. The silicate materials are created as thin sheets called nanoclays.

4.19 Nanoherbicides

Due to their propensity for consuming nutrients that would otherwise be accessible to plants, weeds are thought to pose the greatest threat of damage crops in bigger quantities. The traditional approach to getting rid of weeds takes a long time and a lot of work, involving hand pulling. There are numerous herbicides on the market right now that can potentially eliminate weeds in a field but also harm crops and reduce soil fertility. With no hazardous residues left in the soil, nanoherbicides can be a superior, environmentally friendly substitute for conventional herbicides [260]. A longer time of continuous application of the same pesticide could cause weed resistance. For the traditional herbicide atrazine, poly (epsilon-caprolactone) (PCL) nanocapsules have been employed as a delivery mechanism. Compared to a commercial atrazine formulation, the atrazine-containing nanocapsules had stronger postemergence herbicidal efficacy on mustard plants [250]. By cross-linking diuron disulfide linkages, chitosan nanoparticles were created for regulated administration based on glutathione content [370].

4.20 Nanofungicides

The majority of agricultural damage, including that to important crops like rice, wheat, barley, groundnuts, and cotton, is caused by fungi [128]. These illnesses have an impact on society since they severely reduce crop productivity and the economy. Conventional fungicides can reduce these losses, but they can harm biodiversity because they target a wide variety of living things. As a result, we must consider other strategies to improve management of fungal diseases [259]. The creation of NPs as a successful tactic against fungal diseases is one of the greatest methods. Because of their antibacterial qualities, AgNPs are frequently utilized in disinfection processes [46]. The efficiency of AgNPs and Ag ions against the plant-pathogenic fungus Bipolaris sorokiniana and Magnaporthe grisea was tested in an experiment. When used for three hours, the ionic and nanoparticulate versions dramatically lessened the severity of the condition and were effective against fungus [162]. Both CuNPs and AgNPs have the ability to inhibit the growth of the fungi Alternaria alternata and Botrytis cinerea [252]. Against Alternaria alternate, Rhizopus stolonifer, Fusarium oxysporum, and Mucor plumbeus, ZnO NPs and MgO NPs exhibit antifungal action [353]. Two common pesticides, zineb and mancozeb, which was proven to be a more effective fungicide against A, were enclosed in a MWCNT-g-PCA hybrid material [294].

4.21 Nanofertilizers

Agriculture currently faces a wide range of difficulties, including nutrient deficit, stagnant crop yields, declining soil organic matter, limited water availability, nitrogen deficiency in the soil, reduced land area owing to urbanization and land degradation, and labor shortages [127]. The use of nanoscience and nanotechnology is expanding dramatically, and new processes are always being suggested to create unique and attractive materials for crop management and production. It is one of the most important concepts in the rapidly developing field of precision agriculture, which involves farmers using fertilizers and other inputs in an effective manner. Large-scale production of conventional fertilizers has been driven by the unchecked population growth in an effort to boost crop protection and food production, but this ultimately depletes soil fertility and lowers food quality. As they sit unused, these chemical fertilizers have an impact on human health in addition to aggravating the delicate ecosystem. According to a data provided by the Food and Agricultural Organization of the United Nations, the demand for fertilizers is anticipated to increase globally in the upcoming years. Therefore, it is crucial to use clever techniques for sustainable agriculture, including nanotechnology. Nanofertilizers, sometimes known as "smart fertilizers" or "environmentally friendly fertilizers," have the potential to boost fertilizer application rates and lessen nutrient loss from them, especially phosphorus and nitrogen [90]. It provides a regulated, slow delivery of nutrients to the desired place, preventing environmental and water body contamination [99].

The term "nanofertilizers" refers to substances that either give plants nutrients or improve the results of fertilizers even when used in smaller amounts [279]. Encapsulating fertilizers in nanoform can improve nutrient uptake, reduce nutrient loss, increase crop quality and yield, and lower the risk of environmental degradation. It has also been shown that these micro fertilizers can be applied topically to plants to reduce stress [332]. Based on the nutritional needs of the plants, nanofertilizers can be categorized into three groups: macronutrient nanofertilizers, micronutrient nanofertilizers, and nanoparticulate nanofertilizers (Chhipa, 2017). Magnesium (Mg), calcium (Ca), phosphorus (P), nitrogen (N), potassium (K), and magnesium (Mg) are some of the components that make up macronutrient nanofertilizers (P). There will be a severe need for these fertilizers in the agriculture industry as the overall consumption of macronutrient fertilizers is predicted to rise to 263 million tons (Mt) in 2050. Delfani and his team examined the effects of foliar application of Mg and Fe NPs on black-eyed peas' (Vigna unguiculata) growth and found that it boosted seed weight and photosynthetic capacity, which in turn increased output [84]. The Ca NPs and humic acids increased the growth of peanut seedlings by around 30%, and the dry biomass per plant increased to 5.78 g [199]. The growth rate and yield of soybeans (Glycine max) were greatly improved over control by the produced hydroxyapatite [Ca5 (PO4)3 OH] NPs of size 16 nm. According to these findings, NPs can be used as powerful fertilizers for field crops [196].

In comparison to macronutrients, plant micronutrients such as molybdenum (Mo), copper (Cu), iron (Fe), nickel (Ni), manganese (Mn), and zinc (Zn) are needed in considerably lesser quantities for the proper growth of crops. Composite formulations of micronutrients are more successful at providing enough nutrients and harm the environment less. Comparatively to other traditional ions and salts, micronutrients packaged in NPs increase nutrient bioavailability and fertilizer usage efficiency [90]. ZnO, CuO, and B2O3 have been successfully established as a composite of three micronutrients NPs for agronomic fortification to reduce drought stress in soybean plants [92]. Other NPs found in nanoparticulate fertilizers, like CNTs, TiO_2, and SiO_2, help plants develop more quickly. In soybeans, the combination of TiO_2 and SiO_2 boosted seed germination, growth stimulation, and nitrogen fixation [204].

4.22 Nanofertilizer Uptake, Translocation, and Action: Molecular Mechanism

Due to its smaller size (up to 20 nm) than the pore size of the plant cell wall, which aids in nutrient uptake, NM can be used as nanofertilizer. Nanofertilizers can be applied directly to the ground or to leaves. As a result, there are two important ways to apply nanofertilizer. Foliar entrance comes first, followed by root entry. For foliar entry, spray is applied to the trough cuticle, stromata, hydathodes, lenticels, and wounds. For root entry, spray is applied to root tips, lateral roots, root hairs, and rhizodermis.

Nanofertilizer treatments are preferred or possible when the soil and weather are unfavorable. Additionally, this method of application encourages the direct uptake of nutrients into the plant's leaf system, hence minimizing fertilizer waste. As a result, using nanofertilizer topically increases NUE and hastens the growth of plants or crops [209]. Nanopores, the base of the trichome, and the stromal aperture in leaves promote foliar entry, facilitating NP uptake and their transport inside the leaves. For instance, it has been shown that nanosized particles (43 nm) are deeply buried in the majority of the leaf interior compartment in faba beans (*Vicia faba*) and more (in *C. arabica* below 2.5 nm and *P. cerasus* below 100 nm), which suggests the effectiveness of nanofertilizer to improve nutrient uptake [100].

Additionally, this method of application encourages the direct uptake of nutrients into the plant's leaf system, hence minimizing fertilizer waste. As a result, using nanofertilizer topically increases NUE and hastens the growth of plants or crops [209]. Nanopores, the base of the trichome, and the stromal aperture in leaves promote foliar entry, facilitating NP uptake and their transport inside the leaves. For instance, it has been shown that nanosized particles (43 nm) are deeply buried in the majority of the leaf interior compartment in faba beans (*Vicia faba*) and more (in *C. arabica* below 2.5 nm and *P. cerasus* below 100 nm), which suggests the effectiveness of nanofertilizer to improve nutrient uptake. ZnO and CeO_2 NPs, for example, are demonstrated by following their paths from the ryegrass root cells to the vascular tissues of soybeans (Glycine max) [203]. These particles enter the cell through the apoplastic or symplastic pathway, passing through the plasmodesmata of neighboring cells before arriving in the cytoplasm. They start to disperse to numerous cytoplasmic organelles in the cytoplasm and engage in various cell metabolic pathways there. Additionally, NP absorption in plants grown in hydroponic medium has been found to be higher than in plants cultivated in sand [312]. Therefore, nanofertilizers have the potential to distribute and transport nutrients from cell to cell through a narrower plasmodesmata channel (50–60 nm). Consequently, nanofertilizer results in higher NUE and less nutrient loss during entrance to transport, which leads to increased production of field crops and plants of 6–17% and improved nutritional quality [155].

4.23 System for Sensing with NPs

To reduce environmental damage, care must be taken when using pesticides, fertilizers, and heavy metals in agricultural techniques. Emerging technology must be used to regularly check the condition of the soil and any potential disease outbreaks. For the real-time monitoring of vast field regions, accurate sensing technologies that are convenient and portable must be created. Any biosensor's performance could be improved by nanotechnology to support its actual use in agriculture. The creation of responsive sensors is made possible by remarkable advancements in nanofabrication technology and sensitive analytical procedures. Numerous applications for nanosensors could show their potential, including monitoring soil parameters (such as pH and nutrients), disease detection, and crop cultivation and harvesting, to mention a few. NPs are suitable for better sensing systems because of their distinctive surface chemistry, electrical and thermal properties, enhanced sensitivities, and enhanced detection limits [363].

4.24 Pesticide Residue Detection Using NP-Based Biosensor

High-performance liquid chromatography and GC/MS are two common traditional methods for pesticide detection. While extremely reproducible, both techniques have limited detection limits and involve additional work (sample collection, extraction, and analysis) and well-equipped labs (GC/MS pesticide detection). Various NMs, including metal NPs, graphene, CNTs, and QDs, are used to identify the pesticide, heavy metals, and other pollutants that are present in water and soil [376]. Faster detection, low sample requirement, and excellent sensitivity are further benefits of using NMs. By observing changes in color, electrical potential, or fluorescence, for example, one can measure the binding of the target molecule with that of the biosensor. Another method is array-based, which enables simultaneous detection of numerous analytes by fixing biomolecules to a substrate [121]. Based on a fluoroimmunoassay-based sensing system, NMs such as cadmium telluride (CdTe) QDs were used for the detection of 2,4-dichlorophenoxyactic acid [342].

Additionally, the fluorescent property of QDs can be used to detect other dangerous pesticides as atrazine [78], pyrethroid [175], and methiocarb due to their excellent photostability and size-tunable emission capabilities [149, 341]. Several NMs, including QDs and AuNPs, are used in optically based immunosensors to provide improved signal amplification. QDs coupled with streptavidin are used by Chen *et al.* to identify the pesticide chlorpyrifos in drinking water [70]. Pesticides that are present on-site in water from agricultural fields and other sources can be discovered using AuNPs [193]. For immunochromatographic test strips used to identify various pesticides, AuNP function is made possible. Triazophos and carbofuran can be detected simultaneously using a single-step Au-based, lateral-flow immunoassay strip test, according to the scientists. A critical enzyme for human health called acetylcholinesterase (AChE) is inhibited by the majority of pesticides. Biosensors can be created employing this enzymatic process; for instance, a localized surface plasmon resonance fiber-optic biosensor that measures AChE inhibition was used to determine the quantity of a pesticide [134]. The enzyme, which is mounted on the surface of AuNPs, modifies the light when it is inhibited by a pesticide (such as paraoxon), making it possible to quantify the enzyme's activity.

4.25 NPs for Detecting Plant Pathogens

NMs can be used as a quick diagnostic tool, a biomarker against different plant pathogens, or as an indicator for finding particular illnesses. They can be employed directly or indirectly to detect infections. Using surface plasmon resonance, AuNP-based immunosensors were used to identify the karnal bunt (Tilletia indica) disease in wheat [311]. The bacteria known to cause bacterial spot disease in Solanaceae plants, Xanthomonas axonopodis pv. vesicatoria, was found using fluorescence SiNPs in combination with Ab [364]. Although the topic is young and in its infancy, it has a lot of potential that needs to be realized in practical field applications. Utilizing nanochips, a form of microarray that uses fluorescent oligo probes to detect hybridization, is another crucial tactic [202]. Phytoplasma aurantifolia, which causes witches broom disease of lime, was discovered using QD fluorescence resonance energy transfer-based immunosensors, which demonstrated great sensitivity (100%) and specificity with a detection limit of 5 ca. P. aurantifolia per L. [268]. Plants develop distinctive volatile signatures (plant volatiles) whenever they become infected with any pathogen, and the identification of these signature molecules may aid to confirm the existence of any pathogen infection in plants. TiO_2 and SnO_2 detected p-ethylguaiacol on screen-printed carbon electrodes using an electrochemical sensing method, despite the fact that it was produced by plants when Phytophthora cactorum infected them [109]. Both NPs had good sensitivity (174–188 A cm^2 mM1) and a low limit of detection of 35 to 62 nM [109].

4.26 Smart Plant Sensing System Based on NP

The exact management of expensive agrochemicals (such as pesticides, fertilizer, nutrients, and water) is made possible by the smart plant sensing system, which also reduces major plant stress events in a changing climate. For the purpose of creating plant stress-tolerant cultivars and maximizing the use of finite resources, such as nutrients and water, boosting total crop output necessitates the application of novel and high-throughput smart plant sensing techniques. Smart nanosensors can communicate via wireless and optical signals for real-time monitoring of plant health status. They are used to precisely detect stress and limited resources and to optimize crop growth [126]. Raman and infrared spectroscopy, which produce poor signal-to-noise ratios and require complex, expensive equipment, must be used to regularly monitor the health status of plants and possible onset of any illness [31, 355]. A smart plant sensing system based on nanobiotechnology may directly communicate with hyperspectral imaging cameras and smartphones, and remote sensing tools have recently increased the signal-to-noise ratio for monitoring the health state of individual plants [47, 140, 253, 357].

A smart sensing system based on nanotechnology may monitor plant resource shortages or stressors in real time using key signaling molecules like calcium, ROS, NO, glucose, and sucrose as well as plant hormones like ABA, ethylene, jasmonic acid, and methyl salicylate [123, 126, 173]. NMs are appropriate for more effective in vivo sensing systems for important signaling chemicals due to their distinct surface chemistry, electrical and thermal properties, improved sensitivities, and enhanced detection limits [363]. These newly discovered characteristics of NMs enable the detection of analytes with high spatiotemporal resolution down to the single-molecule level and millisecond time scales, as well as their fluorescence in low or transparent background windows and ultralow photobleaching of living tissue [135, 146].

QDs are fluorescent NM sensors used to measure plant productivity and stress with highly glucose specific (500–1,000 M) emission ranges from the visible to the near-infrared (nIR) [188]. Using SWCNT sensors in a chosen region of the leaf that can report fluorescence intensity with high spatiotemporal resolution (seconds), short-lived signaling chemicals like calcium, ROS, NO, glucose, and nitroaromatics can be tracked in real-time [124, 125, 358]. The humic phase coating of the SWCNT sensor binds with the analytes, changing the nIR fluorescence intensity or shifting the wavelengthFast spatiotemporal pattern variations in waves with time and subcellular locations are caused by an analyte's fluctuating concentration in cells, subcellular space, and the entire plant. Examples include calcium and ROS [176, 375]. Diffusion and stochastic kinetic simulations, in particular, are integrated to clarify the complex data with a beneficial link under various resource deficiency and stress circumstances. Future agricultural devices will need to make decisions and take action based on system biology integration of large amounts of biological data (affinities, the rate constant, sensitivity, selectivity, and dynamic range) through computational modeling.

Wearable sensors for human clothing and skin have been developed in recent years through nanotechnological advancement [142]. Therefore, it opens a new door that provides wireless real-time communication of volatile compounds with low concentrations using wearable nanoelectronic circuits on plant surfaces. Live plants' leaves equipped with graphitic electrodes and SWCNT channels may detect trace quantities of a gaseous substance at concentrations as low as 5 ppm [184]. When copper complex-equipped SWCNT-based sensors are bound to ethylene gas released by plant fruits, the resistance varies in a manner that is proportionate to the volatile compound concentration [105]. Ethylene is a plant hormone and a key predictor of fruit ripening. CNT-based sensors are now commercially accessible for its detection. Additionally, plants embedded with functionalized SWCNTs containing bombolitin peptide are used to monitor explosive nitroaromatics (such as picric acid) in real time using a smartphone or camera of a high-quality laboratory [358]. Similar to this, calcium sensors based on green fluorescent protein can sense plant stress.

For example, delivery of ZnO and cerium oxide NPs to protect the plant from heat and salinity (increasing potassium retention in leaf mesophyll) by scavenging reactive oxygen, or agricultural devices embedded with smartphone, meteorological station, or hyperspectral imaging cameras to alleviate environmental stress. Nanobiotechnology-based sensors have the capacity to monitor chemical signaling in real time [132, 360, 361]. Similar to how nanocrystals lessen cold damage, CuO can protect watermelon (Citrullus lanatus) from fungal root disease [27, 55].

In agriculture field circumstances, such as environmental conditions, plant growth, and how development can affect their performance, smart plant sensors still need to be tested. Plant sensors can only currently communicate with electronic equipment under carefully regulated laboratory circumstances. The sensor can offer real-time information on the need for nutrients, water, pesticides, and fertilizer for particular demands if the field tests are successful.

4.27 NPs for Managing the Postharvest Waste in Agriculture

Nanoinsecticides, nanofertilizers, nanosensors, and crop protection are the most often mentioned applications of nanotechnology in contemporary literature. Agricultural waste contains an abundance of lignocellulose that can be effectively used to create useful NMs including nanocellulose, nanocomposites, and biochar [302]. Value-added goods made from rice and wheat husks help allay worries about how to properly dispose of agricultural trash [169]. Waste cotton fibers and cellulose can be made into goods with a higher value via electrospinning [278]. The biopolymer that makes up the majority of tree trunks and branches is cellulose. It is regarded as the primary element of the paper industry. It is effectively used in the creation and use of nanocellulose and natural fiber [81]. Aerogels, which are extremely hydrophilic nanoscale cellulosic materials, have a wide range of uses. From their bulk parts, their characteristics are significantly different. Due to their nanofibrillar structure, which has a high surface area and excellent adsorbing characteristics, holocellulose aerogels can be used widely as adsorption materials [232]. The rice husk, a renowned great source of nanosilica and microsilica, can be used to extract silica, an inorganic element. Validamycin-conjugated nanosilica is most useful for the regulated distribution of water-soluble pesticides [194]. The use of nanosilica conjugated with methyl methoxy silane as a nanofungicide and plant growth regulator against mildew disease has been proposed [151]. While certain nanopesticides are now on the market, research into how these products interact with the controlled release of agrochemicals is continuing and could take years.

4.28 Application of NP Risk and Health Hazards in Agriculture: Toxicological Impact

Despite extensive research, little progress has been made in comprehending the risk and health risks associated with NM. For the most part, NP implementation in the agrifood industry lacks a clear and precise framework for risk governance across the various sectors. Safety-by-design synthesis methods have been the focus of the creation of unified regulatory frameworks in the field of nanosafety [174, 321]. For information to be shared between various nanotechnology fields, risk assessment, risk management, policymaking, and support for risk communication among the various stakeholders are also necessary (for example, society, industries, researchers, policymakers, insurance companies, and the general public). Most of the time, nanosafety researchers employ a two-way method [56]. Understanding human health toxicity, which primarily targets organisms, tissues, and cells, is the first step (for example, mouse skin and immune cells). Environmental health risk comes in second; communities are mostly worried about other branches of the tree of life (for example, single-cell organisms, plant tissues, invertebrates, and fish). Plant-based nanosafety research has since concentrated on the deliberate or unintentional exposure of food crops to certain NMs (for instance, nanopesticides, nanofertilizers, and nanoherbicides), as well as intentional or unintentional exposure (i.e., spray drifting, leaching, and runoff of NPs).

Nanopesticides and nanofertilizers that are applied directly to plants have the potential to cause ROS, which can damage intracellular organelles' functions. Humans are exposed at work through industrial processes, direct contact with nanopesticides and nanofertilizers, and indirect release (i.e., released through agricultural products as well as food processing and packaging). To increase the sustainability of agriculture nanotechnology, a risk framework for risk assessment, risk management, policymaking, and to facilitate risk communication among the various stakeholders.

Before using NPs in any field, it is important to first understand their physiochemical characteristics because doing so could have a negative impact on their potential for posing environmental and health risks that ultimately limit their usefulness. The two aspects of the NPs that are most crucial from a toxicological standpoint are size and concentration. It can be explained by the relationship between surface area and size, which eventually promotes greater uptake and contact with the plant system and raises the possibility of negative consequences [240]. Ma *et al.* [208] conducted a thorough review of the subject and found that NPs as small as 5 nm can easily pass through cell wall pores and as large as 20 nm can pass through plasmodesmata. As a result, it is reasonable to assume that NPs in the smaller size range can enter plants more quickly and cause greater phytotoxicity even at lower concentrations [208, 286, 345]. The potential transfer from roots to shoots is hampered, as in the case of bigger size graphene oxide (GO), lowering the bioaccumulation [68].

It is important to note that the concentration-dependent effects serve as the primary indicator of whether NPs are beneficial or hazardous. For example, ZnO and CeO_2 nanoparticles exhibit harmful effects on cucumber plants, such as reduced yield (31.6%), when provided at higher concentrations (800 mg/kg) rather than lower quantities (400 mg/kg) [379]. On the contrary, when given to peanut plants at a dosage of 1,000 ppm, ZnO

NPs encourage seed germination and seedling vigor in addition to a higher chlorophyll content [264]. More specifically, metal NPs that shatter ions in the biological environment rely on the creation of excessive ROS as the primary mechanism of NPs toxicity, which results in toxic consequences [362]. Additionally, it disrupts the chloroplast's and mitochondria's electron transport chain. A variety of stressors also have an impact on carbon fixation, which leads to photoinhibition and increased formation of superoxide anion radicals and water [113]. As a result of the excessive ROS production, DNA damage, lipid peroxidation, and protein modification occur in the cells [150, 348]. Additionally, it has advantages since it aids in reducing stress-related problems in plants that highlight the significance of a balance between ROS production and scavenging [306]. In plants under NP stress, the increased ROS has a significant impact on the activity of antioxidant enzymes [106].

Recent studies have also demonstrated that phytohormones are crucial to the signaling of the external insult response. A hormone imbalance in plants may be indicated by the NPs' reduction of auxins, cytokinins, and salicylic acid, which affects overall metabolism [243, 339]. When NP concentrations are extremely high, photosynthesis may be substantially hampered, maybe even leading to plant death. As a result of NP exposure, significant decreases or inhibitions in plant growth have been seen in several instances.

In light of the aforementioned data, it is apparent that a number of parameters must be taken into account before realizing the beneficial exploitation of NPs in the agricultural system. The interaction between plants and soil is what drives agricultural productivity, hence it is crucial to understand the physicochemical properties of the soil system. It is regarded as the NPs' primary sink, therefore how they interact with the soil's constituents could have a significant effect on how they turn out. The risk of toxicity has been found to rise with increasing pH, organic content, and cation exchange capacity [300]. For instance, loamy clay soil with pH 5.5 reported zero toxicity by ZnO NPs even at 2,000 mg/kg, whereas loamy clay soil with pH 7.36 observed toxicity at a dosage of 45.45 mg/kg [97]. In addition, the lowered toxicity of the ZnO NPs when given Alginate at a dosage of 400–800 mg/kg shows that the soil organic matter also plays a key role in altering the safety problems of the NPs. But ZnO NPs by themselves significantly reduce plant biomass [377]. The DNA fingerprinting analysis showed declining taxa of Rhizobiales, Bradyrhizobiaceae, and Bradyrhizobium (the process that fixes nitrogen); however, positive effects were also anticipated on Sphingomonadaceae and Streptomycetaceae. NPs also have a direct impact on the soil microbial community structure in a dose-dependent manner. It is important to observe that NPs are increasing the community's taxa linked to the breakdown of organic contaminants and biopolymers while decreasing the taxa linked to nitrogen fixation [119]. Additionally, Rhizobium and Azotobacter's nitrate-reduction activity was affected by AgNPs in a dose-dependent manner, with Azotobacter's activity increasing by 0.2 ppm [303]. The majority of environmental-based nanosafety research has focused on preventing unintentional NP releases into the environment during the production, use, and disposal of agriculture-related nanoproducts. Green peas (*Pisum sativum* L.) were cultivated, and Mukherjee *et al.* [230] investigated the relative toxicity of NPs Al_2O_2 at ZnO (15 nm), ZnO at KH550 (20 nm), and bare-ZnO (10 nm). Zn levels in seeds and roots were greater in plants treated with Al_2O_3 at ZnO NP (1,000 mg/kg for 65 days) compared to bulk and other Zn NP exposure. With the exception of Al_2O_3 at ZnO NP, photosynthetic pigments (Chl-a and carotenoid) greatly increased, but the quantity of carbohydrates and proteins remained mostly unaltered. In agreement, root elongation and seed germination were unaffected by a relatively high concentration of ZnO NPs (1,000 mg/L). Cu and Ag-NPs, however, prevented the growth of zucchini's roots (Cucurbita pepo). AgNP exposure at 500 mg/L and 100 mg/L drastically reduced plant biomass and transpiration rate by 57% and 41%, respectively [319].

The effects of exposure to 200 and 300 (mg/L) ZnO NP, which reduced Arabidopsis growth by 20 and 80%, respectively, in comparison to the control, were studied by Wang *et al.* According to the findings, the expression of genes involved in photosystem structure and chlorophyll synthesis has decreased. On the other hand, a rise in the expression of genes involved in carotenoid production may help to lessen the photosynthetic efficiency associated with overall phytotoxicity in plants. According to Asli and Neumann [39], cell wall pore size dropped from 6.6 nm to around 3 nm as colloidal suspensions of TiO_2-NPs (30 nm) collected at the cell wall surfaces in the root, impacting the root's ability to transport water and, consequently, the transpiration of intake plants (Zea mays L.). CeO_2 and TiO_2 NP suspensions were tested on the early growth of barley at doses of 0, 500, 1,000, and 2,000 mg/L for 7 days to determine their genotoxic and phytotoxic effects [213]. Genotoxicity [randomly amplified polymorphic DNA (RAPDs) and mitotic index] found changes in chromosomal level and RAPD banding patterns with reduced cell divisions in treated root tip cells compared to control plants. At a concentration of 500 mg/L, both the root and the shoot produced more ROS and contained more ATP. Through the use of ICP-OES and

TEM EDSX microanalysis, the authors also discovered the presence of CeO_2 and TiO_2 NPs in exposed root cells.

Theoretically, agronanotechnologies may directly release NMs into the environment and incorporate them into the human exposure pathway through the food chain. Due to the possible migration of NMs via the food chain, agricultural nanochemical application now faces hurdles; therefore, safety assessment is urgently needed. Human exposure to nanoagrochemicals occurs in a variety of ways, including increased demand for food crops (due, for example, to population growth and rising consumer demand for animal-based goods), significant food product waste in wealthy sectors, and other agricultural operations (preharvesting and postharvesting of food crops). Additionally, because they are predominantly ingested by lower trophic-level creatures, animals, and humans, plants present a tremendous opportunity to assist the transfer of NPs to the various tropic levels and ultimately to the food web.

Understanding the one health idea is crucial, particularly in the context of agricultural nanotechnology. The importance of seeing today's health concerns via a one health concept cannot be understated, according to past studies or reports. Additionally, a lot of peer-reviewed journals use the results of their research as an illustration of the need for a one health approach. The World Health Organization, the World Organization for Animal Health, and the Food and Agriculture Organization of the United Nations must also be commended for their tremendous work in advancing the one health idea and raising awareness of its significance [201]. Here, we suggest a transdisciplinary and comprehensive strategy that supports the idea of one health and is essential for the long-term advancement and application of nanotechnology in policy. All of the species on the world, including people, animals, plants, and related microbiota, depend on one another for their survival in an interconnected ecosystem. The migration of NMs in the food chain is especially exacerbated by the excessive and improper use of agrochemicals as growth boosters of food crops. For the risk assessment of agronanotechnologies, an intelligent testing strategy framework is needed due to the nanotoxicity and associated risks to human health. Along with an initial assessment (document) of the knowledge that is currently known and the knowledge gaps, they classify and integrate data to create an intelligent approach to grouping NMs based on their characteristics (for example, particle size, charge, molecular weight, solubility, and composition) and their biological effects in order to intelligently design next-generation risk assessment strategies and safety to the risks that are associated with this technology.

4.29 Required Qualifications for Selection as Nanobiopesticides

The important factors that need to be taken into consideration in order for pesticides to be safe, simple to use, affordable, and effective in reducing plant pests are first, they must be simple to prepare, efficient at low costs, targeted against a specific pest, possibly effective against a wide variety of insect pests, safer for all living things and environments, nontoxic to farmers, free of any harmful substances, and easily applied. It must not have an impact on food quality, flavor, smell, or texture. It must be accepted on the global market. It must not be combustible, explosive, or corrosive.

4.30 Reasons for Research

Despite the fact that there have been many studies on how harmful nanoparticles are to bacteria, fungus, and animal diseases. Little research has been done to examine the toxicity effect of nanoparticles on pest insects like Termites, Amsacta moorei Butle, Episomus lacerta F, Brachytrypes portentosus Licht, Chrotogonus sp, Gryllulus domesticus Linn, Helicoverpa armigera, Laphygma exigua kb, Myllocerus maculosus Desbr, Pemphere The significant funds and focus in this field could be directed to produce the excellent plant protection research results. It is necessary to identify the plant species that produce insect-repelling secondary metabolites and create complexes with silver or other nanoparticles. It is important to understand how these biologically created nanoparticles affect certain pest insects. This will bring to a close the discussion of a plant species' special insect resistance.

4.31 Important Considerations for Nanobiopesticides

There is a lot of worry about the nanobioinsecticides since they have the potential to have harmful effects on the environment, people, and the soil. Critical research is required to address any potential drawbacks of nanopesticide's beneficial benefits. Other crucial factors include how soil organic matter reacts chemically and how pesticides with nanotechnology interact with the environment. In addition to nanoparticles, it is essential to recognize that certain plant species with unusual photochemicals may be critical for the development of nanoparticles. The nanoparticles that are created in the future are influenced by the plant's inherited secondary metabolites and phytochemical pesticides. Catharanthus roseus L, Datura metal L, Cardiospermum halicacabum L, Argemone mexicana L, Calotropis procera L, Azadirachta indica L, Oscimum species wild, Oscimum sanctum L, Oscimum canum L, and Citrus sinensis L are some examples of plants whose extracts have been used as insecticides.

4.32 Outlook for the Future

Without a doubt, nanotechnology has a significant impact on many areas of the food sector. Although it has many potential uses, it is important to note that most of the information we currently know about it comes from laboratory tests. Understanding the practical application without knowing how it affects environmental toxicity is extremely impossible. The sale of innovative NP-based agrochemicals is further hampered by a lack of regulations, regulatory authorities, and other comparable entities. In order to advance and build futuristic studies based on identified knowledge shortages, it is necessary to carry out extensive trials. We suggest six crucially crucial essential elements in this situation:

1. The open government awareness initiative to inform the general public about food nanotechnology and its uses by creating a sufficient database and supporting documentation to serve as logistical support for both the general public and food producers.
2. To develop a more workable plan, a thorough analysis of upcoming studies is required to ascertain the danger connected with the use of NP-based products in the agricultural sector.
3. To determine the lowest concentration dose of the NP that would be nontoxic and practical for field applications using the concentration-dependent study in the natural soil system.
4. To get a thorough understanding of the bioaccumulation potential of NPs during field applications and its impact on nanotoxicity, as well as to learn more about the transmission of the tropical chain in the ecosystem chain and to determine the appropriate level of safety.
5. Comprehensive knowledge of soil physicochemical interactions with NPs to lessen risk of altering the soil microbiota

Further research into soil rotation is needed to change the fate, transport, and bioavailability of NPs in order to reduce their long-term toxicity and their reliable and beneficial application in agriculture.
6. Lastly, we strongly advise switching from biosynthesized NPs to synthetic ones and conducting extensive research. As a result, employing the ecologically friendly protocol might have negligible or no toxicity, therefore going forward, investigations must especially focus on their functional effectiveness. Pilot tests must also be conducted under real-world circumstances (such as when growing plants in soil) in order to clearly demonstrate how NPs are influenced by their surroundings.

Additionally, a few safety precautions need to be taken

1. In order to prevent misuse at the toxic level, it is essential to harmonize the frameworks of nanosafety laws focused on safety-by-design synthesis techniques.
2. Before being used in any field work, new NP-based agrochemicals should have their physiochemical characteristics thoroughly examined in long-term trials since they may directly affect health and environmental risks, which will restrict their potential to be useful.
3. Environmentally based nanosafety research needs to focus on investigating unintentional NP releases into the environment that occur during the production, use, and disposal of nanoformulated agricultural nanoproducts.

4.33 Conclusion

The many functions that NPs perform in the agriculture and livestock sectors are highlighted in the current review. The application of nanotechnology in contemporary agriculture supports and develops the global economy in a variety of ways. The introduction of NPs makes a significant contribution to addressing the various problems caused by the growing population, plant and animal diseases, and climate change. However, by increasing the use of conventional pesticides and fertilizers to increase crop production, it ultimately harms the environment. In comparison to conventional resources, the effectiveness and agronomic efficiency of NPs have greatly increased. Pesticides can be applied more effectively thanks to varied nanoformulations, and controlled release safeguards the environment. Some nanofertilizers and herbicides are already on the market, and many more are being developed. A crucial factor to take into account when attempting to comprehend the physiological and biochemical reactions imposed by plants is the study of NP–plant interaction. The idea of their transformation and safety features in a complex system is provided by emerging approaches that are designed to find and measure NPs through the plant system. Nanotechnology makes use of NPs' special characteristics to create nanosensors that can measure soil parameters like pH and moisture as well as pesticide residue.

Pest problems are not confined to certain geographic areas, and the existing control methods are exceedingly harmful to both the environment and human health. The choice and availability of safe, efficient, and affordable bio pesticides is a major concern for pest control today. Magnetic nanoparticles could also be used as a substitute for insect repellents that have magnetic properties that are useful in geosensing technologies. The plant with the most pronounced insecticidal activity might be a compound of silver or zinc, and their effectiveness may be increased to control a variety of pests. This would also help reduce the use of the most harmful chemicals now being used as pesticides in fields.

References

1. Abdel Latef, A.A.H., Srivastava, A.K., El-sadek, M.S.A., Kordrostami, M., Tran, L.S.P., Titanium dioxide nanoparticles improve growth and enhance tolerance of broad bean plants under saline soil conditions. *Land. Degrad. Dev.*, 29, 1065–1073, 2018.
2. Abou-Zeid, H. and Ismail, G., The role of priming with biosynthesized silver nanoparticles in the response of Triticum aestivum L to salt stress. *Egypt. J. Bot.*, 58, 73–85, 2018.
3. Acharya, P., Jayaprakasha, G.K., Crosby, K.M., Jifon, J.L., Patil, B.S., Nanoparticle-mediated seed priming improves germination, growth, yield, and quality of watermelons (Citrullus lanatus) at multi-locations in texas. *Sci. Rep.*, 10, 5037, 2020.
4. Adetuyi, B.O., Adebayo, P.F., Olajide, P.A., Atanda, O.O., Oloke, J.K., Involvement of free radicals in the ageing of cutaneous membrane. *World News Nat. Sci.*, 43, 11–37, 2022.
5. Adetuyi, B.O., Adebisi, O.A., Adetuyi, O.A., Ogunlana, O.O., Toloyai, T.Y., Egbuna, C., Uche, C.J., Khan, J., Adumanya, O.C.U., Patrick-Iwuanyanwu, K.C., Ficus exasperata attenuates acetaminophen-induced hepatic damage

via NF-κB Signaling mechanism in experimental rat model. *BioMed. Res. Int.*, Article number 6032511, 2022.

6. Adetuyi, B.O., Adebisi, O.A., Awoyelu, E.H., Adetuyi, O.A., Ogunlana, O.O., Phytochemical and toxicological effect of ethanol extract of *Heliotropium indicum on* liver of male albino rats. *Lett. Appl. NanoBioscience (LIANB)*, 10, 2, 2085–2095, 2020. https://doi.org/10.33263/LIANBS102.20852095.
7. Adetuyi, B.O. and Farombi, E.O., 6 Gingerol, an active constituent of ginger, attenuates lipopolysaccharide-induced oxido-inflammation, cognitive deficits, neuroplasticity and amyloidogenesis in rat. *J. Food Biochem.*, 45, 4, e13660, 2021. https://doi.org/10.1111/jfbc.13660.
8. Adetuyi, B.O., Dairo, J.O., Didunyemi, M.O., Anti-hyperglycemic potency of *Jatropha Gossypiifolia in* alloxan induced diabetes. *Biochem. Pharmacol. (Los Angel)*, 4, 5, 193, 2015.
9. Adetuyi, B.O., Odine, G.O., Olajide, P.A., Adetuyi, O.A., Atanda, O.O., Oloke, J.K., Nutraceuticals: Role in metabolic disease, prevention and treatment. *World News Nat. Sci.*, 42, 1–27, 2022.
10. Adetuyi, B.O., Ogundipe, A.E., Ogunlana, O.O., Egbuna, C., Estella, O.U., Mishra, A.P., Akram, M., Achar, R.R., Banana peel as a source of nutraceuticals, in: *Food and Agricultural Byproducts as Important Source of Valuable Nutraceuticals*, 1st ed, p. 400, Springer, Berlin, Gebunden, ISBN 978-3-030-98759-6, 2022.
11. Adetuyi, B.O., Okeowo, T.O., Adetuyi, O.A., Adebisi, O.A., Ogunlana, O.O., Oretade, O.J., Marraiki, N., Beshbishy, A.M., Welson, N.N., Batiha, G.E.S., Ganoderma lucidum from red mushroom attenuates formaldehyde-induced liver damage in experimental male rat model. *Biology, 9,* 10, 313, 2020. https://doi.org/10.3390/biology9100313.
12. Adetuyi, B.O., Olajide, P.A., Adetuyi, O.A., Oloke, J.K., Preventive phytochemicals of cancer as speed breakers in inflammatory signaling. *Res. J. Pharm. Biol. Chem. Sci.*, 8, 1, 30–61, 2022.
13. Adetuyi, B.O., Olajide, P.A., Awoyelu, E.H., Adetuyi, O.A., Adebisi, O.A., Oloke, J.K., Epidemiology and therapeutic measure for COVID-19; A review. *Afr. J. Reprod. Health (Special Edition COVID-19)*, 24, 2, 142, June 2020. https://www.ajrh.info/index.php/ajrh/article/view/2300.
14. Adetuyi, B.O., Oluwole, E.O., Dairo, J.O., Biochemical effects of shea butter and groundnut oils on white albino rats. *Int. J. Chem. Processes (IJCC)*, 1, 8, 1–17, 2015.
15. Adetuyi, B.O., Oluwole, E.O., Dairo, J.O., Chemoprotective potential of ethanol extract of *Ganoderma Lucidum on* liver and kidney parameters in *Plasmodium Beghei*-induced mice. *Int. J. Chem. Processes (IJCC)*, 1, 8, 29–36, 2015.
16. Adetuyi, B.O., Omolabi, F.K., Olajide, P.A., Oloke, J.K., Pharmacological, biochemical and therapeutic potential of milk thistle (Silymarin): A review. *World News Nat. Sci.*, 37, 75–91, 2021.

17. Adetuyi, B.O., Toloyai, P.Y., Ojugbeli, E.T., Oyebanjo, O.T., Adetuyi, O.A., Uche, C.Z., Olisah, M.C., Adumanya, O.C.U., Jude, C., Chikwendu, J.K., Akram, M., Awuchi, C.G., Egbuna, C., Neurorestorative roles of microgliosis and astrogliosis in neuroinflammation and neurodegeneration. *Scicom J. Med. Appl. Med. Sci.*, 1, 1, 1–5, 2021.
18. Adisa, I.O., Reddy Pullagurala, V.L., Rawat, S., Hernandez-Viezcas, J.A., Dimkpa, C.O., Elmer, W.H., Role of cerium compounds in Fusarium wilt suppression and growth enhancement in tomato (Solanum lycopersicum). *J. Agricul. Food. Chem.*, 66, 5959–5970, 2018.
19. Adrees, M., Khan, Z.S., Ali, S., Hafeez, M., Khalid, S., Ur Rehman, M.Z., Simultaneous mitigation of cadmium and drought stress in wheat by soil application of iron nanoparticles. *Chemosphere*, 238, 124681, 2020.
20. Aghdam, M.T., Mohammadi, H., Ghorbanpour, M., Effects of nanoparticulate anatase titanium dioxide on physiological and biochemical performance of Linum usitatissimum (Linaceae) under well-watered and drought stress conditions. *Braz. J. Bot.*, 39, 139–146, 2016.
21. Ahmad, B., Shabbir, A., Jaleel, H., Khan, M.M.A., Sadiq, Y., Efficacy of titanium dioxide nanoparticles in modulating photosynthesis, peltate glandular trichomes and essential oil production and quality in mentha piperita L. *Curr. Plant Biol.*, 13, 6–15, 2018.
22. Ahmad, F., Aziz, T., Maqsood, M.A., Tahir, M.A., Kanwal, S., Effect of silicon application on wheat (Triticum aestivum L.) growth under water deficiency stress. *J. Food Agric.*, 7, 1–7, 2007.
23. Ahmed, B., Khan, M.S., Musarrat, J., Toxicity assessment of metal oxide nano-pollutants on tomato (Solanum lycopersicon): A study on growth dynamics and plant cell death. *Environ. Poll.*, 240, 802–816, 2018.
24. Ahmed, B., Rizvi, A., Zaidi, A., Khan, M.S., Musarrat, J., Understanding the phyto-interaction of heavy metal oxide bulk and nanoparticles: Evaluation of seed germination, growth, bioaccumulation, and metallothionein production. *RSC Adv.*, 9, 4210–4225, 2019.
25. Akanbi-Gada, M.A., Ogunkunle, C.O., Vishwakarma, V., Viswanathan, K., Fatoba, P.O., Phytotoxicity of nano-zinc oxide to tomato plant (Solanum lycopersicum L.): Zn uptake, stress enzymes response and influence on non-enzymatic antioxidants in fruits. *Environ. Technol. Innov.*, 14, 100325, 2019.
26. Alabdallah, N.M. and Alzahrani, H.S., The potential mitigation effect of ZnO nanoparticles on [Abelmoschus esculentus L. Moench] metabolism under salt stress conditions. *Saudi. J. Biol. Sci.*, 27, 3132–3137, 2020.
27. Alhamid, J.O., Mo, C., Zhang, X., Wang, P., Whiting, M.D., Zhang, Q., Cellulose nanocrystals reduce cold damage to reproductive buds in fruit crops. *Biosys. Eng.*, 172, 124–133, 2018.
28. Ali, M., Kim, B., Belfield, K.D., Norman, D., Brennan, M., Ali, G.S., Inhibition of phytophthora parasitica and P. capsici by silver nanoparticles synthesized

using aqueous extract of artemisia absinthium. *Phytopathology*, 105, 1183–1190, 2015.
29. Alidoust, D. and Isoda, A., Effect of γFe2O3 nanoparticles on photosynthetic characteristic of soybean (Glycine max (L.) Merr.): Foliar spray versus soil amendment. *Acta Physiologiae. Plantarum.*, 35, 3365–3375, 2013.
30. Almutairi, Z.M., Effect of nano-silicon application on the expression of salt tolerance genes in germinating tomato ('Solanum lycopersicum' L.) seedlings under salt stress. *Plant Omics.*, 9, 106, 2016.
31. Altangerel, N., Ariunbold, G.O., Gorman, C., Alkahtani, M.H., Borrego, E.J., Bohlmeyer, D., *In vivo* diagnostics of early abiotic plant stress response via raman spectroscopy. *Proc. Natl. Acad. Sci. U. S. A.*, 114, 3393–3396, 2017.
32. Amenta, V., Aschberger, K., Arena, M., Bouwmeester, H., Moniz, F.B., Brandhoff, P., Regulatory aspects of nanotechnology in the agri/feed/food sector in EU and non-EU countries. *Regul. Toxicol. Pharma.*, 73, 463–476, 2015.
33. An, J., Hu, P., Li, F., Wu, H., Shen, Y., White, J.C., Emerging investigator series: Molecular mechanisms of plant salinity stress tolerance improvement by seed priming with cerium oxide nanoparticles. *Environ. Sci. Nano*, 7, 2214–2228, 2020.
34. Anand, K.V., Anugraga, A.R., Kannan, M., Singaravelu, G., Govindaraju, K., Bio-engineered magnesium oxide nanoparticles as nano-priming agent for enhancing seed germination and seedling vigor of green gram (Vigna radiata L.). *Mater. Lett.*, 8, 127792, 2020.
35. Andersen, C.P., King, G., Plocher, M., Storm, M., Pokhrel, L.R., Johnson, M.G., Germination and early plant development of ten plant species exposed to titanium dioxide and cerium oxide nanoparticles. *Environ. Toxicol. Chem.*, 35, 2223–2229, 2016.
36. Anjali, C.H., Sharma, Y., Mukherjee, A., Chandrasekaran, N., Neem oil (Azadirachta indica) nanoemulsion—A potent larvicidal agent against culex quinquefasciatus. *Pest Manag. Sci.*, 68, 158–163, 2012.
37. Ashkavand, P., Tabari, M., Zarafshar, M., Tomášková, I., Struve, D., Effect of SiO2 nanoparticles on drought resistance in hawthorn seedlings. *Fore. Res. Paper*, 76, 350–359, 2015.
38. Askary, M., Talebi, S.M., Amini, F., Bangan, A.D., Effects of iron nanoparticles on mentha piperita L. under salinity stress. *Biologija*, 63, 65–75, 2017.
39. Asli, S. and Neumann, P.M., Colloidal suspensions of clay or titanium dioxide nanoparticles can inhibit leaf growth and transpiration via physical effects on root water transport. *Plant Cell. Environ.*, 32, 577–584, 2009.
40. Avellan, A., Schwab, F., Masion, A., Chaurand, P., Borschneck, D., Vidal, V., Nanoparticle uptake in plants: Gold nanomaterial localized in roots of arabidopsis thaliana by X-ray computed nanotomography and hyperspectral imaging. *Environ. Sci. Technol.*, 51, 8682–8691, 2017.
41. Avellan, A., Yun, J., Zhang, Y., Spielman-Sun, E., Unrine, J.M., Thieme, J., Nanoparticle size and coating chemistry control foliar uptake pathways,

translocation, and leaf-to-rhizosphere transport in wheat. *ACS Nano*, 13, 5291–5305, 2019.

42. Avestan, S., Ghasemnezhad, M., Esfahani, M., Byrt, C.S., Application of nano-silicon dioxide improves salt stress tolerance in strawberry plants. *Agron.*, 9, 246, 2019.
43. Awoyelu, E.H., Oladipo, E.K., Adetuyi, B.O., Senbadejo, T.Y., Oyawoye, O.M., Oloke, J.K., Phyloevolutionary analysis of SARS-CoV-2 in nigeria. *New Microbes New Infect.*, 36, 100717, 2020. https://doi.org/10.1016/j.nmni.2020.100717.
44. Ayoub, H.A., Khairy, M., Elsaid, S., Rashwan, F.A., Abdel-Hafez, H.F., Pesticidal activity of nanostructured metal oxides for generation of alternative pesticide formulations. *J. Agricul. Food. Chem.*, 66, 5491–5498, 2018.
45. Babajani, A., Iranbakhsh, A., Ardebili, Z.O., Eslami, B., Differential growth, nutrition, physiology, and gene expression in Melissa officinalis mediated by zinc oxide and elemental selenium nanoparticles. *Environ. Sci. Poll. Res.*, 26, 24430–24444, 2019.
46. Baker, C., Pradhan, A., Pakstis, L., Pochan, D.J., Shah, S.I., Synthesis and antibacterial properties of silver nanoparticles. *J. Nanosci. Nanotechnol.*, 5, 244–249, 2005.
47. Baret, F., Houlès, V., Guerif, M., Quantification of plant stress using remote sensing observations and crop models: The case of nitrogen management. *J. Experi. Bot.*, 58, 869–880, 2007.
48. Barik, T.K., Sahu, B., Swain, V., Nanosilica—From medicine to pest control. *Parasitol. Res.*, 103, 253–258, 2008.
49. Baskar, V., Safia, N., Sree Preethy, K., Dhivya, S., Thiruvengadam, M., Sathishkumar, R., A comparative study of phytotoxic effects of metal oxide (CuO, ZnO and NiO) nanoparticles on *in-vitro* grown Abelmoschus esculentus. *Plant Biosys Int. J. Deal. Aspec. Plant Biol.*, 155, 2, 374-383, 2021.
50. Batiha, G.E.B., Awad, D.A., Abdelazeem, M.A., Nyamota, R., Wahed, M.I., Shah, M.A., Amin, M.N., Adetuyi, B.O., Hetta, H.F., Cruz-Marins, N., Koirala, N., Ghosh, A., Sabatier, J., Diary-derived and egg white proteins in enhancing immune system against COVID-19. *Front. Nutr. (Nutritional Immunology)*, 8, 629440, 2021.
51. Batley, G.E., Kirby, J.K., McLaughlin, M.J., Fate and risks of nanomaterials in aquatic and terrestrial environments. *Acc. Chem. Res.*, 46, 854–862, 2013.
52. Behboudi, F., Tahmasebi Sarvestani, Z., Kassaee, M.Z., Modares Sanavi, S.A.M., Sorooshzadeh, A., Ahmadi, S.B., Evaluation of chitosan nanoparticles effects on yield and yield components of barley (Hordeum vulgare L.) under late season drought stress. *J. Water. Environ. Nanotechnol.*, 3, 22–39, 2018.
53. Bharti, A.S., Sharma, S., Shukla, N., Uttam, K.N., Steady state and time resolved laser-induced fluorescence of garlic plants treated with titanium dioxide nanoparticles. *Spectrosco. Lett.*, 51, 45–54, 2018.

54. Bin Hussein, M.Z., Yahaya, A.H., Zainal, Z., Kian, L.H., Nanocomposite-based controlled release formulation of an herbicide, 2, 4-dichlorophenoxyacetate incapsulated in zinc–aluminium-layered double hydroxide. *Sci. Technol. Adv. Mater.*, 6, 956–962, 2005.
55. Borgatta, J., Ma, C., Hudson-Smith, N., Elmer, W., Plaza Pérez, C.D., De La Torre-Roche, R., Copper based nanomaterials suppress root fungal disease in watermelon (Citrullus lanatus): Role of particle morphology, composition and dissolution behavior. *ACS. Sustain. Chem. Eng.*, 61, 4847–14856, 2018.
56. Bos, P.M., Gottardo, S., Scott-Fordsmand, J.J., Van Tongeren, M., Semenzin, E., Fernandes, T.F., The marina risk assessment strategy: A flexible strategy for efficient information collection and risk assessment of nanomaterials. *Int. J. Environ. Res. Public. Health*, 12, 15007–15021, 2015.
57. Boykov, I.N., Shuford, E., Zhang, B., Nanoparticle titanium dioxide affects the growth and microRNA expression of switchgrass (Panicum virgatum). *Genomics*, 111, 450–456, 2019.
58. Bramhanwade, K., Shende, S., Bonde, S., Gade, A., Rai, M., Fungicidal activity of Cu nanoparticles against fusarium causing crop diseases. *Environ. Chem. Lett.*, 14, 229–235, 2016.
59. Buchman, J.T., Elmer, W.H., Ma, C., Landy, K.M., White, J.C., Haynes, C.L., Chitosan-coated mesoporous silica nanoparticle treatment of citrullus lanatus (watermelon): Enhanced fungal disease suppression and modulated expression of stress-related genes. *ACS Sustain. Chem. Eng.*, 7, 19649–19659, 2019.
60. Cai, L., Cai, L., Jia, H., Liu, C., Wang, D., Sun, X., Foliar exposure of Fe3O4 nanoparticles on Nicotiana benthamiana: Evidence for nanoparticles uptake, plant growth promoter and defense response elicitor against plant virus. *J. Hazard. Mater.*, 393, 122415, 2020.
61. Cai, L., Liu, M., Liu, Z., Yang, H., Sun, X., Chen, J., MgONPs can boost plant growth: Evidence from increased seedling growth, morpho-physiological activities, and Mg uptake in tobacco (Nicotiana tabacum L.). *Molecules*, 23, 3375, 2018.
62. Cañas, J.E., Long, M., Nations, S., Vadan, R., Dai, L., Luo, M., Effects of functionalized and nonfunctionalized single-walled carbon nanotubes on root elongation of select crop species. *Environ. Toxicol. Chem.*, 27, 1922–1931, 2008.
63. Carlson, C., Hussain, S.M., Schrand, A.M.K., Braydich-Stolle, L., Hess, K.L., Jones, R.L., Unique cellular interaction of silver nanoparticles: Size-dependent generation of reactive oxygen species. *J. Phys. Chem. B*, 112, 13608–13619, 2008.
64. Carpita, N.C. and Gibeaut, D.M., Structural models of primary cell walls in flowering plants: Consistency of molecular structure with the physical properties of the walls during growth. *Plant J.*, 3, 1–30, 1993.
65. Carpita, N., Sabularse, D., Montezinos, D., Delmer, D.P., Determination of the pore size of cell walls of living plant cells. *Science*, 205, 1144–1147, 1979.

66. Chandra, S., Pradhan, S., Mitra, S., Patra, P., Bhattacharya, A., Pramanik, P., High throughput electron transfer from carbon dots to chloroplast: A rationale of enhanced photosynthesis. *Nanoscale*, 6, 3647–3655, 2014.
67. Chen, H. and Yada, R., Nanotechnologies in agriculture: New tools for sustainable development. *Trends Food Sci. Tech.*, 22, 585–594, 2011.
68. Chen, L., Wang, C., Li, H., Qu, X., Yang, S.T., Chang, X.L., Bioaccumulation and toxicity of 13C-skeleton labeled graphene oxide in wheat. *Environ. Sci. Technol.*, 51, 10146–10153, 2017.
69. Chen, Y.E., Mao, H.T., Wu, N., Mohi Ud Din, A., Khan, A., Zhang, H.Y., Salicylic acid protects photosystem II by alleviating photoinhibition in arabidopsis thaliana under high light. *Int. J. Mol. Sci.*, 21, 1229, 2020.
70. Chen, Y., Ren, H.L., Liu, N., Sai, N., Liu, X., Liu, Z., A fluoroimmunoassay based on quantum dot–streptavidin conjugate for the detection of chlorpyrifos. *J. Agric. Food Chem.*, 58, 8895–8903, 2010.
71. Chhipa, H., Nanofertilizers and nanopesticides for agriculture. *Environ. Chem. Lett.*, 15, 15–22, 2017.
72. Chhipa, H. and Kaushik, N., Development of nano-bio-pesticide using Iron and Eucalyptus plant extract and their application in pest management, in: *Conference Proceeding of Symposium on Recent Advances in Biotechnology for Food and Fuel*, pp. 19–20, New Delhi, TERI, 2015.
73. Chookhongkha, N., Sopondilok, T., Photchanachai, S., Effect of chitosan and chitosan nanoparticles on fungal growth and chilli seed quality, in: *International Conference on Postharvest Pest and Disease Management in Exporting Horticultural Crops-PPDM2012 973*, Bankok, pp. 231–237, 2012.
74. Cifuentes, Z., Custardoy, L., de la Fuente, J.M., Marquina, C., Ibarra, M.R., Rubiales, D., Absorption and translocation to the aerial part of magnetic carbon-coated nanoparticles through the root of different crop plants. *J. Nanobiotechnol.*, 8, 26, 2010.
75. Colebrook, E.H., Thomas, S.G., Phillips, A.L., Hedden, P., The role of gibberellin signalling in plant responses to abiotic stress. *J. Experiment. Boil.*, 217, 67–75, 2014.
76. Collin, B., Auffan, M., Johnson, A.C., Kaur, I., Keller, A.A., Lazareva, A., Environmental release, fate and ecotoxicological effects of manufactured ceria nanomaterials. *Environ. Sci. Nano*, 1, 533–548, 2014.
77. Cromwell, W.A., Yang, J., Starr, J.L., Jo, Y.K., Nematicidal effects of silver nanoparticles on root-knot nematode in bermudagrass. *J. Nematol.*, 46, 261–266, 2014.
78. Cummins, C.M., Koivunen, M.E., Stephanian, A., Gee, S.J., Hammock, B.D., Kennedy, I.M., Application of europium (III) chelate-dyed nanoparticle labels in a competitive atrazine fluoroimmunoassay on an ITO waveguide. *Biosens. Bioelectron.*, 21, 1077–1085, 2006.
79. Cumplido-Nájera, C.F., González-Morales, S., Ortega-Ortíz, H., Cadenas-Pliego, G., Benavides-Mendoza, A., Juárez-Maldonado, A., The application

of copper nanoparticles and potassium silicate stimulate the tolerance to Clavibacter michiganensis in tomato plants. *Sci. Horticult.*, 245, 82–89, 2019.

80. da Costa, M.V.J. and Sharma, P.K., Effect of copper oxide nanoparticles on growth, morphology, photosynthesis, and antioxidant response in Oryza sativa. *Photosynthetica*, 54, 110–119, 2016.
81. Dai, D. and Fan, M., Green modification of natural fibres with nanocellulose. *RSC Adv.*, 3, 4659–4665, 2013.
82. Davis, R.A., Rippner, D.A., Hausner, S.H., Parikh, S.J., McElrone, A.J., Sutcliffe, J.L., *In vivo* tracking of copper-64 radiolabeled nanoparticles in Lactuca sativa. *Environ. Sci. Technol.*, 51, 12537–12546, 2017.
83. de Oliveira, J.L., Campos, E.V., Bakshi, M., Abhilash, P.C., Fraceto, L.F., Application of nanotechnology for the encapsulation of botanical insecticides for sustainable agriculture: Prospects and promises. *Biotech. Adv.*, 32, 1550–1561, 2014.
84. Delfani, M., Baradarn Firouzabadi, M., Farrokhi, N., Makarian, H., Some physiological responses of black-eyed pea to iron and magnesium nanofertilizers. *Commun. Soil Sci. Plant Anal.*, 45, 530–540, 2014.
85. Deng, Y., Petersen, E.J., Challis, K.E., Rabb, S.A., Holbrook, R.D., Ranville, J.F., Multiple method analysis of TiO2 nanoparticle uptake in rice (Oryza sativa L.) plants. *Environ. Sci. Technol.*, 51, 10615–10623, 2017.
86. Dhoke, S.K., Mahajan, P., Kamble, R., Khanna, A., Effect of nanoparticles suspension on the growth of mung (Vigna radiata) seedlings by foliar spray method. *Nanotechnol. Dev.*, 3, e1, 2013.
87. Didunyemi, M.O., Adetuyi, B.O., Oyebanjo, O.O., Morinda lucida attenuates acetaminophen-induced oxidative damage and hepatotoxicity in rats. *J. Biomed. Sci.*, 8, 1-7, 2019. https://www.jbiomeds.com/biomedical-sciences/morinda-lucida-attenuates-acetaminopheninduced-oxidative-damage-and-hepatotoxicity-in-rats.php?aid=24482.
88. Didunyemi, M.O., Adetuyi, B.O., Oyewale, I.A., Inhibition of lipid peroxidation and *in-vitro* antioxidant capacity of aqueous, acetone and methanol leaf extracts of green and red Acalypha wilkesiana Muell Arg. *Int. J. Biol. Med. Res.*, 11, 3, 7089–7094, 20202020.
89. Dietz, K.J. and Herth, S., Plant nanotoxicology. *Trends. Plant Sci.*, 16, 582–589, 2011.
90. Dimkpa, C.O. and Bindraban, P.S., Fortification of micronutrients for efficient agronomic production: A review. *Agro. Sustain Dev.*, 36, 7, 2016.
91. Dimkpa, C.O. and Bindraban, P.S., Nanofertilizers: New products for the industry? *J. Agric. Food Chem.*, 66, 6462–6473, 2017.
92. Dimkpa, C.O., Bindraban, P.S., Fugice, J., Agyin-Birikorang, S., Singh, U., Hellums, D., Composite micronutrient nanoparticles and salts decrease drought stress in soybean. *Agro. Sustain. Dev.*, 37, 5, 2017a.
93. Dimkpa, C.O., White, J.C., Elmer, W.H., Gardea-Torresdey, J., Nanoparticle and ionic Zn promote nutrient loading of sorghum grain under low NPK fertilization. *J. Agricul. Food. Chem.*, 65, 8552–8559, 2017b.

94. Djanaguiraman, M., Belliraj, N., Bossmann, S.H., Prasad, P.V., High-temperature stress alleviation by selenium nanoparticle treatment in grain sorghum. *ACS Omega*, *3*, 2479–2491, 2018a.
95. Djanaguiraman, M., Nair, R., Giraldo, J.P., Prasad, P.V.V., Cerium oxide nanoparticles decrease drought-induced oxidative damage in sorghum leading to higher photosynthesis and grain yield. *ACS Omega*, *3*, 14406–14416, 2018b.
96. Dogaroglu, Z.G. and Köleli, N., TiO2 and ZnO nanoparticles toxicity in barley (Hordeum vulgare L.). *Clean Soil Air Water*, 45, 1700096, 2017.
97. Du, W., Sun, Y., Ji, R., Zhu, J., Wu, J., Guo, H., TiO2 and ZnO nanoparticles negatively affect wheat growth and soil enzyme activities in agricultural soil. *J. Environ. Monit.*, 13, 822–828, 2011.
98. Dutta, P., Reddy, S.G.E., Borthakur, B.K., Effect of neem kernal aqueous extract (NKAE) in tea mosquito bug, helopeltis theivora (Waterhouse, 1886) (Heteroptera: Miridae). *Munis Entomology & Zoology*, 8, 1, 213–218, 2013.
99. Dwivedi, S., Saquib, Q., Al-Khedhairy, A.A., Musarrat, J., Understanding the role of nanomaterials in agriculture, in: *Microbial Inoculants in Sustainable Agricultural Productivity*, pp. 271–288, Springer, New Delhi, 2016.
100. Eichert, T., Kurtz, A., Steiner, U., Goldbach, H.E., Size exclusion limits and lateral heterogeneity of the stomatal foliar uptake pathway for aqueous solutes and water-suspended nanoparticles. *Physiol. Plant*, 134, 151–160, 2008.
101. Elek, N., Hoffman, R., Ravi, U., Resh, R., Ishaaya, I., Magdassia, S., Novaluron nanoparticles: Formation and potential use in controlling agricultural insect pests. *Colloids Surf. A: Physicochem. Eng. Asp.*, 372, 66–72, 2010.
102. Elias, E.H., Flynn, R., Idowu, O.J., Reyes, J., Sanogo, S., Schutte, B.J., Crop vulnerability to weather and climate risk: Analysis of interacting systems and adaptation efficacy for sustainable crop production. *Sustainability*, 11, 6619, 2019.
103. Elmer, W.H. and White, J.C., The use of metallic oxide nanoparticles to enhance growth of tomatoes and eggplants in disease infested soil or soilless medium. *Environ. Sci. Nano*, 3, 1072–1079, 2016.
104. Elsheery, N.I., Helaly, M.N., El-Hoseiny, H.M., Alam-Eldein, S.M., Zinc oxide and silicone nanoparticles to improve the resistance mechanism and annual productivity of salt-stressed mango trees. *Agronomy*, 10, 558, 2020.
105. Esser, B., Schnorr, J.M., Swager, T.M., Selective detection of ethylene gas using carbon nanotube-based devices: Utility in determination of fruit ripeness. *Angew. Chem. Int. Ed.*, 51, 5752–5756, 2012.
106. Faisal, M., Saquib, Q., Alatar, A.A., Al-Khedhairy, A.A., Hegazy, A.K., Musarrat, J., Phytotoxic hazards of NiO-nanoparticles in tomato: A study on mechanism of cell death. *J. Hazard. Mat.*, 250, 318–332, 2013.
107. Falco, W.F., Scherer, M.D., Oliveira, S.L., Wender, H., Colbeck, I., Lawson, T., Phytotoxicity of silver nanoparticles on Vicia faba: Evaluation of particle

size effects on photosynthetic performance and leaf gas exchange. *Sci. Total. Environ.*, 701, 134816, 2020.
108. Fan, P., Gu, Z., Xu, D., Xu, X., Xu, G., Action analysis of drops of emamection-benzoate microemulsion on rice leaf. *Chine. J. Rice Sci.*, 24, 503–508, 2010.
109. Fang, Y., Umasankar, Y., Ramasamy, R.P., Electrochemical detection of p-ethylguaiacol, a fungi infected fruit volatile using metal oxide nanoparticles. *Analyst*, 139, 3804–3810, 2014.
110. Faraji, J. and Sepehri, A., Exogenous nitric oxide improves the protective effects of tio 2 nanoparticles on growth, antioxidant system, and photosynthetic performance of wheat seedlings under drought stress. *J. Soil. Sci. Plant Nutr.*, 20, 703–714, 2020.
111. Farombi, E.O., Abolaji, A.O., Adetuyi, B.O., Awosanya, O.O., Fabusoro, M., Neuroprotective role of 6-Gingerol-rich fraction of Zingiber officinale (Ginger) against acrylonitrile-induced neurotoxicity in male Wistar rats. *J. Basic Clin. Physiol. Pharmacol. (JBCPP), 30,* 1-8, 3, 2018. https://doi.org/10.1515/jbcpp-2018-0114.
112. Feng, Y., Cui, X., He, S., Dong, G., Chen, M., Wang, J., The role of metal nanoparticles in influencing arbuscular mycorrhizal fungi effects on plant growth. *Environ. Sci. Technol.*, 47, 9496–9504, 2013.
113. Foyer, C.H. and Noctor, G., Redox homeostasis and antioxidant signaling: A metabolic interface between stress perception and physiological responses. *Plant Cell.*, 17, 1866–1875, 2005.
114. Fraceto, L.F., Grillo, R., de Medeiros, G.A., Scognamiglio, V., Rea, G., Bartolucci, C., Nanotechnology in agriculture: Which innovation potential does it have? *Front. Environ. Sci.*, 4, 20, 2016.
115. Frazier, T.P., Burklew, C.E., Zhang, B., Titanium dioxide nanoparticles affect the growth and microRNA expression of tobacco (Nicotiana tabacum). *Funct. Integ. Genom.*, 14, 75–83, 2014.
116. Gandhi, V.P. and Namboodiri, N.V., *Economic of BT cotton vis-à-vis non BT cotton in india, A study across for major cotton growing states,* pp. 1–127, Centre for Management in Agriculture Indian Institute of Management, Ahmedabad, India, 2009.
117. Gao, G., Tester, M.A., Julkowska, M.M., The use of high-throughput phenotyping for assessment of heat stress-induced changes in arabidopsis. *Plant Phenom.*, 2020, 14, 2020.
118. Ge, F. and Ding, Y., The population energy dynamics of predacious natural enemies and their pest control activity in different cotton agroecosystems. *Acta Entomologica*, 39, 3, 266–273, 1996.
119. Ge, Y., Schimel, J.P., Holden, P.A., Identification of soil bacteria susceptible to TiO2 and ZnO nanoparticles. *Appl. Environ. Microbial.*, 78, 6749–6758, 2012.
120. Ghafariyan, M.H., Malakouti, M.J., Dadpour, M.R., Stroeve, P., Mahmoudi, M., Effects of magnetite nanoparticles on soybean chlorophyll. *Environ. Sci. Technol.*, 47, 10645–10652, 2013.

121. Ghormade, V., Deshpande, M.V., Paknikar, K.M., Perspectives for nano-biotechnology enabled protection and nutrition of plants. *Biotechnol. Adv.*, 29, 792–803, 2011.
122. Giannousi, K., Avramidis, I., Dendrinou-Samara, C., Synthesis, characterization and evaluation of copper based nanoparticles as agrochemicals against Phytophthora infestans. *RSC Adv.*, 3, 21743–21752, 2013.
123. Gilroy, S., Suzuki, N., Miller, G., Choi, W.G., Toyota, M., Devireddy, A.R., A tidal wave of signals: Calcium and ROS at the forefront of rapid systemic signaling. *Trends. Plant Sci.*, 19, 623–630, 2014.
124. Giraldo, J.P., Landry, M.P., Faltermeier, S.M., McNicholas, T.P., Iverson, N.M., Boghossian, A.A., Plant nanobionics approach to augment photosynthesis and biochemical sensing. *Nat. Mater.*, 13, 400–408, 2014.
125. Giraldo, J.P., Landry, M.P., Kwak, S.Y., Jain, R.M., Wong, M.H., Iverson, N.M., A ratiometric sensor using single chirality near-infrared fluorescent carbon nanotubes: Application to *in vivo* monitoring. *Small*, 11, 3973–3984, 2015.
126. Giraldo, J.P., Wu, H., Newkirk, G.M., Kruss, S., Nanobiotechnology approaches for engineering smart plant sensors. *Nat. Nanotech.*, 14, 541–553, 2019.
127. Godfray, H.C., Beddington, J.R., Crute, I.R., Haddad, L., Lawrence, D., Muir, J.F., Food security: The challenge of feeding 9 billion people. *Science*, 327, 812–818, 2010.
128. Godfray, H.C., Mason-D'Croz, D., Robinson, S., Food system consequences of a fungal disease epidemic in a major crop. *Philos. Trans. R. Soc. Lond. B. Biol. Sci.*, 371, 20150467, 2016.
129. Gohari, G., Mohammadi, A., Akbari, A., Panahirad, S., Dadpour, M.R., Fotopoulos, V., Titanium dioxide nanoparticles (TiO2 NPs) promote growth and ameliorate salinity stress effects on essential oil profile and biochemical attributes of Dracocephalum moldavica. *Sci. Rep.*, 10, 912, 2020.
130. Govea-Alcaide, E., Masunaga, S.H., De Souza, A., Fajardo-Rosabal, L., Effenberger, F.B., Rossi, L.M., Tracking iron oxide nanoparticles in plant organs using magnetic measurements. *J. Nano. Res.*, 18, 305, 2016.
131. Gowayed, M.H., Al-Zahrani, H.S., Metwali, E.M., Improving the salinity tolerance in potato (Solanum tuberosum) by exogenous application of silicon dioxide nanoparticles. *Int. J. Agric. Biol.*, 19, 183–194, 2017.
132. Graham, J.H., Johnson, E.G., Myers, M.E., Young, M., Rajasekaran, P., Das, S., Potential of nano-formulated zinc oxide for control of citrus canker on grapefruit trees. *Plant Dis.*, 100, 2442–2447, 2016.
133. Gui, X., Zhang, Z., Liu, S., Ma, Y., Zhang, P., He, X., Fate and phytotoxicity of CeO2 nanoparticles on lettuce cultured in the potting soil environment. *PLoS One*, 10, e0134261, 2015.
134. Guo, Y.R., Liu, S.Y., Gui, W.J., Zhu, G.N., Gold immunochromatographic assay for simultaneous detection of carbofuran and triazophos in water samples. *Anal. Biochem.*, 389, 32–39, 2009.

135. Guo, Z., Park, S., Yoon, J., Shin, I., Recent progress in the development of near-infrared fluorescent probes for bioimaging applications. *Chem. Soc. Rev.*, 43, 16–29, 2014.
136. Haghighi, M., Abolghasemi, R., da Silva, J.A.T., Low and high temperature stress affect the growth characteristics of tomato in hydroponic culture with Se and nano-Se amendment. *Sci. Horticult.*, 178, 231–240, 2014.
137. Haghighi, M. and Pessarakli, M., Influence of silicon and nano-silicon on salinity tolerance of cherry tomatoes (Solanum lycopersicum L.) at early growth stage. *Sci. Horticult.*, 161, 111–117, 2013.
138. Hao, Y., Yu, F., Lv, R., Ma, C., Zhang, Z., Rui, Y., Carbon nanotubes filled with different ferromagnetic alloys affect the growth and development of rice seedlings by changing the C: N ratio and plant hormones concentrations. *PLoS One*, 11, e0157264, 2016.
139. Hatami, M., Toxicity assessment of multi-walled carbon nanotubes on Cucurbita pepo L. under well-watered and water-stressed conditions. *Ecotoxicol. Environ. Saf.*, 142, 274–283, 2017.
140. Hatfield, J.L., Gitelson, A.A., Schepers, J.S., Walthall, C.L., Application of spectral remote sensing for agronomic decisions. *Agrono. J.*, 100, 117–131, 2008.
141. Hauser, F. and Horie, T., A conserved primary salt tolerance mechanism mediated by HKT transporters: A mechanism for sodium exclusion and maintenance of high K+/Na+ ratio in leaves during salinity stress. *Plant Cell. Environ.*, 33, 552–565, 2010.
142. Heikenfeld, J., Jajack, A., Rogers, J., Gutruf, P., Tian, L., Pan, T., Wearable sensors: Modalities, challenges, and prospects. *Lab. Chip.*, 18, 217–248, 2018.
143. Hernández-Hernández, H., González-Morales, S., Benavides-Mendoza, A., Ortega-Ortiz, H., Cadenas-Pliego, G., Juárez-Maldonado, A., Effects of chitosan–PVA and Cu nanoparticles on the growth and antioxidant capacity of tomato under saline stress. *Molecules*, 23, 178, 2018.
144. Ho, V.A., Le, P.T., Nguyen, T.P., Nguyen, C.K., Nguyen, V.T., Tran, N.Q., Silver core-shell nanoclusters exhibiting strong growth inhibition of plant-pathogenic fungi. *J. Nanomater.*, 2015, 241614, 2015.
145. Hojjat, S.S. and Kamyab, M., The effect of silver nanoparticle on Fenugreek seed germination under salinity levels. *Russian. Agricult. Sci.*, 43, 61–65, 2017.
146. Hong, G., Diao, S., Antaris, A.L., Dai, H., Carbon nanomaterials for biological imaging and nanomedicinal therapy. *Chem. Rev.*, 115, 10816–10906, 2015.
147. Hotze, E.M., Phenrat, T., Lowry, G.V., Nanoparticle aggregation: Challenges to understanding transport and reactivity in the environment. *J. Environ. Qual.*, 39, 1909–1924, 2010.
148. Hu, P., An, J., Faulkner, M.M., Wu, H., Li, Z., Tian, X., Nanoparticle charge and size control foliar delivery efficiency to plant cells and organelles. *ACS. Nano*, 14, 7970–7986, 2020.

149. Hua, X., Qian, G., Yang, J., Hu, B., Fan, J., Qin, N., Development of an immunochromatographic assay for the rapid detection of chlorpyrifos-methyl in water samples. *Biosens. Bioelectron.*, 26, 189–194, 2010.
150. Huang, H., Ullah, F., Zhou, D.X., Yi, M., Zhao, Y., Mechanisms of ROS regulation of plant development and stress responses. *Front. Plant Sci.*, 10, 800, 2019.
151. Huang, S., Wang, L., Liu, L., Hou, Y., Li, L., Nanotechnology in agriculture, livestock, and aquaculture in china. A review. *Agro. Sustain. Dev.*, 35, 369–400, 2015.
152. Huang, Z., Rajasekaran, P., Ozcan, A., Santra, S., Antimicrobial magnesium hydroxide nanoparticles as an alternative to Cu biocide for crop protection. *J. Agricul. Food. Chem.*, 66, 8679–8686, 2018.
153. Imada, K., Sakai, S., Kajihara, H., Tanaka, S., Ito, S., Magnesium oxide nanoparticles induce systemic resistance in tomato against bacterial wilt disease. *Plant Pathol.*, 65, 551–560, 2016.
154. Iqbal, M.A., Nano-fertilizers for sustainable crop production under changing climate: A global perspective, in: *Sustainable Crop Production*, IntechOpen, London, UK, 2019.
155. Iqbal, M., Raja, N.I., Hussain, M., Ejaz, M., Yasmeen, F., Effect of silver nanoparticles on growth of wheat under heat stress. *Ira. J. Sci. Technol. Transac. A Sci.*, 43, 387–395, 2019.
156. Jaberzadeh, A., Moaveni, P., Moghadam, H.R., Zahedi, H., Influence of bulk and nanoparticles titanium foliar application on some agronomic traits, seed gluten and starch contents of wheat subjected to water deficit stress. *Not. Bot. Horti Agrobot. Cluj Napoca*, 41, 201–207, 2013.
157. James-Okoro, Paula-Peace, O., Iheagwam, F.N., Sholeye, M.I., Umoren, I.A., Adetuyi, B.O., Ogundipe, A.E., Braimah, A.A., Adekunbi, T.S., Ogunlana, O.E., Ogunlana, O.O., Phytochemical and *in vitro* antioxidant assessment of Yoyo bitters. *World News Nat. Sci.*, 37, 1–17, 2021.
158. Jampílek, J. and Králová, K., Nanopesticides: Preparation, targeting, and controlled release, in: *New Pesticides and Soil Sensors*, A.M. Grumezescu (Ed.), pp. 81–127, Academic Press, Netherlands, 2017.
159. Jeevanandam, J., Barhoum, A., Chan, Y.S., Dufresne, A., Danquah, M.K., Review on nanoparticles and nanostructured materials: History, sources, toxicity and regulations. *Beilstein J. Nanotech.*, 9, 1050–1074, 2018.
160. Jiang, H.S., Qiu, X.N., Li, G.B., Li, W., Yin, L.Y., Silver nanoparticles induced accumulation of reactive oxygen species and alteration of antioxidant systems in the aquatic plant Spirodela polyrhiza. *Environ. Toxicol. Chem.*, 33, 1398–1405, 2014.
161. Jin, Y., Fan, X., Li, X., Zhang, Z., Sun, L., Fu, Z., Distinct physiological and molecular responses in Arabidopsis thaliana exposed to aluminum oxide nanoparticles and ionic aluminum. *Environ. Pollut.*, 228, 517–527, 2017.
162. Jo, Y.K., Kim, B.H., Jung, G., Antifungal activity of silver ions and nanoparticles on phytopathogenic fungi. *Plant Dis.*, 93, 1037–1043, 2009.

163. Jordan, J.T., Oates, R.P., Subbiah, S., Payton, P.R., Singh, K.P., Shah, S.A., Carbon nanotubes affect early growth, flowering time and phytohormones in tomato. *Chemosphere*, 256, 127042, 2020.
164. Ju, M., Navarreto-Lugo, M., Wickramasinghe, S., Milbrandt, N.B., McWhorter, A., Samia, A.C., Exploring the chelation-based plant strategy for iron oxide nanoparticle uptake in garden cress (Lepidium sativum) using magnetic particle spectrometry. *Nanoscale*, 11, 18582–18594, 2019.
165. Juárez-Maldonado, A., Ortega-Ortíz, H., Morales-Díaz, A.B., González-Morales, S., Morelos-Moreno, Á., Sandoval-Rangel, A., Nanoparticles and nanomaterials as plant biostimulants. *Int. J. Molec. Sci.*, 20, 162, 2019.
166. Kamaraj, C., Abdul Rahuman, A., Bagavan, A., Screening for antifeedant and larvicidal activity of plant extract against Helicoverpa armigera (Hubner), Sylepta derogata (F) and Anopheles stephensi (Liston). *Parasitol. Res.*, 103, 6, 1361–1368, 2020.
167. Kang, Y., Khan, S., Ma, X., Climate change impacts on crop yield, crop water productivity and food security–A review. *Prog. Natural Sci.*, 19, 1665–1674, 2009.
168. Karamian, R., Ghasemlou, F., Amiri, H., Physiological evaluation of drought stress tolerance and recovery in Verbascum sinuatum plants treated with methyl jasmonate, salicylic acid and titanium dioxide nanoparticles. *Plant Biosys. Int. J. Deal. Aspect. Plant Biol.*, 154, 277–287, 2020.
169. Kaur, T., Singh, G.P., Kaur, G., Kaur, S., Gill, P.K., Synthesis of biogenic silicon/silica (Si/SiO2) nanocomposites from rice husks and wheat bran through various microorganisms. *Mater. Res. Expr.*, 3, 085026, 2016.
170. Keller, A.A., Huang, Y., Nelson, J., Detection of nanoparticles in edible plant tissues exposed to nano-copper using single-particle ICP-MS. *J. Nano. Res.*, 20, 101, 2018.
171. Khan, M.N., AlSolami, M.A., Basahi, R.A., Siddiqui, M.H., Al-Huqail, A.A., Abbas, Z.K., Nitric oxide is involved in nano-titanium dioxide-induced activation of antioxidant defense system and accumulation of osmolytes under water-deficit stress in Vicia faba L. *Ecotoxicol. Environ. Saf.*, 190, 110152, 2020.
172. Khodakovskaya, M.V., De Silva, K., Biris, A.S., Dervishi, E., Villagarcia, H., Carbon nanotubes induce growth enhancement of tobacco cells. *ACS Nano*, 6, 2128–2135, 2012.
173. Kim, T.H., Böhmer, M., Hu, H., Nishimura, N., Schroeder, J.I., Guard cell signal transduction network: Advances in understanding abscisic acid, CO2, and Ca2+ signaling. *Ann. Review. Plant Boil.*, 61, 561–591, 2010.
174. Kraegeloh, A., Suarez-Merino, B., Sluijters, T., Micheletti, C., Implementation of safe-by-design for nanomaterial development and safe innovation: Why we need a comprehensive approach. *Nanomaterials*, 8, 239, 2018.
175. Kranthi, K.R., Davis, M., Mayee, C.D., Russell, D.A., Shukla, R.M., Satija, U., Development of a colloidal-gold based lateral-flow immunoassay kit for

'quality-control' assessment of pyrethroid and endosulfan formulations in a novel single strip format. *Crop Prot.*, 28, 428–434, 2009.
176. Kruss, S., Landry, M.P., Vander Ende, E., Lima, B.M., Reuel, N.F., Zhang, J., Neurotransmitter detection using corona phase molecular recognition on fluorescent single-walled carbon nanotube sensors. *J. Amer. Chem. Soc.*, 136, 713–724, 2014.
177. Kumar, R.S., Shiny, P.J., Anjali, C.H., Jerobin, J., Goshen, K.M., Magdassi, S., Distinctive effects of nano-sized permethrin in the environment. *Environ. Sci. Pollut. Res. Int.*, 20, 2593–2602, 2013.
178. Kumbhakar, P., Ray, S.S., Stepanov, A.L., Optical properties of nanoparticles and nanocomposites. *J. Nanomater.*, 2014, 181365, 2014.
179. Lahiani, M.H., Dervishi, E., Chen, J., Nima, Z., Gaume, A., Biris, A.S., Impact of carbon nanotube exposure to seeds of valuable crops. *ACS Appl. Mater. Interfaces.*, 5, 7965–7973, 2013.
180. Larue, C., Castillo-Michel, H., Sobanska, S., Trcera, N., Sorieul, S., Cécillon, L., Fate of pristine TiO2 nanoparticles and aged paint-containing TiO2 nanoparticles in lettuce crop after foliar exposure. *J. Hazard Mat.*, 273, 17–26, 2014.
181. Larue, C., Laurette, J., Herlin-Boime, N., Khodja, H., Fayard, B., Flank, A.M., Accumulation, translocation and impact of TiO2 nanoparticles in wheat (Triticum aestivum spp.): Influence of diameter and crystal phase. *Sci. Total Environ.*, 431, 197–208, 2012a.
182. Larue, C., Pinault, M., Czarny, B., Georgin, D., Jaillard, D., Bendiab, N., Quantitative evaluation of multi-walled carbon nanotube uptake in wheat and rapeseed. *J. Hazard. Mater.*, 227, 155–163, 2012b.
183. Larue, C., Veronesi, G., Flank, A.M., Surble, S., Herlin-Boime, N., Carrière, M., Comparative uptake and impact of TiO2 nanoparticles in wheat and rapeseed. *J. Toxicol. Environ. Health A*, 75, 722–734, 2012c.
184. Lee, K., Park, J., Lee, M.S., Kim, J., Hyun, B.G., Kang, D.J., *In-situ* synthesis of carbon nanotube–graphite electronic devices and their integrations onto surfaces of live plants and insects. *Nano. Lett.*, 14, 2647–2654, 2014.
185. Lehr, P., *BCC Research Global Markets for Bio pesticides, Market Research Reports. Code - CHM029E*, BCC Research Publisher, United State, 2014.
186. Leso, V., Fontana, L., Iavicoli, I., Biomedical nanotechnology: Occupational views. *Nano Today*, 24, 10–14, 2019.
187. Li, J., Hu, J., Ma, C., Wang, Y., Wu, C., Huang, J., Uptake, translocation and physiological effects of magnetic iron oxide (γ-Fe2O3) nanoparticles in corn (Zea mays L.). *Chemosphere*, 159, 326–334, 2016.
188. Li, J., Wu, H., Santana, I., Fahlgren, M., Giraldo, J.P., Standoff optical glucose sensing in photosynthetic organisms by a quantum dot fluorescent probe. *ACS. Appl. Mater. Interf.*, 10, 28279–28289, 2018.
189. Li, R., He, J., Xie, H., Wang, W., Bose, S.K., Sun, Y., Effects of chitosan nanoparticles on seed germination and seedling growth of wheat (Triticum aestivum L.). *Int. J. Boil. Macromolec.*, 126, 91–100, 2019.

190. Li, W.Q., Qing, T., Li, C.C., Li, F., Ge, F., Fei, J.J., Integration of subcellular partitioning and chemical forms to understand silver nanoparticles toxicity to lettuce (Lactuca sativa L.) under different exposure pathways. *Chemosphere*, 258, 127349, 2020.
191. Li, Y., Jin, Q., Yang, D., Cui, J., Molybdenum sulfide induce growth enhancement effect of rice (Oryza sativa L.) through regulating the synthesis of chlorophyll and the expression of aquaporin gene. *J. Agricul. Food. Chem.*, 66, 4013–4021, 2018.
192. Lin, S., Reppert, J., Hu, Q., Hudson, J.S., Reid, M.L., Ratnikova, T.A., Uptake, translocation, and transmission of carbon nanomaterials in rice plants. *Small*, 5, 1128–1132, 2009.
193. Lisha, K.P. and Pradeep, T., Enhanced visual detection of pesticides using gold nanoparticles. *J. Environ. Sci. Health Part B*, 44, 697–705, 2009.
194. Liu, F., Wen, L.X., Li, Z.Z., Yu, W., Sun, H.Y., Chen, J.F., Porous hollow silica nanoparticles as controlled delivery system for water-soluble pesticide. *Mater. Res. Bull.*, 41, 2268–2275, 2006.
195. Liu, M., Feng, S., Ma, Y., Xie, C., He, X., Ding, Y., Influence of surface charge on the phytotoxicity, transformation, and translocation of CeO2 nanoparticles in cucumber plants. *ACS Appl. Mater. Interf.*, 11, 16905–16913, 2019.
196. Liu, R. and Lal, R., Synthetic apatite nanoparticles as a phosphorus fertilizer for soybean (Glycine max). *Sci. Rep.*, 4, 5686, 2015.
197. Liu, R., Zhang, H., Lal, R., Effects of stabilized nanoparticles of copper, zinc, manganese, and iron oxides in low concentrations on lettuce (Lactuca sativa) seed germination: Nanotoxicants or nanonutrients? *Water. Air. Soil. Poll.*, 227, 42, 2016.
198. Liu, X.M., Feng, Z.B., Zhang, F.D., Zhang, S.Q., He, X.S., Preparation and testing of cementing and coating nano-subnanocomposites of slow/controlled-release fertilizer. *Agricult. Sci. China*, 5, 700–706, 2006.
199. Liu, X., Zhang, F., Zhang, S., He, X., Wang, R., Fei, Z., Responses of peanut to nano-calcium carbonate. *Plant Nutri. Ferti. Sci.*, 11, 385–389, 2005.
200. Liu, Y. and Shi, J., Antioxidative nanomaterials and biomedical applications. *Nano Today*, 27, 146–177, 2019.
201. Lombi, E., Donner, E., Dusinska, M., Wickson, F., A one health approach to managing the applications and implications of nanotechnologies in agriculture. *Nat. Nanotechnol.*, 14, 523–531, 2019.
202. López, M.M., Llop, P., Olmos, A., Marco-Noales, E., Cambra, M., Bertolini, E., Are molecular tools solving the challenges posed by detection of plant pathogenic bacteria and viruses? *Curr. Issues Mol. Biol.*, 11, 13–46, 2009.
203. López-Moreno, M.L., de la Rosa, G., Hernández-Viezcas, J. Á., Castillo-Michel, H., Botez, C.E., Peralta-Videa, J.R., Evidence of the differential biotransformation and genotoxicity of ZnO and CeO2 nanoparticles on soybean (Glycine max) plants. *Environ. Sci. Technol.*, 44, 7315–7320, 2010.

204. Lu, C., Zhang, C., Wen, J., Wu, G., Tao, M., Research of the effect of nanometer materials on germination and growth enhancement of Glycine max and its mechanism. *Soybean Sci.*, 21, 168–171, 2002.
205. Luyckx, M., Hausman, J.F., Lutts, S., Guerriero, G., Silicon and plants: Current knowledge and technological perspectives. *Front. Plant Sci.*, 8, 411, 2017.
206. Lv, J., Christie, P., Zhang, S., Uptake, translocation, and transformation of metal-based nanoparticles in plants: Recent advances and methodological challenges. *Environ. Sci. Nano*, 6, 41–59, 2019.
207. Ma, C., Rui, Y., Liu, S., Li, X., Xing, B., Liu, L., Phytotoxic mechanism of nanoparticles: Destruction of chloroplasts and vascular bundles and alteration of nutrient absorption. *Sci. Rep.*, 5, 11618, 2015.
208. Ma, X., Geiser-Lee, J., Deng, Y., Kolmakov, A., Interactions between engineered nanoparticles (ENPs) and plants: Phytotoxicity, uptake and accumulation. *Sci. Total. Environ.*, 408, 3053-3061, 2010.
209. Mahil, T., Baburai, N., Aravinda, K., Foliar application of nanofertilizers in agricultural crops- A review. *J. Farm Sci.*, 32, 239–249, 2019.
210. Mahmoodzadeh, H., Nabavi, M., Kashefi, H., Effect of nanoscale titanium dioxide particles on the germination and growth of canola (Brassica napus). *J. Ornamen. Horticult. Plants.*, 3, 25–32, 2013.
211. Mahmoud, L.M., Dutt, M., Shalan, A.M., El-Kady, M.E., El-Boray, M.S., Shabana, Y.M., Silicon nanoparticles mitigate oxidative stress of *in vitro*-derived banana (Musa acuminata 'Grand Nain') under simulated water deficit or salinity stress. *South Afr. J. Bot.*, 132, 155–163, 2021.
212. Malik, M.F., Khan, A.G., Hussiany, S.W., Rehman, D.U., Amin, M., Scouting and control of Helicoverpa armigera by synthetic pheromones technology in apple. *Asian J. Plant Sci.*, 3, 663–664, 2002.
213. Mattiello, A., Filippi, A., Pošćić, F., Musetti, R., Salvatici, M.C., Giordano, C., Evidence of phytotoxicity and genotoxicity in Hordeum vulgare L. exposed to CeO2 and TiO2 nanoparticles. *Frontiers. Plant Sci.*, 6, 1043, 2015.
214. Mattos, B.D. and Magalhães, W.L., Biogenic nanosilica blended by nanofibrillated cellulose as support for slow-release of tebuconazole. *J. Nano Res.*, 18, 274, 2016.
215. McGehee, D.L., Alimohammadi, M., Khodakovskaya, M.V., Carbon-based nanomaterials as stimulators of production of pharmaceutically active alkaloids in cell culture of Catharanthus roseus. *Nanotechnology*, 30, 275102, 2019.
216. Meena, R.S., Kumar, S., Datta, R., Lal, R., Vijayakumar, V., Brtnicky, M., Impact of agrochemicals on soil microbiota and management: A review. *Land*, 9, 34, 2020.
217. Meychik, N.R., Nikolaeva, J.I., Yermakov, I.P., Ion exchange properties of the root cell walls isolated from the halophyte plants (Suaeda altissima L.) grown under conditions of different salinity. *Plant Soil*, 277, 163, 2005.
218. Milenković, I., Mitrović, A., Algarra, M., Lázaro-Martínez, J.M., Rodríguez-Castellón, E., Maksimović, V., Interaction of carbohydrate coated

cerium-oxide nanoparticles with wheat and pea: Stress induction potential and effect on development. *Plant*, 8, 478, 2019.
219. Milewska-Hendel, A., Zubko, M., Stróz, D., Kurczyńska, E.U., Effect of nanoparticles surface charge on the Arabidopsis thaliana (L.) roots development and their movement into the root cells and protoplasts. *Int. J. Molec. Sci.*, 20, 1650, 2019.
220. Mingyu, S., Chao, L., Chunxiang, Q., Lei, Z., Liang, C., Hao, H., Nano-anatase relieves the inhibition of electron transport caused by linolenic acid in chloroplasts of spinach. *Biol. Trace Ele. Res.*, 122, 73–81, 2008.
221. Mishra, S., Singh, B.R., Singh, A., Keswani, C., Naqvi, A.H., Singh, H.B., Biofabricated silver nanoparticles act as a strong fungicide against Bipolaris sorokiniana causing spot blotch disease in wheat. *PLoS One*, 9, e97881, 2014.
222. Mohamed, A.K.S., Qayyum, M.F., Abdel-Hadi, A.M., Rehman, R.A., Ali, S., Rizwan, M., Interactive effect of salinity and silver nanoparticles on photosynthetic and biochemical parameters of wheat. *Arch. Agro. Soil Sci.*, 63, 1736–1747, 2017.
223. Mohammadi, R., Maali-Amiri, R., Abbasi, A., Effect of TiO2 nanoparticles on chickpea response to cold stress. *Biolo. Trace Elem. Res.*, 152, 403–410, 2013.
224. Møller, I.M., Jensen, P.E., Hansson, A., Oxidative modifications to cellular components in plants. *Annu. Rev. Plant Biol.*, 58, 459–481, 2007.
225. Mondal, K.K. and Mani, C., Investigation of the antibacterial properties of nanocopper against Xanthomonas axonopodis pv. punicae, the incitant of pomegranate bacterial blight. *Annal. Microbiol*, 62, 889–893, 2012.
226. Moradbeygi, H., Jamei, R., Heidari, R., Darvishzadeh, R., Investigating the enzymatic and non-enzymatic antioxidant defense by applying iron oxide nanoparticles in Dracocephalum moldavica L. plant under salinity stress. *Sci. Horticult.*, 272, 109537, 2020.
227. Mousavi Kouhi, S.M., Lahouti, M., Ganjeali, A., Entezari, M.H., Comparative phytotoxicity of ZnO nanoparticles, ZnO microparticles, and Zn2+ on rapeseed (Brassica napus L.): Investigating a wide range of concentrations. *Toxicol. Environ. Chem.*, 96, 861–868, 2014.
228. Mozafari, A., Havas, F., Ghaderi, N., Application of iron nanoparticles and salicylic acid in *in vitro* culture of strawberries (Fragaria× ananassa Duch.) to cope with drought stress. *Plant Cell. Tiss.Organ. Cult.*, 132, 511–523, 2018.
229. Mozhayeva, D. and Engelhard, C., A critical review of single particle inductively coupled plasma mass spectrometry—A step towards an ideal method for nanomaterial characterization. *J. Anal. Atom. Spectrom.*, 43, 5290–5294, 2020.
230. Mukherjee, A., Sun, Y., Morelius, E., Tamez, C., Bandyopadhyay, S., Niu, G., Differential toxicity of bare and hybrid ZnO nanoparticles in green pea (Pisum sativum L.): A life cycle study. *Front. Plant Sci.*, 6, 1242, 2016.
231. Munir, N., Hasnain, M., Waqif, H., Adetuyi, B.O., Egbuna, C., Olisah, M.C., Gelling agents, micro and nanogels in food system applications, in:

Application of Nanotechnology in Food Science, Processing and Packaging, pp. 153–167, 2022.
232. Muñoz-García, R.O., Hernández, M.E., Ortiz, G.G., Fernández, V.V., Arellano, M.R., Sánchez-Díaz, J.C., A novel polyacrylamide-based hydrogel crosslinked with cellulose acetate and prepared by precipitation polymerization. *Química Nova*, 38, 1031–1036, 2015.
233. Murugan, K., Sivaramakrishnan, S., Senthilkumar, N., Jayabalan, D., Nathan, S.S., Synergistic interaction of botanicals and biocides nuclear polyhedrosis virus on pest control. *J. Sci. Ind. Res.*, 57, 732–739, 1998.
234. Mustafa, G. and Komatsu, S., Insights into the response of soybean mitochondrial proteins to various sizes of aluminum oxide nanoparticles under flooding stress. *J. Proteo. Res.*, 15, 4464–4475, 2016.
235. Nadiminti, P.P., Dong, Y.D., Sayer, C., Hay, P., Rookes, J.E., Boyd, B.J., Nanostructured liquid crystalline particles as an alternative delivery vehicle for plant agrochemicals. *ACS Appl. Mat. Interf.*, 5, 1818–1826, 2013.
236. Narayanan, K.B. and Park, H.H., Antifungal activity of silver nanoparticles synthesized using turnip leaf extract (Brassica rapa L.) against wood rotting pathogens. *Eur. J. Plant Pathol.*, 140, 185–192, 2014.
237. Nath, J., Dror, I., Landa, P., Vanek, T., Kaplan-Ashiri, I., Berkowitz, B., Synthesis and characterization of isotopically-labeled silver, copper and zinc oxide nanoparticles for tracing studies in plants. *Environ. Pollut.*, 242, 1827–1837, 2018.
238. Navarro, D.A., Bisson, M.A., Aga, D.S., Investigating uptake of water-dispersible CdSe/ZnS quantum dot nanoparticles by Arabidopsis thaliana plants. *J. Hazard. Mater.*, 211, 427–435, 2012.
239. Nazir, A., Itrat, N., Shahid, A., Mushtaq, Z., Abdulrahman, S.A., Egbuna, C., Adetuyi, B.O., Khan, J., Uche, C.Z., Toloyai, P.Y., Orange peel as a source of nutraceuticals, in: *Food and Agricultural Byproducts as Important Source of Valuable Nutraceuticals*, 1st ed, p. 400, Springer, Berlin, 2022.
240. Nel, A., Xia, T., Mädler, L., Li, N., Toxic potential of materials at the nanolevel. *Science*, 311, 622–627, 2006.
241. Nievola, C.C., Carvalho, C.P., Carvalho, V., Rodrigues, E., Rapid responses of plants to temperature changes. *Temperature*, 4, 371–405, 2017.
242. Nuruzzaman, M.D., Rahman, M.M., Liu, Y., Naidu, R., Nanoencapsulation, nano-guard for pesticides: A new window for safe application. *J. Agric. Food Chem.*, 64, 1447–1483, 2016.
243. O'Brien, J.A. and Benková, E., Cytokinin cross-talking during biotic and abiotic stress responses. *Front. Plant Sci.*, 4, 451, 2013.
244. Ocsoy, I., Paret, M.L., Ocsoy, M.A., Kunwar, S., Chen, T., You, M., Nanotechnology in plant disease management: DNA-directed silver nanoparticles on graphene oxide as an antibacterial against Xanthomonas perforans. *ACS Nano.*, 7, 8972–8980, 2013.
245. Ogunlana, O.O., Adetuyi, B.O., Adekunbi, T.S., Adegboye, B.E., Iheagwam, F.N., Ogunlana, O.E., Ruzu bitters ameliorates high–fat diet induced

non-alcoholic fatty liver disease in male Wistar rats. *J. Pharm. Pharmacogn. Res.*, 9, 3, 251–260, 2021.
246. Ogunlana, O.O., Adetuyi, B.O., Esalomi, E.F., Rotimi, M.I., Popoola, J.O., Ogunlana, O.E., Adetuyi, O.A., Antidiabetic and antioxidant activities of the twigs of Andrograhis paniculata on Streptozotocin-induced diabetic male rats. *BioChem.*, 1, 3, 238–249, 2021.
247. Ogunlana, O.O., Adetuyi, B.O., Rotimi, M., Esalomi, I., Adeyemi, A., Akinyemi, J., Ogunlana, O., Adetuyi, O., Adebisi, O., Okpata, E., Baty, R., Batiha, G., Hypoglycemic activities of ethanol seed extract of *Hunteria umbellate* (Hallier F.) on Streptozotocin-induced diabetic rats. *Clin. Phytoscience*, 7, 1, 1–9, 2021.
248. Ogunlana, O.O., Ogunlana, O.E., Adekunbi, T.S., Adetuyi, B.O., Adegboye, B.E., Iheagwam, F.N., Anti-inflammatory mechanism of ruzu bitters on diet-induced nonalcoholic fatty liver disease in male wistar rats. *J. Evid. Based Complement. Altern. Med.*, 2020, 1, 1–8, 2020.
249. Ogunlana, O.O., Ogunlana, O.E., Popoola, J.O., Adetuyi, B.O., Adekunbi, T.S., David, O.L., Adeleye, O.J., Udeogu, S.A., Adeyemi, A.O., Twigs of Andrographis paniculata (Burn. F) Nees attenuates carbon tetrachloride (CCl4) induced liver damage in Wistar albino rats. *RASAYAN J.Chem.*, 14, 4, 2598–2603, 2021.
250. Oliveira, H.C., Stolf-Moreira, R., Martinez, C.B., Grillo, R., de Jesus, M.B., Fraceto, L.F., Nanoencapsulation enhances the post-emergence herbicidal activity of atrazine against mustard plants. *PLoS One*, 10, e0132971, 2015.
251. Othman, S.H., Bio-nanocomposite materials for food packaging applications: Types of biopolymer and nano-sized filler. *Agric. Agricul. Sci. Proc.*, 2, 296–303, 2014.
252. Ouda, S.M., Antifungal activity of silver and copper nanoparticles on two plant pathogens, Alternaria alternata and Botrytis cinerea. *Res. J. Microbiol.*, 9, 34–42, 2014.
253. Padilla, F.M., Gallardo, M., Peña-Fleitas, M.T., De Souza, R., Thompson, R.B., Proximal optical sensors for nitrogen management of vegetable crops: A review. *Sensors*, 18, 2083, 2018.
254. Palchoudhury, S., Jungjohann, K.L., Weerasena, L., Arabshahi, A., Gharge, U., Albattah, A., Enhanced legume root growth with pre-soaking in α-Fe2O3 nanoparticle fertilizer. *RSC Adv.*, 8, 24075–24083, 2018.
255. Palmqvist, N.M., Seisenbaeva, G.A., Svedlindh, P., Kessler, V.G., Maghemite nanoparticles acts as nanozymes, improving growth and abiotic stress tolerance in Brassica napus. *Nanoscale Res. Lett.*, 12, 631, 2017.
256. Palocci, C., Valletta, A., Chronopoulou, L., Donati, L., Bramosanti, M., Brasili, E., Endocytic pathways involved in PLGA nanoparticle uptake by grapevine cells and role of cell wall and membrane in size selection. *Plant Cell Rep.*, 36, 1917–1928, 2017.

257. Pandey, K., Anas, M., Hicks, V.K., Green, M.J., Khodakovskaya, M.V., Improvement of commercially valuable traits of industrial crops by application of carbon-based nanomaterials. *Sci. Rep.*, 9, 19358, 2019.
258. Paret, M.L., Vallad, G.E., Averett, D.R., Jones, J.B., Olson, S.M., Photocatalysis: Effect of light-activated nanoscale formulations of TiO2 on Xanthomonas perforans and control of bacterial spot of tomato. *Phytopathology*, 103, 228–236, 2013.
259. Patel, N., Desai, P., Patel, N., Jha, A., Gautam, H.K., Agronanotechnology for plant fungal disease management: A review. *Int. J. Curr. Microbiol. App. Sci.*, 3, 71–84, 2014.
260. Pérez-de-Luque, A., Interaction of nanomaterials with plants: What do we need for real applications in agriculture. *Front. Environ. Sci.*, 5, 12, 2017.
261. Pokhrel, L.R. and Dubey, B., Evaluation of developmental responses of two crop plants exposed to silver and zinc oxide nanoparticles. *Sci. Total Environ.*, 452, 321–332, 2013.
262. Pouratashi, M. and Iravani, H., Farmers' knowledge of integrated pest management and learning style preferences: Implications for information delivery. *Int. J. Pest Manage*, 58, 347–353, 2012.
263. Pradhan, S., Patra, P., Das, S., Chandra, S., Mitra, S., Dey, K.K., Photochemical modulation of biosafe manganese nanoparticles on Vigna radiata: A detailed molecular, biochemical, and biophysical study. *Environ. Sci. Technol.*, 47, 13122–13131, 2013.
264. Prasad, T. N. V. K.V., Sudhakar, P., Sreenivasulu, Y., Latha, P., Munaswamy, V., Reddy, K.R., Effect of nanoscale zinc oxide particles on the germination, growth and yield of peanut. *J. Plant Nutt.*, 35, 905–927, 2012.
265. Pulido-Reyes, G., Rodea-Palomares, I., Das, S., Sakthivel, T.S., Leganes, F., Rosal, R., Untangling the biological effects of cerium oxide nanoparticles: The role of surface valence states. *Sci. Rep.*, 5, 15613, 2015.
266. Qian, K., Shi, T., Tang, T., Zhang, S., Liu, X., Cao, Y., Preparation and characterization of nano-sized calcium carbonate as controlled release pesticide carrier for validamycin against Rhizoctonia solani. *Microchim. Acta*, 173, 51–57, 2011.
267. Qi-liang, H.U., Wen-ji, Z.H., Feng-min, L.I., Dong-mei, S.H., Chun-hua, Z.H., Xiao-li, B.U., Solubilization of chlorpyrifos in the mixed system of surfactants and the bioactivity evaluation. *J. Chin. J. Pestic. Sci.*, 8, 71–76, 2006.
268. Rad, F., Mohsenifar, A., Tabatabaei, M., Safarnejad, M.R., Shahryari, F., Safarpour, H., Detection of candidatus phytoplasma aurantifolia with a quantum dots FRET-based biosensor. *J. Plant Pathol.*, 1, 525–534, 2012.
269. Rafi, M.M., Epstein, E., Falk, R.H., Silicon deprivation causes physical abnormalities in wheat (Triticum aestivum L). *J. Plant Physiol.*, 151, 497–501, 1997.
270. Ragaei, M. and Sabry, A.H., Nanotechnology for insect pest control. *Int. J. Sci. Environ. Technol.*, 3, 2, 528–545, 2014.

271. Rahman, A.K.M.Z., Haque, M.H., Alam, S.N., Mahmudunnabi, M., Dutta, N.K., Efficacy of botanicals against Helicoverpa armigera (Hubner) in tomato. *Agriculturists*, 12, 1, 131–139, 2014.
272. Rahman, M.S., Chakraborty, A., Mazumdar, S., Nandi, N.C., Bhuiyan, M.N., Alauddin, S.M., Effects of poly (vinylpyrrolidone) protected platinum nanoparticles on seed germination and growth performance of Pisum sativum. *NanoStruc. NanoObj.*, 21, 100408, 2020.
273. Raja, K., Sowmya, R., Sudhagar, R., Moorthy, P.S., Govindaraju, K., Subramanian, K.S., Biogenic ZnO and Cu nanoparticles to improve seed germination quality in blackgram (Vigna mungo). *Mater. Lett.*, 235, 164–167, 2019.
274. Rajaee Behbahani, S., Iranbakhsh, A., Ebadi, M., Majd, A., Ardebili, Z.O., Red elemental selenium nanoparticles mediated substantial variations in growth, tissue differentiation, metabolism, gene transcription, epigenetic cytosine DNA methylation, and callogenesis in bittermelon (Momordica charantia): An *in vitro* experiment. *PLoS One*, 15, e0235556, 2020.
275. Rajesh, S., Raja, D.P., Rathi, J.M., Sahayaraj, K., Biosynthesis of silver nanoparticles using Ulva fasciata (Delile) ethyl acetate extract and its activity against Xanthomonas campestris pv. malvacearum. *J. Biopesti.*, 5, 119, 2012.
276. Rajput, V., Minkina, T., Fedorenko, A., Sushkova, S., Mandzhieva, S., Lysenko, V., Toxicity of copper oxide nanoparticles on spring barley (Hordeum sativum distichum). *Sci. Total. Environ.*, 645, 1103–1113, 2018.
277. Raliya, R., Franke, C., Chavalmane, S., Nair, R., Reed, N., Biswas, P., Quantitative understanding of nanoparticle uptake in watermelon plants. *Front. Plant Sci.*, 7, 1288, 2016.
278. Ramakrishna, S., Fujihara, K., Teo, W.E., Yong, T., Ma, Z., Ramaseshan, R., Electrospun nanofibers: Solving global issues. *Mater. Today*, 9, 40–50, 2006.
279. Rameshaiah, G.N., Pallavi, J., Shabnam, S., Nano fertilizers and nano sensors—An attempt for developing smart agriculture. *Int. J. Engineer. Res. Gen. Sci.*, 3, 314–320, 2015.
280. Ramya, S., Rajasekaran, C., Sundararajan, G., Alaguchamy, N., Jayakumararaj, R., Antifeedant activity of leaf aqueous extracts of selected medicinla plants on VI instar larva of Helicoverpa armigera. *Ethnobotanical Leaflets*, 12, 938–943, 2008.
281. Rastogi, A., Tripathi, D.K., Yadav, S., Chauhan, D.K., Živčák, M., Ghorbanpour, M., Application of silicon nanoparticles in agriculture. *Biotech*, 9, 90, 2019.
282. Rathore, I. and Tarafdar, J.C., Perspectives of biosynthesized magnesium nanoparticles in foliar application of wheat plant. *J. Bionanosci.*, 9, 209–214, 2015.
283. Raza, A., Mehmood, S.S., Tabassum, J., Batool, R., Targeting plant hormones to develop abiotic stress resistance in wheat, in: *Wheat Production in Changing Environments*, pp. 557–577, Springer, Singapore, 2019.
284. Rejeb, I.B., Pastor, V., Mauch-Mani, B., Plant responses to simultaneous biotic and abiotic stress: Molecular mechanisms. *Plants*, 3, 458–475, 2014.

285. Rice, J.C. and Garcia, S.M., Fisheries, food security, climate change, and biodiversity: Characteristics of the sector and perspectives on emerging issues. *ICES J. Marine Sci.*, 68, 1343–1353, 2011.
286. Rico, C.M., Majumdar, S., Duarte-Gardea, M., Peralta-Videa, J.R., Gardea-Torresdey, J.L., Interaction of nanoparticles with edible plants and their possible implications in the food chain. *J. Agricul. Food. Chem.*, 59, 3485–3498, 2011.
287. Riederer, M. and Schreiber, L., Protecting against water loss: Analysis of the barrier properties of plant cuticles. *J. Exp. Bot.*, 52, 2023–2032, 2001.
288. Roberts, A. and Oparka, K.J., Plasmodesmata and the control of symplastic transport. *Plant Cell Environ.*, 26, 103–124, 2003.
289. Rossi, L., Zhang, W., Lombardini, L., Ma, X., The impact of cerium oxide nanoparticles on the salt stress responses of Brassica napus L. *Environ. Poll.*, 219, 28–36, 2016.
290. Rui, M., Ma, C., Hao, Y., Guo, J., Rui, Y., Tang, X., Iron oxide nanoparticles as a potential iron fertilizer for peanut (Arachis hypogaea). *Front. Plant Sci.*, 7, 815, 2016.
291. Saharan, V., Kumaraswamy, R.V., Choudhary, R.C., Kumari, S., Pal, A., Raliya, R., Cu-chitosan nanoparticle mediated sustainable approach to enhance seedling growth in maize by mobilizing reserved food. *J. Agricul. Food. Chem.*, 64, 6148–6155, 2016.
292. Saharan, V., Mehrotra, A., Khatik, R., Rawal, P., Sharma, S.S., Pal, A., Synthesis of chitosan based nanoparticles and their *in vitro* evaluation against phytopathogenic fungi. *Int. J. Boilog. Macromole*, 62, 677–683, 2013.
293. Santiago, M., Pagay, V., Stroock, A.D., Impact of electroviscosity on the hydraulic conductance of the bordered pit membrane: A theoretical investigation. *Plant Physiol.*, 163, 999–1011, 2013.
294. Sarlak, N., Taherifar, A., Salehi, F., Synthesis of nanopesticides by encapsulating pesticide nanoparticles using functionalized carbon nanotubes and application of new nanocomposite for plant disease treatment. *J. Agric. Food Chem.*, 62, 4833–4838, 2014.
295. Sasson, Y., Levy-Ruso, G., Toledano, O., Ishaaya, I., Nanosuspensions: Emerging novel agrochemical formulations, in: *Insecticides Design Using Advanced Technologies*, pp. 1–39, Springer, Berlin, Heidelberg, 2007.
296. Sattelmacher, B., The apoplast and its significance for plant mineral nutrition. *N. Phytol.*, 149, 167–192, 2001.
297. Saxena, R., Tomar, R.S., Kumar, M., Exploring nanobiotechnology to mitigate abiotic stress in crop plants. *J. Pharma. Sci. Res.*, 8, 974, 2016.
298. Schwab, F., Zhai, G., Kern, M., Turner, A., Schnoor, J.L., Wiesner, M.R., Barriers, pathways and processes for uptake, translocation and accumulation of nanomaterials in plants–critical review. *Nanotoxicology*, 10, 257–278, 2016.

299. Sedghi, M., Hadi, M., Toluie, S.G., Effect of nano zinc oxide on the germination parameters of soybean seeds under drought stress. *Annal. West Univ. Timisoara Ser. Biol.*, 16, 73–78, 2013.
300. Servin, A.D. and White, J.C., Nanotechnology in agriculture: Next steps for understanding engineered nanoparticle exposure and risk. *NanoImpact*, 1, 9–12, 2016.
301. Shah, M.A. and Khan, A.A., Use of diatomaceous earth for the management of stored-product pests. *Int. J. Pest Manage.*, 60, 100–113, 2014.
302. Shahabi-Ghahafarrokhi, I., Khodaiyan, F., Mousavi, M., Yousefi, H., Preparation and characterization of nanocellulose from beer industrial residues using acid hydrolysis/ultrasound. *Fibers Polym.*, 16, 529–536, 2015.
303. Shahrokh, S., Hosseinkhani, B., Emtiazi, G., The impact of nano-silver on bacterial aerobic nitrate reductase. *J. Bioproces. Biotechn.*, 4, 4, 2014.
304. Shallan, M.A., Hassan, H.M., Namich, A.A., Ibrahim, A.A., Biochemical and physiological effects of TiO2 and SiO2 nanoparticles on cotton plant under drought stress. *Res. J. Pharma. Bio. Chem. Sci.*, 7, 1540–1551, 2016.
305. Sharifi, M., Faryabi, K., Talaei, A.J., Shekha, M.S., Ale-Ebrahim, M., Salihi, A., Antioxidant properties of gold nanozyme: A review. *J. Mol. Liq.*, 297, 112004, 2020.
306. Sharma, P., Jha, A.B., Dubey, R.S., Pessarakli, M., Reactive oxygen species, oxidative damage, and antioxidative defense mechanism in plants under stressful conditions. *J. Bot.*, 2012, 26, 2012.
307. Shen, C.X., Zhang, Q.F., Li, J., Bi, F.C., Yao, N., Induction of programmed cell death in arabidopsis and rice by single-wall carbon nanotubes. *Am. J. Bot.*, 97, 1602–1609, 2010.
308. Siddiqui, M.H., Al-Whaibi, M.H., Faisal, M., Al Sahli, A.A., Nano-silicon dioxide mitigates the adverse effects of salt stress on Cucurbita pepo L. *Environ. Toxicol. Chem.*, 33, 2429–2437, 2014.
309. Siddiqui, M.H. and Al-Whaibi, M.H., Role of nano-SiO2 in germination of tomato (Lycopersicum esculentum seeds Mill.). *Saudi. J. Biol. Sci.*, 21, 13–17, 2014.
310. Singh, A., Singh, N.B., Hussain, I., Singh, H., Effect of biologically synthesized copper oxide nanoparticles on metabolism and antioxidant activity to the crop plants Solanum lycopersicum and brassica oleracea var. botrytis. *J. Biotechnol.*, 262, 11–27, 2017.
311. Singh, S., Singh, M., Agrawal, V.V., Kumar, A., An attempt to develop surface plasmon resonance based immunosensor for Karnal bunt (Tilletia indica) diagnosis based on the experience of nano-gold based lateral flow immunodipstick test. *Thin Solid Films*, 519, 1156–1159, 2010.
312. Solanki, P., Bhargava, A., Chhipa, H., Jain, N., Panwar, J., Nano-fertilizers and their smart delivery system, in: *Nanotechnologies in Food and Agriculture*, M. Rai (Ed.), pp. 81–101, Springer, Cham, 2015.
313. Soleymanzadeh, R., Iranbakhsh, A., Habibi, G., Ardebili, Z.O., Selenium nanoparticle protected strawberry against salt stress through modifications

in salicylic acid, ion homeostasis, antioxidant machinery, and photosynthesis performance. *Acta Biolog. Cracoviensia. Bot.*, 62, 33–42, 2020.

314. Sotoodehnia-Korani, S., Iranbakhsh, A., Ebadi, M., Majd, A., Ardebili, Z.O., Selenium nanoparticles induced variations in growth, morphology, anatomy, biochemistry, gene expression, and epigenetic DNA methylation in Capsicum annuum: An *in vitro* study. *Environ. Poll.*, 265, 114727, 2020.
315. Souza, L.R.R., Bernardes, L.E., Barbetta, M.F.S., da Veiga, M.A.M.S., Iron oxide nanoparticle phytotoxicity to the aquatic plant Lemna minor: Effect on reactive oxygen species (ROS) production and chlorophyll a/chlorophyll b ratio. *Environ. Sci. Poll. Res.*, 26, 24121–24131, 2019.
316. Spielman-Sun, E., Avellan, A., Bland, G.D., Tappero, R.V., Acerbo, A.S., Unrine, J.M., Nanoparticle surface charge influences translocation and leaf distribution in vascular plants with contrasting anatomy. *Environ. Sci. Nano*, 6, 2508–2519, 2019.
317. Stadler, T., Buteler, M., Weaver, D.K., Novel use of nanostructured alumina as an insecticide. *Pest Manag. Sci.*, 66, 577–579, 2010.
318. Stadler, T., Buteler, M., Valdez, S.R., Gitto, J.G., Particulate nanoinsecticides: A new concept in insect pest management. *Insectic. Agricul. Toxicol.*, 83, 72448, 2018.
319. Stampoulis, D., Sinha, S.K., White, J.C., Assay-dependent phytotoxicity of nanoparticles to plants. *Environ. Sci. Technol.*, 43, 9473–9479, 2009.
320. Stefan, L. and Monchaud, D., Applications of guanine quartets in nanotechnology and chemical biology. *Nat. Rev. Chem.*, 3, 650–668, 2019.
321. Stone, V., Führ, M., Feindt, P.H., Bouwmeester, H., Linkov, I., Sabella, S., The essential elements of a risk governance framework for current and future nanotechnologies. *Risk. Anal.*, 38, 1321–1331, 2018.
322. Su, Y., Ashworth, V.E., Geitner, N.K., Wiesner, M.R., Ginnan, N., Rolshausen, P., Delivery, fate, and mobility of silver nanoparticles in citrus trees. *ACS Nano*, 14, 2966–2981, 2020.
323. Su, Y., Ashworth, V., Kim, C., Adeleye, A.S., Rolshausen, P., Roper, C., Delivery, uptake, fate, and transport of engineered nanoparticles in plants: A critical review and data analysis. *Environ. Sci. Nano*, 6, 2311–2331, 2019.
324. Subbaiah, L.V., Prasad, T. N. V. K. V., Krishna, T.G., Sudhakar, P., Reddy, B.R., Pradeep, T., Novel effects of nanoparticulate delivery of zinc on growth, productivity, and zinc biofortification in maize (Zea mays L.). *J. Agricul. Food. Chem.*, 64, 3778–3788, 2016.
325. Sullivan, M. and Molet, T., *CPHST Pest Datasheet for Helicoverpa armigera*, USDA-APHIS-PPQ-CPHST, Chicago, 2007.
326. Sun, D., Hussain, H.I., Yi, Z., Rookes, J.E., Kong, L., Cahill, D.M., Mesoporous silica nanoparticles enhance seedling growth and photosynthesis in wheat and lupin. *Chemosphere*, 152, 81–91, 2016.
327. Sun, D., Hussain, H.I., Yi, Z., Rookes, J.E., Kong, L., Cahill, D.M., Delivery of abscisic acid to plants using glutathione responsive mesoporous silica nanoparticles. *J. Nanosci. Nanotech.*, 18, 1615–1625, 2018.

328. Sun, D., Hussain, H., Yi, Z., Siegele, R., Cresswell, T., Kong, L., Uptake and cellular distribution, in four plant species, of fluorescently labeled mesoporous silica nanoparticles. *Plant Cell. Rep.*, 33, 1389–1402, 2014.
329. Sundararajan, G. and Kumuthakalavalli, R., Antifeedant activity of aqueous extract of Gnidia glauca Gilg. and Toddalia asiatica Lam. on gram pod borer, Helicoverpa armigera (Hbn). *J. Environ. Biol.*, 22, 1, 11–14, 2001.
330. Suriyaprabha, R., Karunakaran, G., Yuvakkumar, R., Rajendran, V., Kannan, N., Silica nanoparticles for increased silica availability in maize (Zea mays. L) seeds under hydroponic conditions. *Curr. Nanosci.*, 8, 902–908, 2012.
331. Tantawy, A.S., Salama, Y.A.M., El-Nemr, M.A., Abdel-Mawgoud, A.M.R., Nano silicon application improves salinity tolerance of sweet pepper plants. *Int. J. Chem.Tech. Res.*, 8, 11–17, 2015.
332. Tarafdar, J.C., Xiong, Y., Wang, W.N., Quinl, D., Biswas, P., Standardization of size, shape and concentration of nanoparticle for plant application. *Appl. Biol. Res.*, 14, 138–144, 2012.
333. Tepfer, M. and Taylor, I.E., The permeability of plant cell walls as measured by gel filtration chromatography. *Science*, 213, 761–763, 1981.
334. Torabian, S., Zahedi, M., Khoshgoftar, A.H., Effects of foliar spray of nanoparticles of FeSO4 on the growth and ion content of sunflower under saline condition. *J. Plant Nutr.*, 40, 615–623, 2017.
335. Torney, F., Trewyn, B.G., Lin, V.S.Y., Wang, K., Mesoporous silica nanoparticles deliver DNA and chemicals into plants. *Nat. Nanotech.*, 2, 295–300, 2007.
336. Tripathi, D.K., Singh, S., Singh, V.P., Prasad, S.M., Dubey, N.K., Chauhan, D.K., Silicon nanoparticles more effectively alleviated UV-B stress than silicon in wheat (Triticum aestivum) seedlings. *Plant Physiol. Biochem.*, 110, 70–81, 2017.
337. Usman, M., Farooq, M., Wakeel, A., Nawaz, A., Cheema, S.A., ur Rehman, H. Nanotechnology in agriculture: Current status, challenges and future opportunities. *Sci. Total. Environ.*, 721, 137778, 2020.
338. Van Ha, C., Van Nguyen, D., Nguyen, H.M., Le, N.T., Nguyen, K.H., Le, H.M., Copper nanoparticle application enhances plant growth and grain yield in maize under drought stress conditions. *bioRxiv*, 41, 364-375, 2020.
339. Vankova, R., Landa, P., Podlipna, R., Dobrev, P.I., Prerostova, S., Langhansova, L., ZnO nanoparticle effects on hormonal pools in Arabidopsis thaliana. *Sci. Total. Environ.*, 593, 535–542, 2017.
340. Verma, V., Ravindran, P., Kumar, P.P., Plant hormone-mediated regulation of stress responses. *BMC Plant Boil.*, 16, 86, 2016.
341. Vinayaka, A.C. and Thakur, M.S., Focus on quantum dots as potential fluorescent probes for monitoring food toxicants and foodborne pathogens. *Anal. Bioanal. Chem.*, 397, 1445–1455, 2010.
342. Vinayaka, A.C., Basheer, S., Thakur, M.S., Bioconjugation of CdTe quantum dot for the detection of 2, 4-dichlorophenoxyacetic acid by competitive fluoroimmunoassay based biosensor. *Biosens. Bioelectron.*, 24, 1615–1620, 2009.

343. Wakeel, A., Xu, M., Gan, Y., Chromium-induced reactive oxygen species accumulation by altering the enzymatic antioxidant system and associated cytotoxic, genotoxic, ultrastructural, and photosynthetic changes in plants. *Int. J. Molec. Sci.*, 21, 728, 2020.
344. Wang, H., Kou, X., Pei, Z., Xiao, J.Q., Shan, X., Xing, B., Physiological effects of magnetite (Fe3O4) nanoparticles on perennial ryegrass (Lolium perenne L.) and pumpkin (Cucurbita mixta) plants. *Nanotoxicology*, 5, 30–42, 2011.
345. Wang, J., Koo, Y., Alexander, A., Yang, Y., Westerhof, S., Zhang, Q., Phytostimulation of poplars and Arabidopsis exposed to silver nanoparticles and Ag+ at sublethal concentrations. *Environ. Sci. Technol.*, 47, 5442–5449, 2013.
346. Wang, P., Lombi, E., Zhao, F.J., Kopittke, P.M., Nanotechnology: A new opportunity in plant sciences. *Trends Plant Sci.*, 21, 699–712, 2016.
347. Wang, Q., Ma, X., Zhang, W., Pei, H., Chen, Y., The impact of cerium oxide nanoparticles on tomato (Solanum lycopersicum L.) and its implications for food safety. *Metallomics*, 4, 1105–1112, 2012.
348. Wang, S., Wu, X.M., Liu, C.H., Shang, J.Y., Gao, F., Guo, H.S., Verticillium dahliae chromatin remodeling facilitates the DNA damage repair in response to plant ROS stress. *PLoS Pathog.*, 16, e1008481, 2020.
349. Wang, X., Liu, X., Chen, J., Han, H., Yuan, Z., Evaluation and mechanism of antifungal effects of carbon nanomaterials in controlling plant fungal pathogen. *Carbon*, 68, 798–806, 2014.
350. Wang, X., Yang, X., Chen, S., Li, Q., Wang, W., Hou, C., Zinc oxide nanoparticles affect biomass accumulation and photosynthesis in Arabidopsis. *Front. Plant Sci.*, 6, 1243, 2016.
351. Wang, Y., Jiang, F., Ma, C., Rui, Y., Tsang, D.C., Xing, B., Effect of metal oxide nanoparticles on amino acids in wheat grains (Triticum aestivum) in a life cycle study. *J. Environ. Manage*, 241, 319–327, 2019.
352. Wang, Z., Xie, X., Zhao, J., Liu, X., Feng, W., White, J.C., Xylem and phloem-based transport of CuO nanoparticles in maize (Zea mays L.). *Environ. Sci. Technol.*, 46, 4434–4441, 2012.
353. Wani, A.H. and Shah, M.A., A unique and profound effect of MgO and ZnO nanoparticles on some plant pathogenic fungi. *J. Appl. Pharma. Sci.*, 2, 4, 2012.
354. Wild, E. and Jones, K.C., Novel method for the direct visualization of *in vivo* nanomaterials and chemical interactions in plants. *Environ. Sci. Technol.*, 43, 5290–5294, 2009.
355. Wilson, R.H., Nadeau, K.P., Jaworski, F.B., Tromberg, B.J., Durkin, A.J., Review of short-wave infrared spectroscopy and imaging methods for biological tissue characterization. *J. Biomed. Optics.*, 20, 030901, 2015.
356. Wojcieszek, J., Jiménez-Lamana, J., Bierla, K., Asztemborska, M., Ruzik, L., Jarosz, M., Elucidation of the fate of zinc in model plants using single particle ICP-MS and ESI tandem MS. *J. Anal. Atom. Spectrom.*, 34, 683–693, 2019.

357. Wolfert, S., Ge, L., Verdouw, C., Bogaardt, M.J., Big data in smart farming—A review. *Agricult. Syst.*, 153, 69–80, 2017.
358. Wong, M.H., Giraldo, J.P., Kwak, S.Y., Koman, V.B., Sinclair, R., Lew, T.T.S., Nitroaromatic detection and infrared communication from wild-type plants using plant nanobionics. *Nat. Mat.*, 16, 264–272, 2017.
359. Wu, H., Santana, I., Dansie, J., Giraldo, J.P., *In vivo* delivery of nanoparticles into plant leaves. *Curr. Protoc. Chem. Biol.*, 9, 269–284, 2017a.
360. Wu, H., Shabala, L., Shabala, S., Giraldo, J.P., Hydroxyl radical scavenging by cerium oxide nanoparticles improves arabidopsis salinity tolerance by enhancing leaf mesophyll potassium retention. *Environ. Sci. Nano*, 5, 1567–1583, 2018.
361. Wu, H., Tito, N., Giraldo, J.P., Anionic cerium oxide nanoparticles protect plant photosynthesis from abiotic stress by scavenging reactive oxygen species. *ACS Nano*, 11, 11283–11297, 2017b.
362. Yan, A. and Chen, Z., Impacts of silver nanoparticles on plants: A focus on the phytotoxicity and underlying mechanism. *Int. J. Mole. Sci.*, 20, 1003, 2019.
363. Yao, J., Yang, M., Duan, Y., Chemistry, biology, and medicine of fluorescent nanomaterials and related systems: New insights into biosensing, bioimaging, genomics, diagnostics, and therapy. *Chem. Rev.*, 114, 6130–6178, 2014.
364. Yao, K.S., Li, S.J., Tzeng, K.C., Cheng, T.C., Chang, C.Y., Chiu, C.Y., Fluorescence silica nano probe as a biomarker for rapid detection of plant pathogens. *Adv. Mater. Res.*, 79, 513–516, 2009.
365. Yaqoob, A.A., Parveen, T., Umar, K., Mohamad, M.N., Role of nanomaterials in the treatment of wastewater: A review. *Water*, 12, 495, 2020.
366. Ye, Y., Cota-Ruiz, K., Hernández-Viezcas, J.A., Valdés, C., Medina-Velo, I.A., Turley, R.S., Manganese nanoparticles control salinity-modulated molecular responses in Capsicum annuum L. through priming: A sustainable approach for agriculture. *ACS. Sustain. Chem. Eng.*, 8, 1427–1436, 2020.
367. Yearla, S.R. and Padmasree, K., Exploitation of subabul stem lignin as a matrix in controlled release agrochemical nanoformulations: A case study with herbicide diuron. *Environ. Sci. Poll. Res.*, 23, 18085–18098, 2016.
368. Yoon, K.Y., Byeon, J.H., Park, J.H., Hwang, J., Susceptibility constants of Escherichia coli and Bacillus subtilis to silver and copper nanoparticles. *Sci. Total. Environ.*, 373, 572–575, 2007.
369. Yousefi, S., Kartoolinejad, D., Naghdi, R., Effects of priming with multi-walled carbon nanotubes on seed physiological characteristics of Hopbush (Dodonaeaviscosa L.) under drought stress. *Int. J. Environ. Stud.*, 74, 528–539, 2017.
370. Yu, Z., Sun, X., Song, H., Wang, W., Ye, Z., Shi, L., Glutathione-responsive carboxymethyl chitosan nanoparticles for controlled release of herbicides. *Mater. Sci. Appl.*, 6, 591, 2015.

371. Yuan, J., Chen, Y., Li, H., Lu, J., Zhao, H., Liu, M., New insights into the cellular responses to iron nanoparticles in Capsicum annuum. *Sci. Rep.*, 8, 3228, 2018.
372. Yusefi-Tanha, E., Fallah, S., Rostamnejadi, A., Pokhrel, L.R., Particle size and concentration dependent toxicity of copper oxide nanoparticles (CuONPs) on seed yield and antioxidant defense system in soil grown soybean (Glycine max cv. Kowsar). *Sci. Total. Environ.*, 715, 136994, 2020.
373. Ze, Y., Liu, C., Wang, L., Hong, M., Hong, F., The regulation of TiO 2 nanoparticles on the expression of light-harvesting complex II and photosynthesis of chloroplasts of Arabidopsis thaliana. *Biol. Trace. Elem. Res.*, 143, 1131–1141, 2011.
374. Zeng, Y., Himmel, M.E., Ding, S.Y., Visualizing chemical functionality in plant cell walls. *Biotechnol. Biofuel.*, 10, 1–16, 2017.
375. Zhang, J., Landry, M.P., Barone, P.W., Kim, J.H., Lin, S., Ulissi, Z.W., Molecular recognition using corona phase complexes made of synthetic polymers adsorbed on carbon nanotubes. *Nat. Nanotechnol.*, 8, 959–968, 2013.
376. Zhang, L. and Fang, M., Nanomaterials in pollution trace detection and environmental improvement. *Nano Today*, 5, 128–142, 2010.
377. Zhao, L., Hernandez-Viezcas, J.A., Peralta-Videa, J.R., Bandyopadhyay, S., Peng, B., Munoz, B., ZnO nanoparticle fate in soil and zinc bioaccumulation in corn plants (Zea mays) influenced by alginate. *Environ. Sci. Proces. Impact.*, 15, 260–266, 2013a.
378. Zhao, L., Lu, L., Wang, A., Zhang, H., Huang, M., Wu, H., Nano-biotechnology in agriculture: Use of nanomaterials to promote plant growth and stress tolerance. *J. Agricul. Food. Chem.*, 68, 1935–1947, 2020.
379. Zhao, L., Sun, Y., Hernandez-Viezcas, J.A., Servin, A.D., Hong, J., Niu, G., Influence of CeO2 and ZnO nanoparticles on cucumber physiological markers and bioaccumulation of Ce and Zn: A life cycle study. *J. Agricul. Food. Chem.*, 61, 11945–11951, 2013b.
380. Zhu, H., Han, J., Xiao, J.Q., Jin, Y., Uptake, translocation, and accumulation of manufactured iron oxide nanoparticles by pumpkin plants. *J. Environ. Monit.*, 10, 713–717, 2008.
381. Zhu, Z.J., Wang, H., Yan, B., Zheng, H., Jiang, Y., Miranda, O.R., Effect of surface charge on the uptake and distribution of gold nanoparticles in four plant species. *Environ. Sci. Technol.*, 46, 12391–12398, 2012.
382. Adetunji, C.O., Titilayo, O., Mathew, J.T., Inobeme, A., Singh, K.R.B., Olaniyan, O.T., Singh, J., Singh, R.P., *Carbon Dots: Next-generation materials for biomedical applications*, Bristol, England, pp. 8–12, IOP Publishing, 2022.
383. Farid, A., Muzammal, M., Ghazanfar, S.G., Ismail, S., Ullah, A., Dauda, W.P., Adetunji, C.O., Shah, M., Dadrasnia, A., Maryam Idrees, S. *et al.*, Green cleaning activity of *Bacillus salmalaya* 139Sl: A novel strain for removing common household stains. *Biomass Conv. Bioref.*, 1, 1–8, 2022. https://doi.org/10.1007/s13399-022-03147-z.

384. Wali, S., Zahra, M., Okla, M.K., Wahidah, H.A., Tauseef, I., Haleem, K.S., Farid, A., Maryam, A., AbdElgawad, H., Adetunji, C.O., Akhtar, N., Akbar, S., Rehman, W., Yasir, H., Shakira, G., Brassica oleracea L.(Acephala Group) based zinc oxide nanoparticles and their efficacy as antibacterial agent. *Braz. J. Biol.*, 84, 1–7, 2022. https://doi.org/10.1590/1519-6984.259351.
385. Adetunji, C.O., Mathew, J.T., Singh, K.R.B., Inobeme, A., Nayak, V., Olaniyan, O.T., Singh, J., Singh, R.P., Zero-dimensional carbon nanomaterials for targeted drug delivery and photothermal therapy applications, in: *Zero-dimensional Carbon Nanomaterials: Fundamentals and applications*, pp. 1–10, IOP Publishing, Bristol, England, 2022.
386. Adetunji, C.O., Mathew, J.T., Singh, K.R.B., Inobeme, A., Nayak, V., Olaniyan, O.T., Singh, J., Singh, R.P., Zero-dimensional carbon nanomaterials for targeted drug delivery and photothermal therapy applications, in: *Zero-Dimensional Carbon Nanomaterials: Fundamentals and Applications*, pp. 1–12, IOP Publishing, Bristol, England, 2022.
387. Adetunji, C.O., Michael, O.S., Nwankwo, W., Anani, O.A., Adetunji, J.B., Olayinka, A.S., Akram, M., Biogenic nanoparticles based drugs derived from medicinal plants: A sustainable panacea for the treatment of malaria, in: *Green Synthesis in Nanomedicine and Human Health*, pp. 103–122, CRC Press, Boca Raton, Florida, United State, 2021.
388. Adetunji, O., Michael, O.S., Akram, M., Oseni, K., Anani, O.A., Nwankwo, W., Anwar, H., Adetunji, J.B., Olayinka, A.S., Recent advances in the utilization of bioengineered plant-based nanoparticles: A sustainable nanobiotechnology for the management of extensively drug-resistant tuberculosis, in: *Green Synthesis in Nanomedicine and Human Health*, pp. 149–166, CRC Press, Boca Raton, Florida, United State, 2021.
389. Adetunji, C.O., Michael, O.S., Akram, M., Oseni, K., Temidayo, A.K., Anani, O.A., Olayinka, A.S., Nwankwo, W., Ghaffar, I., Adetunji, J.B., *Green Synthesis in Nanomedicine and Human Health*, pp. 81–100, CRC Press, Boca Raton, Florida, United State, 2021.
390. Islam, S., Thangadurai, D., Adetunji, C.O., Michael, O.S., Nwankwo, W., Kadiri, O., Anani, O.A., Makinde, S., Adetunji, J.B., Nanomaterials: Applications in biomedicine and biotechnology, in: *Handbook of Nanomaterials and Nanocomposites for Energy and Environmental Applications*, O.V. Kharissova, L.M. Torres-Martínez, B.I. Kharisov (Eds.), Springer, Cham, 2021, https://doi.org/10.1007/978-3-030-36268-3_4.
391. Thangadurai, D., Prakash, L., Sangeetha, J., Mohammad Said Al-Tawaha, A.R., David, M., Islam, S., Adetunji, C.O., Adetunji, J.B., Sustainable synthesis of greener nanomaterials: Principles, processes, and products, in: *Handbook of Nanomaterials and Nanocomposites for Energy and Environmental Applications*, O.V. Kharissova, L.M.T. Martínez, B.I. Kharisov (Eds.), Springer, Cham, 2021, https://doi.org/10.1007/978-3-030-11155-7_30-1.

392. Adetunji, C.O. and Varma, A., Biotechnological application of *Trichoderma*: A powerful fungal isolate with diverse potentials for the attainment of food safety, management of pest and diseases, healthy planet, and sustainable agriculture, in: *Trichoderma: Agricultural Applications and Beyond. Soil Biology*, vol. 61, C. Manoharachary, H.B. Singh, A. Varma (Eds.), Springer, Cham, 2020, https://doi.org/10.1007/978-3-030-54758-5_12.

5

Application of Plant-Based Nanobiopesticides as Sanitizers

Babatunde Oluwafemi Adetuyi[1]*, Oluwakemi Semiloore Omowumi[1], Peace Abiodun Olajide[1] and Charles Oluwaseun Adetunji[2]

[1]*Department of Natural Sciences, Faculty of Pure and Applied Sciences, Precious Cornerstone University, Ibadan, Nigeria*

[2]*Applied Microbiology, Biotechnology and Nanotechnology Laboratory, Department of Microbiology, Edo University Iyamho, Auchi, Edo State, Nigeria*

Abstract

Nanotechnology involves the synthesis, evaluation, composition, and incorporation of several research fields to create useful substances or composite materials for a number of applications. The use of nanotechnology in biopesticides is one example of how this technology is being put to good use. Nanobiopesticides are substance compounds with nanoparticles that have natural beginnings and are used as insect poisons. However, nanobiopesticide can also be used as a sanitizing agent as a result of its properties, such as rapid elimination of vegetative microscopic organisms, yeasts, and molds by uniform, wide range microbial annihilation characteristics. Due to the fact that plant-based nanobiopesticides can kill pests or insects when they are sprayed on the field, they can also be used as a sanitizing agent to clean insect or pest-infested area. However, these nanobiopesticides need to be commercialized and used safely as cleaning agents.

***Keywords*:** Sanitizer, nanopesticides, nanotechnology, cleaning agents

5.1 Introduction

The world's population is currently 7 billion, and by 2050, it is expected to expand by 30% to 9.2 billion (70 million per annum). The explosive rise in

**Corresponding author*: badetuyi@gmail.com

Charles Oluwaseun Adetunji and Julius Kola Oloke (eds.) Handbook of Agricultural Biotechnology: Volume I Nanopesticides, (131–170)

the world's population is predicted to lead to a 70% increase in the demand for food production, largely as a result of dietary shifts toward higher-quality foods in developing nations, such as increased in meat consumption & dairy products & the growing utilization of cereals as livestock feed [47]. However, there is a limited amount of extra agricultural field that can be used to meet the rising food demand. Any growth will primarily come at the price of forests and other natural areas that are habitat of animals, plants that are related to crops in the wild, and pests that naturally attack crops. Nevertheless, there is a limited amount of extra agricultural area that can be to help with the ever-increasing hunger problem. The reduction of present output losses brought on by pests is a significant issue that needs to be managed in agricultural production. Therefore, it is crucial to comprehend both the pest and the pesticide scenarios from an agricultural standpoint [3–5].

An invasive species is one that has the potential to cause significant economic, environmental, and social harm to humans and their communities. A pesticide is defined as a substance that kills pests [85]. When pesticides are sprayed on any surface, they release toxins which have the ability to inhibit or stop the existence of any named pest. Depending on the sprayed targets, pesticides can be referred to in various ways. Herbicides are some of the plant pesticides (tree killers, weed killers, and defoliants,), fungicides (for yeast, mold, mushrooms, mildew, etc.), and Algicides (for algae). There are various pesticides used on animals and they are labeled according to their intended application. For example, against bacteria (bactericides) [6, 7], birds (avicides) [8, 9], namatodes (nematicides) [10], insects (insecticides) [12–14], rodents (rodenticides) [18], fish (piscicides) [21–23], viruses (viricides) [24–26], and mollusks (molluscicides) [27]. According to Randall *et al.* [118], they can also be employed as disinfectants, microbial killers, and repellents. According to one study, herbicides make up over 80% of all pesticides applied (Grace Communications Foundation, 2018), which are weed-controlling plant protection chemicals (sometimes called crop protection products) that can also affect fungus and insects. Although pesticides provide advantages, they also have disadvantages from the bottom up, such as the possibility of harming people & animals up the food chain.

Over the last two decades, pesticide use has greatly increased globally. Around 2 million tons of pesticides are used every year around the globe, with Europe (45%) and the United States (25%) using the most of them, followed by the rest of the world (25%). India's consumption (3.75%) is comparably too low. In Japan and Korea, pesticide usage is 12.0 and 6.6 kg/ha, respectively, while it is barely 0.5 kg/ha in India [29, 39]. Crops have been harmed by the indiscriminate use of pesticides, typically as a result of their toxicity & lack of biodegradability [30, 41, 59]. Due to their widespread accessibility and indiscriminate use, a vast variety of chemical pesticides are used indiscriminately, breast cancer, reproductive toxicity, endocrine disruption, and cytotoxicity are just some of the devastating health impacts that can result from exposure to these chemicals [33, 34, 100]. Soil fertility, soil microbial flora, soil structure, soil mineral cycles, and insect repellency have all been connected to the usage of synthetic pesticides [37, 58, 82, 118]. Every of these negative consequences necessitate the development of flexible, secure, and environmentally friendly techniques to take the place of dangerous chemical pesticides. Biological sources, or biopesticides, have been used as an alternate strategy to control pests on crop plants [60].

5.2 Nanotechnology and Nanoscience

The terms "nanoscience" and "nanotechnology" were developed based on our knowledge of, and allusions to, the study of nanoscale (1–100 nm) materials and their manipulation at the atomic and molecular levels, whose properties differ substantially from those at higher scales [133, 134]. These uses are commonly referred to as "nanoscience." Nanotechnology involves the synthesis, evaluation, composition, & incorporation of several research fields to create useful substances or composite materials for a number of applications [139–149].

5.2.1 Definitions Used in Nanoscience

Numerous explanations of shape, state, nature, size, and attributes are included in nanolevel studies. Understanding these words' definitions in nanoscience is crucial for drawing inferences about their applications and usage [42, 43]. A handful of these expressions are discussed below along with their definitions [38, 39]. One or more sizes which are 100 nm or fewer in numbers are referred to as nanoscale [49–52]. Nanoscience refers to research and manipulation of materials and phenomena on the molecular, macromolecular, and atomic levels [44, 45]. Nanotechnology is the discipline that deals with designing, characterizing, creating, and exploiting systems, materials, and devices by modifying their size and shape on the nanoscale. A substance with different exterior structure or an interior dimension is referred to as a nanomaterial and may display unique properties when compared to similar substances without nanoscale characteristics. Particles with one or more dimensions at the nanoscale are referred to as nanoparticles [46]. If at least each of the components has a minimum of one nanoscale dimension, the composite is referred to as a nanocomposite. A structure at the nanoscale is referred to as being nanostructured [53, 54]. The British Standards Institution has released and made public further information about the terminology of nanoparticles and their requirements [31, 55]. Any event's history helps us assimilate and comprehend the changes that take place in methods, instruments, discoveries, technologies, advancements, and applicability across time. Thus, knowing the background of nanotechnology will shed light on certain potential use cases in every area of science. The development of nanotechnology has seen numerous fascinating occurrences that have altered the possibilities for its use in a variety of industries.

5.2.2 History, Significant Moments, and Landmark Nanotechnology Events

The crucial developments in nanoscience and nanotechnology may be traced to the years 1959–1960, when Feynman was the first to discuss the prenanoscience subject by asserting that Quite a Bit of Space Down Below. When Professor Norio Taniguchi of the University of Tokyo in Japan first introduced the term "nanotechnology" in 1974, he was referring to a specific type of nanoscale technology [66, 130]. In 1981–1982, Gerd Binnig and Heinrich Rohrer created the scanning tunneling microscope (STM). They developed the atomic force microscope (AFM) in the 1980s, which enabled researchers to observe nanostructures and individual atoms, and so won the Nobel Prize in Physics in 1986. By using Coulomb blockade, the first observation of quantized conductance was made the same year, resulting in the construction of a single electron transistor. In the 1980s, electron beam lithography made progress, and in 1991, carbon nanotubes were found by Iijima [68]. Fullerene, a unique form of the carbon cluster C60, was invented by Smalley, Curl, and Kroto in 1985; for this work, they were given the Nobel Prize in Chemistry in 1996. The chemical and physical characteristics of nanocomposite substances had already been studied [7, 76, 121], and in subsequent years, experiments for the development of nanoparticules, composites, and colloids multiplied. According to Schwerdtfeger [75, 77, 122], nanoparticles (NPs) have existed in nature long before humans arrived on the scene [78, 79]. In nature, NPs can be found naturally or can be produced through photochemical and volcanic processes, as well as by plants and algae [81, 85]. They can also be produced through combustion and air pollution. Naturally, nanoscale or nanoscience impressions are produced by nanostructures as visualized in plants (lotus leaf) and insects (super-hydrophobic carbon basis) [71, 74]. Additionally, although the synthesis of nanoparticles or nanostructures may be accomplished through physical, chemical, or biological means, the dimensions of the generated nanoparticles and the chemical reactions that produce them may be of critical importance to their eventual usage [68–70].

5.3 Demand for Biopesticides Worldwide

Rising demand in both Asia and North America has contributed to a surge in the worldwide interest in biopesticides. Overall, the market for pesticides and biopesticides increased from $54.8 to $61.8 billion in 2013–14, and it is predicted that this growth will continue through 2019 to reach $83.7 billion [67, 81]. By 2022, the biopesticides industry is expected to expand from $3.22 billion in 2017 to $6.6 billion, with a compound annual growth rate (CAGR) of 15.43%. This market trend is moving upward due to things like integrated pest management (IPM) practices, pest resistance, rising organic food demand, government agency promotion, significant crop loss from pest infestations, cheaper raw material costs, and quicker approvals from relevant regulatory bodies [85–87]. Utilizing biological pesticides, especially for those insects that are resistant to conventional insecticides, also provides significant societal benefits [72, 73, 84, 91, 113]. As a result of their production using natural ingredients like bacteria and plants, they pose less of a threat to both ecosystems and human health. Biopesticides, which are used to kill pests, have additional advantages because it takes less time and money to create and register them than synthetic pesticides, which take ten years and $200 million to develop. Biological pesticides take 3 to 6 years and $15 to $20 million to develop [94, 119]. A product from natural items should not, however, turn into a pest in and of itself, just as natural enemies should not turn into pests which infest agricultural products [83, 92, 93, 135] or such products should not have an adverse effects on the ecosystem [79]. The requirements state that every biopesticide must not hurt agriculturists, accumulate in the soil, have no impact on the vegetables, fruits or food produced, and must also have a cheap cost of production [38, 95]. Biological pesticides can be broken down into several categories, including plant secondary metabolites, biochemical pesticides, microbiological pesticides, and plant-incorporated protectants (PIPs). Microorganisms like fungus or bacteria that kill and infect pests are referred to as microbial biopesticides. The *Bacillus thuringiensis*, or Bt. bacterium, is a well-known microbial insecticide. Different Bt strains each produce a unique combination of toxins which has the ability to kill larvae or insects. Specific insect species are killed by these various proteins [39, 72, 106].

For the past two decades, phytochemicals derived from plants have been heralded as safe, effective alternatives to industrial pesticides for the control of pests [73]. Plants' secondary metabolites have found utility as biopesticides, most often as insect toxins or antifeedants for controlling insect pests [74, 76, 104, 115]. That nature stores potentially beneficial bioactive chemicals in its vast array of species is a key reason why biodiversity is so crucial. Thus, essential oils and plant secondary metabolites form the backbone of the biopesticides industry. Biopesticides could potentially include items like plant-incorporated protectants (PIPs), which have been modified through genetic engineering to produce a targeted pesticide. The genes of Bacillus thuringiensis that code for pesticide proteins that are incorporated into plants are the best-studied example of this. Communicating plants make their own insecticides, called quality explicit proteins [38].

Semiochemicals, frequently known as pheromones, are a class of natural substances that are delivered by living animals and impact the way of behaving of different creatures. Pheromones are utilized by ranchers or pesticide organizations to tempt bugs into a snare and catch them for expulsion from the field. As a general rule, bug pheromones are critical pieces of innovations utilized for checking and overseeing vermin that influence horticultural harvests. They have recently gone through an exhaustive assessment. Some immediate nuisance the executives strategies that depend on the utilization of pheromones incorporate mating interruption, mass catching, entice, murder, and yank and push and tug [97, 99, 132]. Along these lines, obviously biopesticides are more harmless to the ecosystem than ordinary manufactured pesticides. In any case, to safeguard center materials from debasement, various variables should be considered, including application techniques, conveyance frameworks, and more powerful gathering of the dynamic compound in a network. Thus, during the beyond 10 or so years, the utilization of nanotechnology in agrarian data sources has become progressively significant [101–103].

5.3.1 A Nanobiopesticide: What Is It?

The formation of nanobiopesticides can be achieved in one of two ways. In the main classification, nanobiopesticides are dynamic pesticide synthetic substances or mixtures created naturally that have been joined with nanoparticles and a reasonable polymer [108–111]. In the subsequent kind, a metallic salt is diminished by a functioning pesticide compound to create metallic nanoparticles (NPs), they are combined to form a polymer that works. Advantages, such as slow and controlled delivery, enhanced mobility, and biodegradability of the pesticide ingredient and polymer, motivated the development of APC with nanoparticles and mixes with viable vectors such as polymers, micelles, nanofibers, nanospheres, and liposomes. Additionally, numerous nanocomposites could be created by utilizing various polymers. Through exact conveyance of dynamic pesticide fixings to kill bugs for plant insurance, broad review into the use of nanobiopesticides would diminish how much synthetics showered [112, 113, 116].

We should characterize a nanobiopesticide first. Nanobiopesticides are substance compounds with nanoparticles that have natural beginnings and are used as insect poisons. Polymers, metal oxides, dynamic particles blended in with micelles, and different substances are utilized to build the proficiency and adequacy of nanobiopesticides. Target-situated continuous arrival of dynamic fixings, improved solvency of ineffectively dissolvable dynamic fixings, and dynamic synthetic substances that should not debase too early are a couple of the principal characteristics that confirmed nanobiopesticides should have [116]. Regardless of being utilized in little amounts, nanobiopesticides should not have lessened viability. Furthermore, by combining nanobiopesticides as opposed to utilizing information on the existence cycle and conduct of the microorganism and additionally bug, the objective explicitness and its outcome can be expanded. The qualities and destiny of its dynamic parts are fundamentally affected by nanoformulation. Hence, before their use, it is important to recognize their size reach, shape, or nature, as well as their surface characteristics, adjuvants, and trademark discharge after some time under practical settings [116]. Nanoparticles containing either pesticide-dynamic fixings or other little planned structures with reasonable pesticidal qualities are remembered for the class of "nanobiopesticides" [20, 22, 24, 27]. *Anopheles subpictus* Grassi, a vector for jungle fever, and *Culex quinquefasciatus* Say, a vector for filariasis, both during the fourth instar of development, were used as examples in these evaluations (Diptera: Culicidae) [27, 29, 37, 85]. As indicated by Rajakumar and Rahuman [117], the size of the nanoparticles essentially affected the adequacy of the items. In a rough fluid setup, Ag NP of 35 and 60 nm size showed the greatest efficacy against *C. quinquefasciatus* (LC50 = 27.49 and 4.56 mg/L; LC90 = 70.38 and 13.14 mg/L) and *A. subpictus* (LC50 = 27.85 and 4.56 mg/L; LC90 = 70.38 and 13.14 mg/L). The leaves of the fig trees *Ficus religiosa* (FR) and *Ficus benghalensis* (FB) have been used to expanded spreadability, faster degradability in soil, and more slow degradability in plants with buildup levels below the administrative standards in staples are obtained by replacing the nanobiopesticide with biocomposites or a polymer transporter development displaying properties like solidity, vulnerability, crystallinity, warm dependability, dissolvability, and biodegradability [26, 30, 64]. For example, the hatchlings of the Egyptian cotton leafworm, *Spodoptera littoralis*, are harmed by the pyrethrin-based nanoparticles (30–100 nm) of the water-insoluble insect spray novaluron. Notwithstanding, it has been laid out that these nanoitems are earth hurtful, making removal a test. This clearly raises worries about ecological well-being, which ought to be considered while making NP-based definitions [46].

Chitosan nanoparticles have remarkable physical and synthetic properties and hold extraordinary commitment for use in organic applications like medication conveyance frameworks and harvest security in agribusiness. The level of chitin deacetylation, polymerization, and subatomic weight all influence the bioactivity of chitosan-based nanomaterials [35, 58]. Chitosan nanoparticles (ChNP) are more powerful against microorganisms than conventional chitosan definitions [124]. ChNPs stacked with silver metal particles display worked on antibacterial exercises, with a MIC of only 3 g/ml when dealing with E. coli and 6 g/ml when dealing with *S. aureus* [44]. When compared to regular chitosan, chitosan nanoparticles were found to reduce mycelial growth of *Macrophomina phaseolina* and *Fusarium oxysporum* and to inhibit the development of microsclerotia [56, 61, 66, 124]. Like this, under field settings, a base grouping of 0.06% of Cu-chitosan nanocomposites decreased the rate of the post blossoming tail decay (PFSR) sickness of maize brought about by *F. verticillioides* by 34% contrasted with an untreated control [35, 67, 74–76]. Greenly created metal nanoparticles are becoming turning out to be more huge in the treatment of illness. One of the natural systems that has been extensively studied and is used for the creation of metal nanoparticles is growths due to their ability to survive high metal fixations and their ability to bioaccumulate them [8, 9, 11, 115]. When compared to microorganisms and actinomycetes, organisms are able to release a greater quantity of catalysts needed for the creation of nanoparticles [12, 14, 28, 66]. Trichoderma is a versatile organism that can be used to treat a wide range of plant diseases, from those that affect the leaves and roots to those that affect the medicinal properties of plants and even invertebrate parasites like nematodes. The biocatalytic potential of Trichoderma and its capacity for metal nanoparticle biosynthesis have been evaluated [91]. Kamaraj *et al.* [66] utilized *T. viride* in a concentrate to create titanium dioxide nanoparticles (TDNPs), which were then tested for their bioefficacy against *Helicoverpa armigera* eggs and nymphs. In H. armigera I, II, and III instar hatchlings, TDNPs showed 100% larval death at a concentration of 100 ppm, in contrast to contagious without cell removal (16–52%), $TiO(OH)_2$ (2–8%), and synthetically produced TiO_2 (33–74%). Trichoderma-biosynthesized TDNPs displayed the highest antifeedant activity (100%) and pupal mortality compared to infectious sans cell extricate, $TiO(OH)_2$, and azadirachtin (100%). When *H. armigera* embryos were given Trichoderma viride NPs (TDNPs), the production of stomach enzymes that detoxify metals (-glycosidase and carboxyl esterase) was significantly reduced [66]. Additional research was directed by Mishra [91], using *T. viride* and *Hypocrea lixii* to biosynthesize gold nanoparticles and assess their antibacterial efficacy. Utilizing a sans cell concentrate of *T. viride*, fast biosynthesis of gold nanoparticles happened in the span of 10 minutes at a response temperature of 30°C, while no union was seen utilizing a without cell concentrate of H. lixii at a similar temperature. The development of *E. coli*, *Shigella sonnei*, and *Pseudomonas syringae* was essentially smothered by the antibacterial movement of biogenic gold nanoparticles, which was worked on by 53%, 47%, and 55%, individually. Utilizing the microbes *Serratia* sp. BHU-S4, silver nanoparticles (AgNP) were created extracellularly, and they meaningfully affected the growth *Bipolaris sorokini* ana. While misleadingly vaccinating parasitic conidia onto wheat leaves, Biogenic AgNPs totally hindered their germination.

Comparable outcomes were found in a glass house setting, where the development of the growth *B. sorokiniana* on wheat plants was completely forestalled, as well as improving plant development [4, 6, 7, 92]. Plant illnesses can be overseen in an earth harmless way by utilizing botanicals, and their nanoformulations are fundamentally more powerful than their conventional partners [76, 77, 86]. The adequacy of various plant natural balms in nanoemulsions against plant sicknesses has been researched. Nanoemulsions based on neem and citronella EO showed significant antifungal activity against *Rhizoctonia solani* and *Sclerotium rolfsii. In vitro* tests confirmed that nanoemulsions made from neem and citronella were highly toxic to the two bacteria, with ED50 values much below the threshold for killing [19]. In an alternate report, making nanoemulsions from an EO combination of clove and lemon grass expanded the fungicidal viability. Contrasted with their unformulated mix of rejuvenating ointments, these nanoemulsions had a 1.2 to 1.7 overlay more noteworthy fungistatic movement against *F. oxysporum f.* sp. lycopersici [126]. Noveriza [101] tracked down that utilizing nanoemulsion of citronella EO brought about a 8% to 23% reduction in the rate of vascular streak dieback in cocoa.

5.3.2 Structures for Nanobiopesticides

With their interesting elements, the different nanocarrier shapes act as conveyance vehicles [85]. The most common nanostructure shapes used in the detailing of nanobiopesticides utilizing plant optional metabolites are nanospheres (totals with consistently conveyed dynamic mixtures in a polymeric framework), nanocapsules (totals with concentrated dynamic mixtures close deeply, lined by the lattice polymer), nanogels (hydrophilic, regularly cross-connected polymers that can retain high volumes of water), and micelles (totals). Micelles, which are made up of both hydrophilic and hydrophobic molecular totals, form themselves in fluid configurations, are another example [116]. There have been efforts to build, investigate, and even secure these nanobiopesticide structures. The use of *Azadirachta indica* oil and concentrates, biopolymeric nanoparticles, and powder microparticles in a patented delivery system is just one example [52]. It has been established that *Plutella xylostella* can be killed by a different method of creating nano/microparticles that is loaded with *Azadirachta indica* concentrates [51]. Important criteria for selecting a biopolymer include its biodegradability, viability, and innocuousness. Biodegradability, practicality, and safety are three of the most important criteria for selecting a biopolymer. In order to make a stable nanopesticide (bifenthrin), Liu *et al.* [87] used polymer stabilizers such as poly (acrylic corrosive)-b-poly (butylacrylate), polyvinylpyrrolidone (PVP), and polyvinyl liquor (PVOH).

5.3.3 Nanobiopesticide Synthesis

Biopesticides are natural substances or organic material blends that are utilized to forestall, kill, dispense with, control, or diminish pests. Herbal pesticides, sometimes known as pesticide plants, are made normally from plants. But the executives have been polished for a really long time utilizing plants that produce pesticides. They act as a nuisance safeguard instrument that has developed normally north of millions of years. Many plant types produce bother repulsing compounds. Pesticide-delivering plants have been used for millennia. Before the advancement of compound pesticides during the 1940s, they were normally used in business farming. Substance pesticide utilize caused ecological harm, medical issues like malignant growth, safe framework issues, neurological problems, and metabolic infections like diabetes, endocrine framework unsettling influence, and barrenness. The development of synthetic pesticides has decreased the piece of the pie of biopesticides, and that implies that the ongoing ecological development has made a stage for the improvement of biopesticides.

A developing discipline, nanotechnology, manages materials as little as 10 to 9 m. Ranchers have been utilizing this innovation to analyze plant infections, animate solid plant advancement, and oversee bothers [89, 90, 93]. Substance pesticides have a scope of results. Two hundred passings are believed to be caused every year in the USA by bug spray harming. As per the WHO, there are 25 million episodes of pesticide harming every year, and 20,000 individuals bite the dust thus in unfortunate countries [129]. One hundred thirty compartments of new grapes delivered from India were dismissed by the Netherlands because of the great openness to substance pesticides [112]. The World Wildlife Fund requested that DDT assembling and use be totally prohibited by 2007 [136].

5.4 Biopesticides in Light of Nanoparticles

One class of pesticides known as "biopesticides" is gotten from regular sources such plants, creatures, organisms, and certain minerals. They comprise of neem extricate, microscopic organisms like Bacillus sp., and growth like *Beauveria* sp. baking pop and canola oil are additionally considered biopesticides since they can be utilized to control bugs. The two plants and microorganisms are utilized in the blend of nanoparticles from the previously mentioned sources. For example, silver, nickel, cobalt, zinc, and copper nanoparticles that are made from plants are utilized to control bothers [13]. Biopesticides are earth harmless and less harming. They are utilized to target explicitly specific bugs and break down more rapidly. It shows a lower openness and helps in contamination control. They are both reasonable and degradable. A biopesticide is used for nanoparticle creation and assumes a critical part in horticulture for the government assistance of people and the climate by considering these advantages. Nanoparticles (NPs) with a size somewhere in the range of 10 and 1,000 nm or under 100 nm are utilized in nanotechnology. Metals, including gold, silver, copper, and platinum, are the most widely recognized materials used to make metallic nanoparticles. In contemporary review, nanotechnology is a creating region. Life sciences, especially biomedical advancements and biotechnology, are developing day to day [95, 102, 104, 105]. There are a few purposes for nanotechnology in the synthetic, farming, restorative, organic, and beauty care products enterprises. The water-solvent insect poison mesoporous silica nanoparticles (MSNs) were, at first, used to battle validamycin [45, 108]). *Sitophilus oryzae* L. what is more, *Rhyzopertha dominica*, two huge bug bugs in food capacity, were both effectively killed utilizing alumina nanoparticles. The bundling of food, materials, sanitizers, and various home apparatuses all utilization silver nanoparticles. They are vital in the avoidance of various medical issues, as well as the therapy of sicknesses. The plant world is a rich inventory of natural particles, as per current logical examinations. Most of the plants are known to initiate visualizations and are utilized as insect sprays and drugs [78]. Notwithstanding, there is an information hole in this review. Subsequently, we should focus on plant species whose applications have been portrayed in anthropological and ethnobotanical writing. This study audits proof of plant species used to control bothers from a few papers and rapidly makes sense of some of it. Future investigations to help the climate and mankind can be enlivened by their perceptions, insights, and thoughts.

5.4.1 Advantages of Nanobiopesticides

Due to their nanosize, high surface region/volume proportion, security, upgraded adequacy, better dissolvability, portability, and low poisonousness, nanobiopesticides are ideal [120]. The modification of an unsafe synthetic or pesticide utilized in nanodefinitions is finished to forestall nanopesticide overflow in the air, water, and soil. Be that as it may, the poison might develop in the established pecking order, which is a critical ecological and well-being concern. Nanopesticide particles are incorporated into conveyance materials biopolymers, micelles, or composites to hold and work on the insecticidal worth while likewise limiting natural development to forestall such events. A natural trigger, like temperature, mugginess, wetness, or light, causes the arrival of these nanobiopesticides. Earth nanotubes (halloysite), planned as pesticide transporters, are delivered by the ecological trigger to cooperate better with bugs, like how biopolymers do. Like nanopesticides, nanofertilizers are encased in a specific transporter to permit controlled discharge for explicit impacts [27, 90, 110]. To accomplish more desirable characteristics including firmness, vulnerability, crystallinity, warm dependability, dissolvability, and biodegradability, the nanobiopesticides are supported into biopolymers [15, 30, 64]. As per Shah and Belozerova [125], nanomaterials affected soil microorganisms, and the phototoxicity of powdered Ag nanoparticles was stifled by covering them with biocompatible polyvinyl pyrrole, a nanocovering [88]. The probability of utilizing nanoparticles in plant germination and development is expanded by examining such biocompatible covers to neutralize the poisonousness of nanomaterials. One more opportunities for natural cultivating, as indicated by Oancea, [103, 116], is the controlled arrival of dynamic plant development triggers and different mixtures encased in nanocomposites built of layered twofold hydroxides (anionic dirts). Nanoemulsions, nanotypifies, nanoholders, and nanoconfines have as of late been portrayed as a few nanopesticide conveyance methods with different functionalities for plant security [6, 27, 30]. Along these lines, the ongoing part centers around understanding how to make the nanostructure of nanobiopesticides utilizing electrospinning strategies (NBP). Also, the collection of NBP in plants, bugs, and other living things is analyzed, as well as the capability of NBP *in vitro* and *in vivo*. At last, subjects connected with quality, security, human well-being, and morals are covered [111].

Due to their potential benefits over customary pesticides, like their cheap blend, viability, and eco-similarity, nanobiopesticides have replaced synthetic bug control, or phytochemicals. Be that as it may, this size-subordinate nanoparticles/nanostructures amalgamation requires exact arrangement, hardware, and procedure.

5.4.2 Formulation of Nanobiopesticides

There are various synthetic cycles that can be utilized to make nanoparticles, yet they are in some cases very costly and incorporate hazardous substances. Since they are more secure than synthetic methodologies, researchers are attracted to the utilization of organically determined nanoparticles that include microorganisms and plant extricates [97]. Furthermore, organic beginning responses happen at room temperature and require little energy; they are additionally conservative and recyclable [8, 61]. A captivating area of nanotechnology is the development of nanoparticles from normally happening plant materials for the making of nanobiopesticides (NBP). A polymer made of protein or sugars contains the nanoparticles and the pesticide in addition to nanoparticles or (APC) diminished nanoparticles [21]. With the advantage of cross-connecting, biodegradable polymers like chitosan, polyethylene glycol (PEG), or alginate are utilized to entangle NBP and help in their managed discharge. The medicinal balms (EO) and polymers are likewise complexed in the multifunctional nanoparticles to add further cell reinforcement, antibacterial, and smell veiling qualities. This group of three design, which comprises of a cross-connecting polymer, dynamic biopesticides, and nanoparticles, helps with the directed arrival of the dynamic element for a supported impact of pesticidal action [116]. Foreseeing the capability of nanobiopesticides or nanoparticle is helped by relative investigation of unrefined concentrate, plant separate based metals, nanoparticles, biopolymers, and the combined nanobiopesticides, which could be characterized by deducting the capability of the rough concentrate from the nanobiopesticides. Potential nanobiopesticide plans utilizing silver salt ($AgNO_3$) diminished by aloe vera remove and integrated into a specific biopolymer (chitosan).

5.5 Methods for Synthesis of Nanobiopesticides

Numerous options exist for the formulation of nanobiopesticides. Blending nanoparticles with APC or APC-decreased nanoparticles is the simplest method; this blending can be done in a magnetic stirrer, an RT, an autoclave, a microwave, or through homogenization, high-pressure homogenization, ultrahigh-pressure homogenization, and so on. More perplexing strategies incorporate electro-splashing, electro-turning, This article just examines the electrospinning technique for NBP creation and cross-connecting in polymers.

5.5.1 Electroturning

As of late, the utilization of electroturning processes in various ventures, including medication, wound care, farming, and drugs, has been explored. As indicated by this rule, electrospinning transforms the leading polymer that contains NPs into nanofibers. Under high voltage, the polymer is liquefied and constrained through needles to shape drops. At the point when a high voltage is utilized, the dissolving dissolvable made at the gatherer vanishes because of electrostatic power until it arrives at the authority and structures nanofibers [10]. (1) The needle siphon, which allows the answer for travel through the spinneret at a steady and controlled rate, is one of the central pieces of electroturning. It interfaces the metallic needle of the spinneret to the needle into which the directing polymer is taken care of. (2) The high-voltage arrangement: The essential power drawing nanofibers is high-voltage DC power. The positive terminal and the needle of the needle (provided with polymer) are associated with the high-voltage DC power source, and the negative cathode is associated with a metallic gatherer shrouded in aluminum foil: ordinarily, 30 to 50 kV is utilized. (3) The charged filaments are gathered by the gatherer, which is aluminum or one more leading material, and are then saved on the metal to make a nanofiber web. Gatherers can be static or rotatable and might in different structures, including wire at any point network, pin frameworks, equal or gridded bars, pivoting poles, and wheel fluid showers.

5.5.1.1 Technique

The guide polymer is broken down in a predetermined dissolvable after the orchestrated NBPs have been joined with it. The dissolvable is infused, the NBP and dissolvable containing polymer are consolidated, current is sent through the gadget, and strands are gathered in the gatherer. It is feasible to make NBP + polymer nanofibers; Electro-turning the NBP + NP caught in the directing polymer brings about a high surface/volume proportion, lessening the requirement for synthetics [114].

5.5.2 Potential Nanostructures

Various countries plan, examination, and, surprisingly, patent the designs of nanobiopesticides. In this study, we look into these nanoparticles' capabilities as nanostructure and APC carriers. Nanospheres, in which the active ingredient is distributed uniformly throughout the polymeric scaffolding, (1) nanocapsules, (2) a total where the active component is concentrated near the middle center and is lined by the lattice polymer. (3) Nanogels, which are hydrophilic polymers that are cross-linked and capable of retaining massive amounts of water; micelles, as described by Ragaei and Sabry [116], are aggregates of hydrophilic and hydrophobic atoms. (4) Liposomes: tiny, water-filled, phospholipid-particle sacks, often used in fraudulent drug delivery to tissues; dendrimers are extremely dispersive macromolecules. (5) Hydrogel, a macromolecular polymer gel composed of a network of cross-linked polymer chains, such as hydrophilic monomers with an useful cross-linker to stimulate network formation. (6) Biopolymers with several layers; mesoporous silica (7) and nanofibers (8) are used in catalysis, drug delivery, and imaging, respectively; and nanofibers are delivered from several top-tier polymers and have diameters in the nanometer range.

5.5.3 Nanobiopesticides' Impact on Living Things

By and large, nanoparticles show cytotoxicity, epigenicity, and genotoxicity (mutagenicity, clastogenicity, and aneugenicity) (apoptosis, autophagy, mitoptosis). The nanoparticle shows harmfulness and can possibly cause various sicknesses. Human cells and tissues are impacted by NPs in the mind, nose, lungs, skin, and stomach. As with other organs, NPs can travel through the bloodstream to reach the bone marrow, spleen, endothelium, liver, heart, and placenta. NPs have been linked to the onset of neurological conditions like Alzheimer's disease and Parkinson's disease. Also, breathing in NP can cause disease, emphysema, bronchitis, asthma, and bronchitis [48]. After entering the circulatory framework, NPs incite lethal heart ailments, for example, arthrosclerosis, vasoconstriction, clots, hypertension, and arrhythmia. Moreover, NPs have unidentified infections that have a renal or liver beginning that influence different organs. NPs can cause immune system problems, like dermatitis and podocomiosis, in the skin, as well as Kaposi sarcoma and podocomiosis in the lymphatic framework. NPs likewise affect the intestinal system, causing Crohn's sickness and colon malignant growth [25].

5.5.4 Quality Assurance of Nanobiopesticides

The quality control and well-being checks for the nanomaterials or nanoproducts are regularly represented by the public government specialists of the singular countries. Quality, well-being, harmfulness, and control testing are utilized by numerous offices to choose when nanobased items ought to be acquainted with the market. Nanomaterial-based agrifood varieties are considered for natural guidelines by the top overall food natural certifiers, including the Biological Farmers of Australia (BFA) and the Soil Association, UK [123]. Conflictingly, bunches situated in Germany, including Natural and the International Federation of Organic Agriculture Movements (IFOAM), restricted the natural naming of food sources delivered with fake nanomaterials [60, 98]. Subsequently, there is some conflict among global sanitation experts on the necessities of security for agriproducts made utilizing nanotechnology. This agreement must be accomplished through broad field testing and concentrated review into the most squeezing rural issues in the areas of natural science, green science, science, material science, and nanotechnology.

5.5.5 Nanobiopesticide Safety Issues

The utilization of normal and natural reactants covers or diminishes the security concern related with delivered nanobiopesticides. Green dynamic pesticide mixtures, NPs, and polymer are the components required for NBP; for this situation, green pesticide synthetic compounds kill bothers through retention in the stomach of bugs and staying there; as well as being green in nature, they are biodegradable and innocuous [15]. In nature, NPs assume the part of killing bugs by various cycles or courses, including fingernail skin retention, creation of protein, metabolic chemical, or DNA irregularities, and enlistment of apoptosis. The last segment comprises of polymers used to deliver dynamic NBPs gradually, with dependable antimicrobial action that is solvent in water and soil. Such NBPs are protected partially and significantly affect individuals and plants when they are encased in biopolymers. NBPs are hence protected; by and by, care should be taken while applying them in fields to forestall administrator retention of the NBPs during splashing.

5.6 Potential Human Health Issues

Agribusiness, food, the environment, and human health are all concerned about the biosafety of nanopesticides because of their dermal retention (due to their nanosize, NPs can go through cell films and injure skin tissues). In view of their little weight, NPs can be breathed in and cause respiratory issues. They can likewise infiltrate the blood-cerebrum boundary and enter the mind. An NP's drawn out movement relies upon its steadiness [63]. Yet, these elevated degrees of action for long haul aggregation in the dirt, air, water, and at last in the pecking order are components that should be dealt with cautiously. Some nanoparticles' extraordinary strength or reactivity presents ecological worries. One more issue that requires examination through collaboration in research between organizations, confidential associations, and legislative organizations is the scarcity of concentrates on produced nanomaterials and their openness to the climate [116].

5.7 Morality and Potential Hazards

It totally depends on the kind of nanomaterial utilized in the nanobiopesticides whether they are phytotoxic and harm plants [94, 99]. Given its antimicrobial properties, Ag is used in more than 100 different pesticides. However, the question of whether or not nanosilver is toxic to living organisms, including humans and other animals, remains contentious [26]. Nair *et al.* [96] addressed the unknown consequences of NPs, including (1) unexpected responses of nanomaterials to various plants, (2) phytotoxicity prompted by higher concentrations, and (3) decreased uptake and photosynthesis of plants caused by larger nanoparticles in a 2010 study [127]. It is yet hazy how nanobiopesticides influence individuals and creatures, and further examination is expected to decide what they mean for living things. Such disclosures ought to be dependent upon public conversation at gatherings among analysts and researchers, as well as among the overall population. The nanobased substance, item, or its fixings should be unequivocally distinguished. The expression "nanoscale" suggests that genomic material like DNA or RNA might make fear of negative repercussions; this desolate demeanor should be perceived because of possible dangers to human well-being. The essential impulse behind the improvement of the ethic for the craving to carry on with better and longer lives is this worry for one's well-being [83, 107]. As per Gogoi [54], Syngenta's Banner MAXXÔ, and Nano-5 are instances of plant-safeguarding nanoformulations that have been delivered and are right now available. This shows that general society, government, organizations, and people support nanoitems morally.

5.8 Sanitizers

A sanitizer is an item that brings down microbial impurities on lifeless surfaces to sums that are considered protected according to the viewpoint of general well-being, yet it does not guarantee to dispose of them. Vegetative cells can be successfully killed with a sanitizer. The EPA manages sanitizers, which are dependent upon thorough research center testing necessities and accreditation. They fall into two classifications: no-wash surface sanitizers for food contact and non-contact surface sanitizers. Disinfecting washes for apparatuses, utensils, and holders utilized in dairy handling plants, food handling and refreshment production lines, and eating and drinking places are instances of food-contact sanitizers [5].

5.8.1 Optimal Sanitizer Characteristics

The ideal sanitizer ought to have the accompanying attributes: Rapid eliminate of vegetative microscopic organisms, yeasts, and molds by uniform, wide range microbial annihilation characteristics.

Great cleaning characteristics; ecological obstruction (compelling within the sight of natural matter (soil load), cleanser and cleanser buildups, and water hardness and pH varieties); Nontoxic and aggravation qualities.

Accessibility of water in all sums.

Satisfactory smell or no scent, solidness in concentrated and use weakening, convenience, prepared accessibility, reasonableness, and simplicity of estimation being used arrangement are positive characteristics.

No single universal cleaner is adequate for all circumstances. The substance chosen as a sanitizer should be able to pass the Chambers test, often known as the sanitizer proficiency test: Sanitizers, when used properly at 20°C for 30 seconds, should kill 99.999% of between 75 and 125 million

Escherichia coli and *Staphylococcus aureus* bacteria. The effectiveness of the disinfectant may change depending on the pH at which the fixer is applied. The specialist that kills the germs is typically used to categorize substance deodorizers.

5.9 Cleaning Strategies

5.9.1 Thermal

Because of the energy utilized, warm cleaning is exceptionally wasteful. Its viability is impacted by the imperative mugginess, temperature, and length of a specific temperature. On the off chance that the thing is warmed for an adequate measure of time and the apportioning strategy, application plan, gear, and establish configuration permit the intensity to arrive at all aspects of the item, microorganisms can be killed. To guarantee effective disinfecting, temperature ought to be checked utilizing exact thermometers put at the power source pipes. Steam and high temp water are the two primary energy hotspots for warm disinfection. Steam sterilization is costly and habitually inadequate. Since laborers much of the time mistake water fume for steam, the temperature is habitually inadequate to sanitize the article being cleaned. Industry experience has shown that steam is not reasonable for persistent transport sterilization. Cleaning techniques have been made additional troublesome by buildup from this activity and other steam applications. Warm Water Another warm technique for sanitization includes lowering little parts (like blades, little parts, eating utensils, and little compartments) in water that has been warmed to 80°C or higher. A portion of the cell's protein parts are remembered to have gone through denaturation because of the microbicidal impact. Because of the hardships of keeping a water temperature sufficiently high to empower adequate cleansing, pouring "hot" water into the holders is certainly not a practical sanitizing approach. For food-contact surfaces, high temp water is a proficient, nonselective cleaning approach; notwithstanding, spores might endure for over an hour at 100°C. Eating utensils and plate heat exchangers are typically made of boiling water. How much openness time expected to accomplish sanitization relies upon the water's temperature. Mixes utilized for various plants that utilization 15 minutes of openness time at 85°C or 20 minutes at 80°C are an illustration of a period temperature relationship. A higher temperature is required for a more limited time frame. How much water present and the rate at which it streams will likewise influence what amount of time it requires for the part parts to accomplish the ideal temperature. On the off chance that the water hardness is more prominent than 60 mg/L, scaling of the surfaces being cleaned is normal except if the water is mellowed. Boiling water is effectively available and safe. Disinfecting should be possible by either lowering hardware in water or siphoning water through collected gear.

5.9.2 Radiation

Microorganisms are obliterated by radiation with a frequency of around 2,500, for example, UV light, high-energy cathode beams, or gamma beams. For example, low-pressure mercury fume lights have been utilized to produce bright light to kill microorganisms in homes and medical clinics. UV movement does not appear to be impacted by temperature or pH and leaves treated water bland or unscented. Scarcely any, undesirable results as well as almost no mutagenesis movement or halogenated side-effects have been accounted for to be created. To sanitize drinking and food handling waters, bright light lights are as of now frequently used in Europe and are being underlying the US. Regardless of the way that bright light's activity is autonomous of pH and temperature, its viable dispensing with range for microbes is too little to ever be helpful in food handling. To treat food items, ionizing radiation is accessible from three unique sources. They are X-beams, gamma beams, and electron radiates (e-bar). Gamma and X-beams can infiltrate at least one meters, while e-bar radiation has an infiltrating scope of around 7.5 cm [138]. The dangerous openness not entirely set in stone by bacterial obstruction. The microorganisms should really be presented to the light beams. Radiation does not enter really, thus it ought to simply be utilized to kill microbes that are on sources, in the air, or in straightforward fluids. Refreshment plant water, salt water arrangements, vegetable item move water, cleaning set up wash water, warming and cooling water, cheddar curd farm water, and wastewater effluents are among the fluids that can be treated with UV light. Dust, flimsy oil movies, and hazy or turbid fluids can all ingest them. Likewise, radiation forestalls bug pervasions at all phases of their life cycles. The proficiency of UV lights relies upon the range properties of the light, the term of openness, the separation from the light source, and any meddling substances that block the light, for example, the fouling of the UV response chamber's inner sources or the safeguard that encompasses the bulb. A defensive obstruction is fundamental since the strategy includes the utilization of glass bulbs and quartz response chambers, which represent a gamble of breakage. Diminish the distance of the material or surface being treated from the light since UV force diminishes with distance from the light source [106].

Since openness to UV radiation can bring about serious eye harm and skin disturbance, well-being is a major concern. Besides, since there is no left-over anti-microbial movement, bacterial regrowth may occur. Bacterial cells that had been hurt by UV light can recuperate themselves when presented to noticeable light. Food varieties are exposed to high strain (which shifts relying upon application) for right around five minutes utilizing the high hydrostatic tension (HHP) strategy. Food varieties can be fluid or strong, matured or unaged, bundled or unpackaged. HHP can be applied to a wide assortment of dishes, including crude and cooked meats, fish, shellfish, natural product, vegetable items, cheddar, mixed greens, plunges, oats and grain items, as well as fluids can imagine squeezes, sauces, and soups. Since force is controlled similarly from all sides, the high tension guarantees that the food is not demolished. Microorganisms that are available on the food's outside and inside are delivered idle. By adjusting the sub-atomic construction of synthetic particles expected for the microorganisms' metabolic interaction, inactivation is accomplished. HHP has shown some adequacy in treating bacterial spores, which are impervious to a few biocidal handling methods. It is similarly powerful on molds, microbes, infections, and parasites. In numerous food varieties that have dynamic catalyst frameworks of their own, the breakdown of proteins and lipids (which might be brought about by chemicals of microbial defilement) adds to the development of ruining at chilled temperatures. Certain compounds that cause the crumbling of food have been delivered latent because of HHP.

5.9.3 Vacuum/Steam/Vacuum

A device was developed that subjected robust food items to a vacuum, steam, and then another vacuum [23, 120]. To exploit the critical idle intensity of buildup contrasted with the reasonable intensity sent attributable to temperature distinction in cooling very warmed steam, soaked steam is utilized. The strategy seems to have the ability to dispose of pathogenic microbes in new meat and poultry, handled meats, fish, natural products, and vegetables, in spite of the way that it has not yet been totally researched [117, 118].

5.9.4 Disinfecting With Synthetic Compounds

The synthetic cosmetics and action of the compound sanitizers utilized in food handling and foodservice activities differs relying upon the conditions. By and large, a sanitizer's action is faster and more successful the more thought it is. To pick the best synthetic sanitizer for a specific application, it is important to comprehend and know about every substance sanitizer's one of a kind properties. Microorganisms that are available in breaks, fissure, pockets, and mineral soils may not be totally wiped out by synthetic sanitizers due to their unfortunate entrance limit. At the point when utilized related to cleaning compounds, sanitizers work best while the cleaning arrangement is 55°C or less and the dirt is light. Physical-compound components like the accompanying affect the viability of sanitizers, especially substance sanitizers: Time of openness: Studies have shown that microbial populace passing follows a logarithmic example, intending that if 90% of a populace is cleared out in a given measure of time, the leftover 10% will in any case be alive. In the accompanying unit of time, 90% of the survivors are destroyed, leaving only 1% of the first populace. The amount of microorganisms present and the number of inhabitants in cells with differing sanitizer susceptibilities because of maturing, spore improvement, and other physiological boundaries influence how it requires for the sanitizer to work. You ought to observe the EPA-enlisted mark for the suggested contact time. The greatest focus permitted on the EPA item mark ought to be utilized as a no-wash food-contact surface sanitizer when a sanitizer is applied by a focal sanitizer framework or shower application, which is frequently used to disinfect outside hardware surfaces or for ecological cleaning.

This technique is meant to make up for a lack of manual cleaning, especially in hard-to-clean areas, and to make up for any regular weakening that may occur due to the presence of buildup or additional flush water from cleaning. Temperature: As the temperature increases, microorganisms will develop all the more rapidly and bite the dust all the more regularly from compound openness. A more prominent temperature commonly brings about changes that might help bactericidal activity, including diminished surface pressure, higher pH, diminished consistency, and different alterations. Iodophors that dissipate above 50°C are an exemption. As the temperature climbs, these mixtures become more harming to surfaces, especially elastomers and gasketing materials. Synthetic sanitizers ought to in this way be utilized at room temperature, in a perfect world somewhere in the range of 21 and 38°C. Generally speaking, the degree of sterilization much offsets the speed of bacterial turn of events, in this manner the last effect of raising the temperature is to speed up the pace of microorganism disposal. Focus: Increasing the centralization of sanitizer speeds up the rate at which microscopic organisms are killed. pH: Relatively minor changes in the medium's pH can essentially affect the movement of antimicrobial specialists that exist as different species inside a pH range. Compounds containing chlorine and iodine ordinarily lose adequacy as pH rises. Hardware neatness: Organic soil materials that poor person been eliminated from gear and different surfaces can respond with hypochlorites, other chlorine compounds, iodine compounds, and other sanitizers [123, 125].

A sanitizer's strength can be decreased by ill-advised surface cleaning. Oxidizing synthetic compounds collaborate with natural materials like soils and debilitate their enemy of microorganism impacts. Water hardness: A sanitizer is influenced by the water's substance cosmetics, which could deliver it idle synthetically or support the pH and lessen its viability. Quaternary ammonium mixtures ought not be used in water that contains in excess of 200 sections for every million (ppm) of calcium or without a sequestering or chelating specialist since they are contradictory with calcium and magnesium salts. These sanitizers become less powerful as water hardness rises. Microbial populace: Not each sanitizer actually battles each microorganism. More safe than vegetative and uninhibitedly suspended cells are those that are in a biofilm or spore condition. Liquid milk plants, which are principally worried about psychrotrophic ruining microscopic organisms, may require an alternate sanitizer than refreshment plants, which frequently have yeasts and molds as their essential poisons. The more microbes that are at first present, the more probable it is that some of them will endure in light of the fact that a sanitizer can bring down the quantity of microscopic organisms. The sanitizer can be overpowered by huge amounts. Bacterial connection: Le Chevallier *et al.* [85] shown that the connection of a microorganisms to a strong surface expands the protection from chlorine. Different circumstances, like supplement inadequacy (rigid reaction), additionally make this difference, and connection builds the subsequent protection from chlororine. Since the vast majority of these factors are associated, one is commonly ready to compensate for another. It may be possible to achieve viability comparable to warm temperature in a shorter time period or at a lower fixation if, for instance, the optimum sanitizer must be prepared in cold water.

5.9.5 Different Hand Sanitizer

As a rule, sanitizers can be categorized as one of two classes: liquor based or liquor free. To kill microorganisms and briefly stop their development, an ABHS might contain at least one sorts of liquor, alongside or without other excipients and humectants, for application to the hands [42, 131, 132]. Without water or towel drying, ABHS can productively and immediately obliterate microorganisms across a wide germicidal reach.

Notwithstanding, there are critical downsides to the viability of ABHS, including its transient antimicrobial effect and low movement against different non-wrapped (non-lipophilic) infections, bacterial spores, and protozoa [18, 133].

The liquor free sanitizer, then again, utilizes substances with germicide characteristics to convey the antibacterial impacts. As per their synthetic utilitarian gatherings, these mixtures have different methods of activity and capabilities [42, 80]. They are by and large more secure to use around kids than ABHS on the grounds that they are non-combustible and as often as possible used at low amounts. There are a few dosing types of ABHS, including gel, fluid, and froth. Since each kind has particular characteristics, a review was finished to decide the effect on tangible highlights that can impact a client's acknowledgment of the item and at last impact utilization and consistence with hand cleanliness [70]. Generally, the discoveries demonstrated that gels and froths are more regularly utilized than fluid, especially concerning handleability, despite the fact that the last option left a more elevated level of tidiness and dried quicker [53]. The rundown of OK sterilizers utilized in non-solution (normally known as over-the-counter or OTC) prescriptions has been given by the United States Food and Drug Administration (USFDA) [138, 141]. This rundown is exceptionally useful in picking the right germicide dynamic elements for utilization by clinical experts in emergency clinics or different settings where patients get care beyond emergency clinics [126, 128].

5.9.6 Ingredients in Medicine and Their Purpose

Either ethanol, isopropanol, or n-propanol are available in ABHS. As indicated by studies, the best bactericidal movement is accounted for to happen somewhere in the range of 60% and 95% by volume of liquor [31]. Alcohols' ability to breakdown microorganisms lipid films and denature their protein constituents is believed to be the reason for their antimicrobial impacts. Most of vegetative types of microscopic organisms, including Mycobacterium tuberculosis, parasites, and wrapped infections are powerless to the expansive range antimicrobial impacts of liquor (human immunodeficiency infection [HIV] and herpes simplex infection) [143, 144]. They are inadequate, by and by, against the bacterial spores that are most often tracked down in unrefined substances. This issue may be settled by adding hydrogen peroxide (3%); notwithstanding, because of its destructive nature, care should be taken while dealing with it during fabricate [137]. Various cleaning agents have replaced liquor as the essential dynamic component in items that do exclude liquor. Chlorhexidine Similar to liquor, chlorhexidine accelerates the items in cells by disturbing how the cytoplasmic layers are coordinated [84]. It has a restricted measure of movement against Gram-negative microscopic organisms, wrapped infections, and is generally effective against Gram-positive microbes [17]. Since anionic emulsifying synthetic substances could make the cationic chlorhexidine encourage or become dormant, they ought to be stayed away from while utilizing chlorhexidine-containing items with regular cleansers and hand creams [110, 147]. The Covid is most likely powerless to similar antiviral impacts against other wrapped infections as chlorhexidine gluconate 0.12% [49].

5.9.7 Chloroxylenol

Chloroxylenol is regularly utilized as an antibacterial specialist in cleanser or as an additive in beauty care products. Chloroxylenol's antibacterial properties are because of its ability to impede the arrangement of cell walls in organisms and deactivate compound frameworks. Albeit less successful against Pseudomonas aeruginosa, it is powerful at killing microorganisms and encompassed infections [1].

5.9.7.1 Iodine and Iodophors

Iodine was beforehand a strong disinfectant used to clean the skin. It can enter the microbial cell wall and join with unsaturated fats or amino acids to create intensifies that forestall the arrangement of natural parts. Iodophors have in any case taken over as the dynamic fixing in germicides to supplant iodine because of its capacity to cause skin aggravation and discolouration. Any fluid synthetic sterilant or significant level sterilization containing iodophors as the essential dynamic part has not been supported by the FDA [33], Iodine, iodide, or triiodide are joined with a high subatomic weight polymer transporter, for example, polyvinyl pyrrolidone, to shape iodophors. Iodine's solvency is upgraded, its drawn out discharge is improved, and skin disturbance is diminished thanks to its transporter [131]. relies upon the amount of free iodine contained in the construction, the degree of antibacterial movement. The extent of free iodine will in general ascent following weakening, thus definitions with lower iodophor content may by the by have huge antibacterial action [16]. Both iodine and iodophors have a germicidal effect on a wide variety of bacteria, viruses, fungi, protozoa, and spore-forming microorganisms [13, 146]. However, the iodophor fixation often found in cleaning products (such povidone-iodine 5–10%) is not enough to cause sporicidal effects. Nose detailing with povidone-iodine has been demonstrated to be tolerable and to have an excellent risk/benefit ratio for reducing the spread of COVID-19 during surgery [40].

5.9.7.2 Quaternary Ammonium Compounds, Segment

Four alkyl bunches are joined to a focal nitrogen iota in quaternary ammonium compounds. Models that ring a bell are cetyl peridium chloride, benzethonium chloride, and benzalkonium chloride. They work by sticking to the cytoplasmic layer and permitting the parts to get away. They are more successful against lipophilic infections and Gram-positive microorganisms. The action is very low against mycobacteria, parasites, and Gram-negative microscopic organisms [62].

5.9.7.3 Triclosan

Because of its adverse consequences on bacterial compounds engaged with the amalgamation of unsaturated fats from cell walls and layers, triclosan is bacteriostatic at low fixations. Triclosan breaks the bacterial layer at high fixations, killing the organic entity [65]. It is viable against methicillin-safe Staphylococcus aureus, Candida species, and mycobacteria, which are all Gram-positive microbes. The pH, the use of emollients, and the ionic person of some skin definitions may all affect triclosan's viability [2]. Numerous sanitizers additionally contain humectants in their detailing, like glycerine, to reduce the probability of dry skin brought about by the utilization of liquor based medicines since liquor can strip away sebum that keeps the skin damp. In spite of the fact that fragrance and colorants may be added to upgrade feel, doing so is normally not educated because of the chance concerning unfavorable responses [43, 149].

5.9.7.4 Hand Skin Physiology

A surface epidermis (50–100 m), a center dermis (2 mm), and a deepest hypodermis make up the three essential layers of the skin (1–2 mm). It functions as the body's most memorable line of safeguard against obtrusive microorganisms, protecting the body from actual mischief and restricting extreme water misfortune. The layer corneum, the highest epidermal layer, plays out the pivotal obstruction capability of the skin (SC). The SC is comprised of layers of corneocytes, which have gone through terminal separation from the keratinocytes that make up the epidermis' basal layer [57]. To work on the cohesiveness of the SC, layer intersections known as corneodesmosomes join the adjoining corneocytes [3]. The lipids that are created during the terminal separation of keratinocytes from lamellar bodies will fill the holes between the corneocytes and assist with keeping up with the epidermal boundary capability [50]. The keratinized defined epidermis is the name of the layer that lies underneath the SC. It is comprised of melanocytes, which make melanin, a shade that gives skin its tone and safeguards it from bright light. What is more, this layer contains Merkel cells, which are accountable for light touch sensation, and Langerhan's cells, which are associated with the immunological reaction [36]. Albeit the skin goes about as a hindrance to keep out destructive organisms, it likewise harbors different supportive microorganisms, including Corynebacterium spp., Propionibacterium spp., Micrococcus spp., and Staphylococcus epidermis [34]. By either rivaling them for sustenance or helping the skin's guard systems, these microbes might support the counteraction of the colonization of destructive microorganisms. They have a low pathogenicity under normal circumstances. In any case, they might turn destructive when the skin greenery appropriation is upset, for example on account of delayed utilization of effective anti-microbials or successive hand washing [55]. The microbial equilibrium is reestablished and kept up with through continuous skin recovery to bring down the gamble of disease. From the time that the basal epithelium goes through mitotic division until desquamation, the whole cycle goes on something like 28 days. The microscopic organisms that have colonized the skin surface are killed when the dead keratinocytes in the SC are shed. This continuous interaction effectively accomplishes a decent development among the microbial networks while extraordinarily lessening the intrusion of microscopic organisms.

5.10 Application of Plant-Based Nanobiopesticides as Sanitizers

Plant-based nanobiopesticides possess the properties needed to be used as a sanitizing agent. This properties include: Rapid elimination of vegetative microscopic organisms, yeasts, and molds by uniform, wide range microbial annihilation characteristics; Great cleaning characteristics; ecological obstruction (compelling within the sight of natural matter (soil load), cleanser and cleanser buildups, and water hardness and pH varieties); Nontoxic and aggravation qualities; Accessibility of water in all sums; Satisfactory smell or no scent, solidness in concentrated and use weakening, convenience, prepared accessibility, reasonableness, and simplicity of estimation being used arrangement are positive characteristics. Due to the fact that plant-based nanobiopesticides can kill pests or insects when they are sprayed on the field, they can also be used as a sanitizing agent to clean insect or pest infested area. However, further research needs to be done on nanobiopesticides in order for it to be commercialized and used safely as cleaning agents.

5.11 Conclusion

The utilization of plant-based nanobiopesticides is not only limited to farm products but can also be used as a sanitizing agent. However, further research needs to be done to improve its properties in order to be used safely on both human being and environment without any adverse effects.

References

1. Adeleke, D.A., Olajide, P.A., Omowumi, O.S., Okunlola, D.D., Taiwo, A.M., Adetuyi, B.O., Effect of monosodium glutamate on the body system. *World News Nat. Sci.*, *44*, 1–23, 2022.
2. Adetuyi, B.O., Oluwole, E.O., Dairo, J.O., Chemoprotective potential of ethanol extract of *Ganoderma Lucidum* on liver and kidney parameters in *Plasmodium beghei*-induced mice. *Int. J. Chem. (IJCC)*, 1, 8, 29–36, 2015.
3. Adetuyi, B.O., Oluwole, E.O., Dairo, J.O., Biochemical effects of shea butter and groundnut oils on white albino rats. *Int. J. Chem. (IJCC)*, 1, 8, 1–17, 2015.
4. Adetuyi, B.O., Dairo, J.O., Didunyemi, M.O., Anti-hyperglycemic potency of *Jatropha Gossypiifolia* in alloxan induced diabetes. *Biochem. Pharmacol. (Los Angel)*, 4, 5, 193, 2015b.
5. Adetuyi, B.O., Omolabi, F.K., Olajide, P.A., Oloke, J.K., Pharmacological, biochemical and therapeutic potential of milk thistle (Silymarin): A review. *World News Nat. Sci.*, 37, 75–91, 2021.
6. Adetuyi, B.O. and Farombi, E.O., 6-Gingerol, an active constituent of ginger, attenuates lipopolysaccharide-induced oxidation, inflammation, cognitive deficits, neuroplasticity, and amyloidogenesis in rat. *J. Food Biochem.*, 45, 4, e13660, 2021.
7. Adetuyi, B.O., Adebayo, P.F., Olajide, P.A., Atanda, O.O., Oloke, J.K., Involvement of free radicals in the ageing of cutaneous membrane. *World News Nat. Sci.*, 43, 11–37, 2022.
8. Adetuyi, B.O., Adebisi, O.A., Adetuyi, O.A., Ogunlana, O.O., Toloyai, P.-B., Egbuna, C., Uche, C.Z., Khan, J., Adumanya, O.C., Patrick-Iwuanyanwu, K.C., Ficus exasperata attenuates acetaminophen-induced hepatic damage

via NF-κB signaling mechanism in experimental rat model. *BioMed. Res. Int.*, 2022, 1–10, 2022.
9. Adetuyi, B.O., Adebisi, O.A., Awoyelu, E.H., Adetuyi, O.A., Ogunlana, O.O., Phytochemical and toxicological effect of ethanol extract of *Heliotropium indicum* on liver of male albino rats. *Lett. Appl. NanoBioscience (LIANB)*, 10, 2, 2085–2095, 2020. https://doi.org/10.33263/LIANBS102.20852095.
10. Adetuyi, B.O., Odine, G.O., Olajide, P.A., Adetuyi, O.A., Atanda, O.O., Oloke, J.K., Nutraceuticals: Role in metabolic disease, prevention and treatment. *World News Nat. Sci.*, 42, 1–27, 2022.
11. Adetuyi, B.O., Ogundipe, A.E., Ogunlana, O.O., Egbuna, C., Estella, O.U., Mishra, A.P., Akram, M., Achar, R.R., Banana peel as a source of nutraceuticals, in: *Food and Agricultural Byproducts as Important Source of Valuable Nutraceuticals*, 1st ed, p. 400, Springer, Berlin, 2022.
12. Adetuyi, B.O., Okeowo, T.O., Adetuyi, O.A., Adebisi, O.A., Ogunlana, O., Oretade, O.J., Marraiki, N., Beshbishy, A.M., Welson, N.N., Batiha, G.E., Ganoderma lucidum from red mushroom attenuates formaldehyde-induced liver damage in experimental male rat model. *Biology*, 9, 10, 313, 2020. https://doi.org/10.3390/biology9100313.
13. Adetuyi, B.O., Olajide, P.A., Awoyelu, E.H., Adetuyi, O.A., Adebisi, O.A., Oloke, J.K., Epidemiology and therapeutic measure for COVID-19: A review. *Afr. J. Reprod. Health (Special Edition COVID-19)*, 24, 2, 142, June 2020. https://www.ajrh.info/index.php/ajrh/article/view/2300.
14. Adetuyi, B.O., Olajide, P.A., Omowumi, O.S., Odine, G.O., Okunlola, D.D., Taiwo, A.M., Opayinka, O.D., Blockage of Alzheimer's gene: Breakthrough effect of Apolipoprotein E4. *Afr. J. Adv. Pure Appl. Sci. (AJAPAS)*, 1, 26–33, 2022.
15. Adetuyi, B.O., Toloyai, P.Y., Ojugbeli, E.T., Oyebanjo, O.T., Adetuyi, O.A., Uche, C.Z., Olisah, M.C., Adumanya, O.C., Jude, C., Chikwendu, J.K., Akram, M., Awuchi, C.G., Egbuna, C., Neurorestorative roles of microgliosis and astrogliosis in neuroinflammation and neurodegeneration. *Scicom J. Med. Appl. Med. Sci.*, 1, 1, 1–5, 2021.
16. Adetuyi, B.O., Olajide, P.A., Oluwatosin, A., Oloke, J.K., Preventive phytochemicals of cancer as speed breakers in inflammatory signaling. *Res. J. Life Sci. Bioinform. Pharm. Chem. Sci.*, 8, 1, 30–61, 2022.
17. Adewale, G.G., Olajide, P.A., Omowumi, O.S., Okunlola, D.D., Taiwo, A.M., Adetuyi, B.O., Toxicological significance of the occurrence of selenium in foods. *World News Nat. Sci.*, *44*, 63–88, 2022.
18. Ahmed, S., Saifullah Ahmad, M., Swami, B., Ikram, S., Green synthesis of silver nanoparticles using Azadirachta indica aqueous leaf extract. *J. Radiat. Res. Appl. Sci.*, 9, 1–7, 2016. http://dx.doi.org/10.1016/j.jrras.2015.06.006.
19. Ali, E.O., Shakil, N.A., Rana, V.S., Sarkar, D.J., Majumder, S., Kaushik, P., Singh, B.B., Kumar, J., Antifungal activity of nano emulsions of neem and citronella oils against phytopathogenic fungi, Rhizoctonia solani and Sclerotium rolfsii. *Ind. Crop Prod.*, 108, 379–387, 2017.

20. Arizona State University, *Ask a biologist*, Arizona State University Publisher Unites State of America, 2018, https://askabiologist.asu.edu/building-right-nanoparticles.
21. Atef, M.M.S. and Behle, R.W., Comparing formulations for a mixed-microbial biopesticide with Bacillus thuringiensis var. kurstaki and Beauveria bassiana blastospores. *J. Arch. Phytopathol. Plant Prot.*, 50, 15–16, 2017.
22. Awoyelu, E.H., Oladipo, E.K., Adetuyi, B.O., Senbadejo, T.Y., Oyawoye, O.M., Oloke, J.K., Phyloevolutionary analysis of SARS-CoV-2 in Nigeria. *New Microbes New Infect.*, 36, 100717, 2020. https://doi.org/10.1016/j.nmni.2020.100717.
23. Batiha, G.B., Awad, D.A., Algamma, A.M., Nyamota, R., Wahed, M.I., Shah, M.A., Amin, M.N., Adetuyi, B.O., Hetta, H.F., Cruz-Marins, N., Koirala, N., Ghosh, A., Sabatier, J., Diary-derived and egg white proteins in enhancing immune system against COVID-19. *Front. Nutr. (Nutritional Immunology)*, 8, 629440, 2021.
24. Benelli, G., Mode of action of nanoparticles against insects. *Environ. Sci. Pollut. Res.*, 25, 13, 12329–12341, 2018.
25. Benelli, G., Plant-mediated biosynthesis of nanoparticles as an emerging tool against mosquitoes of medical and veterinary importance: A review. *J. Parasitol. Res.*, 115, 23–34, 2016.
26. Bergeson, L.L., Nanosilver: US EPA's pesticide office considers how best to proceed. *Environ. Qual. Manag.*, 19, 3, 79–85, 2010.
27. Bergeson, L.L., Nanosilver pesticide products: What does the future hold? *Environ. Qual. Manag.*, 19, 4, 73–82, 2010.
28. Bhattacharyya, A., Duraisamy, P., Govindarajan, M., Buhroo, A.A., Prasad, R., Nanobiofungicides: Emerging trend in insect pest control, in: *Advances and Applications Through Fungal Nanobiotechnology*, R. Prasad (Ed.), pp. 307–319, Springer International Publishing, Cham, 2016.
29. Bordes, P., Pollet, E., Avérous, L., Nano-biocomposites: Biodegradable polyester/nanoclay systems. *Prog. Polym. Sci.*, 34, 125–155, 2009.
30. Bouwmeester, H., Dekkers, S., Noordam, M.Y., Hagens, W.I., Bulder, A.S., de Heer, C., ten Voorde, S.E.C.G.S., Buzea, C., Blandino, I.I.P., Robbie, K., Nanomaterials and nanoparticles: Sources and toxicity. *Biointerfaces*, 2, 4, 17–172, 2007.
31. BSI, British Standards Institution, *Vocabulary - nanoparticles, publicly available specification, PAS*, p. 71, BSI, London, 2005.
32. Centers for Disease Control and Prevention, *Chemical Disinfectants—Guideline for Disinfection and Sterilization in Healthcare Facilities*, Centers for Disease Control and Prevention, Atlanta, GA, USA, 2016.
33. Berríos-Torres, S. I., Umscheid, C. A., Bratzler, D. W., Leas, B., Stone, E. C., Kelz, R. R., & Healthcare Infection Control Practices Advisory Committee. Centers for disease control and prevention guideline for the prevention of surgical site infection. *JAMA Surgery*, 152, 8, 784-791, 2017.

34. Chiller, K., Selkin, B.A., Murakawa, G.J., Skin microflora and bacterial infections of the skin. *J. Investig. Dermatol. Symp. Proc.*, 6, 170–174, 2001.
35. Choudhary, M.K., Joshi, A., Sharma, S.S., Saharan, V., Effect of laboratory synthesized Cu-Chitosan nanocomposites on control of PFSR disease of Maize caused by Fusarium verticillioids. *Int. J. Curr. Microbiol. Appl. Sci.*, 6, 1656–1664, 2017.
36. Clayton, K., Vallejo, A.F., Davies, J., Sirvent, S., Polak, M.E., Langerhans cells-programmed by the epidermis. *Front. Immunol.*, 8, 1676, 2017.
37. Costa, M.N., Veigas, B., Jacob, J.M., Santos, D.S., Gomes, J., Baptista, P.V., Martins, R., Inácio, J., Fortunato, E., A low cost, safe, disposable, rapid and self-sustainable paper-based platform for diagnostic testing: Lab-on-paper. *Nanotechnology*, 25, 094006, 2014.
38. An, C., Sun, C., Li, N., Huang, B., Jiang, J., Shen, Y., ... Wang, Y. Nanomaterials and nanotechnology for the delivery of agrochemicals: Strategies towards sustainable agriculture. *J. Nanobiotechnol.*, 20, 1, 1–19, 2022.
39. De, A., Bose, R., Kumar, A., Mozumdar, S., Targeted delivery of pesticides using biodegradable polymeric nanoparticles, in: *Springer Briefs in Molecular Science*, pp. 5–6, Springer, New Delhi, 2014.
40. Dexter, F., Parra, M.C., Brown, J.R., Loftus, R.W., Perioperative COVID-19 defense: An evidence-based approach for optimization of infection control and operating room management. *Anesth. Analg.*, 131, 1, 37–42, 2020.
41. Didunyemi, M.O., Adetuyi, B.O., Oyewale, I.A., Inhibition of lipid peroxidation and *in-vitro* antioxidant capacity of aqueous, acetone and methanol leaf extracts of green and red Acalypha wilkesiana Muell Arg. *Int. J. Biol. Med. Res.*, 11, 3, 7089–7094, 2020.
42. Didunyemi, M.O., Adetuyi, B.O., Oyebanjo, O.O., Morinda lucida attenuates acetaminophen-induced oxidative damage and hepatotoxicity in rats. *J. Biomed. Sci.*, 8, 1–7, 2019. https://www.jbiomeds.com/biomedical-sciences/morinda-lucida-attenuates-acetaminopheninduced-oxidative-damage-and-hepatotoxicity-in-rats.php?aid=24482.
43. Dimetry, N.Z. and Hussein, H.M., Role of nanotechnology in agriculture with special reference to pest control. *Int. J. Pharmtech Res.*, 9, 121–144, 2016.
44. Du, W.L., Xu, Z.R., Han, X.Y., Xu, Y.L., Miao, Z.G., Preparation, characterization and adsorption properties of chitosan nanoparticles for eosin Y as a model anionic dye. *J. Hazard Mater.*, 153, 152–156, 2008.
45. Dureja, P. and Johnson, S., Photodegradation of azadirachtin-A: A neem-based pesticide. *Curr. Sci.*, 79, 1700–1703, 2000.
46. Elek, N., Hoffman, R., Ravi, U., Resh, R., Ishaaya, I., Magdassia, S., Novaluron nanoparticles: Formation and potential use in controlling agricultural insect pests. *Colloids Surf. A Physicochem. Eng. Asp.*, 372, 66–72, 2010.
47. FAO, Feeding the world in 2050, in: *World Agricultural Summit on Food Security*, Rome, 16–18 November 2009, Food and Agriculture Organization of the United Nations, 2009.

48. Farombi, E.O., Abolaji, A.O., Adetuyi, B.O., Awosanya, O., Fabusoro, M., Neuroprotective role of 6-gingerol-rich fraction of Zingiber officinale (Ginger) against acrylonitrile-induced neurotoxicity in male Wistar rats. *J. Basic Clin. Physiol. Pharmacol.*, 30, 1–8, 3, 2019.
49. Fehr, A.R. and Perlman, S., Coronaviruses: An overview of their replication and pathogenesis, in: *Coronaviruses: Methods and Protocols*, pp. 1–23, Springer, New York, NY, USA, 2015.
50. Feingold, K.R., Lamellar bodies: The key to cutaneous barrier function. *J. Investig. Dermatol.*, 132, 1951–1953, 2012.
51. Forim, M.R., Costa, E.S., da Silva, M.F., Fernandes, J.B., Mondego, J.M., Boiça Junior, A.L., Development of a new method to prepare nano-/microparticles loaded with extracts of Azadirachta indica, their characterization and use in controlling Plutella xylostella. *J. Agric. Food Chem.*, 61, 9131–9139, 2013.
52. Forim, M.R., Fernandes Da Silva, M.F.F.D.G., Fernandes, J.B., Vieira, P.C., *Process for obtaining biopolymeric nanoparticles containing Azadirachta indica A. Juss. (neem.) oil and extracts, biopolymeric nanoparticles, and powder microparticles*, pp. 1–23, Patent Application Publication, Pub. No.: US 2015/0320036A1, United States, 2015.
53. Gawade, V.V., Gavade, N.L., Shinde, H.M., Babar, S.B., Kadam, A.N., Garadkar, K.M., Green synthesis of ZnO nanoparticles by using Calotropis procera leaves for the photodegradation of methyl orange. *J. Mater. Sci. Mater. Electron.*, 28, 14033–14039, 2017. https://doi.org/10.1007/s10854-017-7254-2.
54. Gogoi, R., Dureja, P., Singh, P.K., Nanoformulations a safer and effective option for agrochemicals. *Indian Farming*, 59, 7–12, 2009.
55. Grice, E.A. and Segre, J.A., The skin microbiome. *Nat. Rev. Microbiol.*, 9, 244–253, 2011.
56. Halden, R.U., Van Den Wymelenberg, K. *et al.*, Antimicrobial chemicals associate with microbial function and antibiotic resistance indoors. *Am. Soc. Microbiol.*, 3, e00200-18, 2018.
57. Honari, G. and Maibach, H., Skin Structure and function, in: *Applied Dermatotoxicology*, pp. 1–10, Elsevier, Amsterdam, The Netherlands, 2014.
58. Huang, B., Chen, F., Shen, Y., Qian, K., Wang, Y., Sun, C., Zhao, X., Cui, B., Gao, F., Zeng, Z., Cui, H., Advances in targeted pesticides with environmentally responsive controlled release by nanotechnology. *Nanomaterials (Basel)*, 8, 102, 2018.
59. IFOAM, PositionPaperNanotech2011_Approved.pdf IFOAM Organic Clinic. publisher, United kingdom, 1, 1-13, (Accessed March 1, 2012).
60. Khot, L.R., Sankaran, S., Maja, J.M., Ehsani, R., Schuster, E.W., Applications of nanomaterials in agricultural production and crop protection: A review. *Crop Protection*, 35, 64–70, 2012.
61. Ingale, A.G. and Chaudhari, A.N., Biogenic synthesis of nanoparticles and potential applications: An eco-friendly approach. *J. Nanomed. Nanotechol.*, 4, 165, 1–7, 2013.

62. Ionidis, G., Hübscher, J., Jack, T., Becker, B., Bischoff, B., Todt, D., Hodasa, V., Brill, F.H.H., Steinmann, E., Steinmann, J., Development and virucidal activity of a novel alcohol-based hand disinfectant supplemented with urea and citric acid. *BMC Infect. Dis.*, 16, 77, 2016.
63. James-Okoro, P.O., Iheagwam, F.N., Sholeye, M.I., Umoren, I.A., Adetuyi, B.O., Ogundipe, A.E., Braimah, A.A., Adekunbi, T.S., Ogunlana, O.E., Ogunlana, O.O., Phytochemical and *in vitro* antioxidant assessment of Yoyo bitters. *World News Nat. Sci.*, 37, 1–17, 2021.
64. Jianhui, Y., Kelong, H., Yuelong, W., Suqin, L., Study on anti-pollution nano-preparation of dimethomorph and its performance. *Chin. Sci. Bull.*, 50, 2, 108–112, 2005.
65. Jones, R.D., Jampani, H.B., Newman, J.L., Lee, A.S., Triclosan: A review of effectiveness and safety in health care settings. *Am. J. Infect. Control*, 28, 184–196, 2000.
66. Kamaraj, C., Balasubramani, G., Deepak, P., Aiswarya, D., Arul, D., Amutha, V., Karthi, S., Perumal, P., Bio-pesticidal effects of Trichoderma viride formulated titanium dioxide nanoparticle and their physiological and biochemical changes on Helicoverpa armigera (Hub.). *Pestic. Biochem. Physiol.*, 149, 26–36, 2018.
67. Kamaraj, C., Rajakumar, G., Rahuman, A.A., Velayutham, K., Bagavan, A., Zahir, A.A., Elango, G., Feeding deterrent activity of synthesized silver nanoparticles using Manilkara zapota leaf extract against the house fly, Musca domestica (Diptera: Muscidae). *Parasitol. Res.*, 111, 2439–2448, 2012.
68. Elnahal, A.S., El-Saadony, M.T., Saad, A.M., Desoky, E.S.M., El-Tahan, A.M., Rady, M.M., El-Tarabily, K.A., The use of microbial inoculants for biological control, plant growth promotion, and sustainable agriculture: A review. *Eur. J. Plant Pathol.*, 162, 4, 759–792, 2022.
69. Kataria, B., Shyam, V., Kaushik, B., Vasoya, J., Joseph, J., Savaliya, C., Kumar, S., Parikh, S., Thakar, C., Pandya, D., Ravalia, A., Markna, J., Shah, N., Green synthesis of silver nanoparticle using Bambusa arundinacea leaves. *Proceedings of the International Conference on Functional Oxides and/ Nanomaterials*, 2017, https://doi.org/10.1063/1.4982127.
70. Khan, S., Ismail, M., Khan, T., Hussain, F., Majid, A., Iqbal, T., Tasleem, F., Wahab, Z., Fayaz, M., Rahim, T., Khan, Y., Green synthesis of Brassica campestris mediated silver nanoparticles, their antibacterial and antioxidant activities. *Int. J. Chem. Studies*, 6, 4, 1943–1949, 2018.
71. Komarneni, S., Nanocomposites. *J. Mater. Chem.*, 2, 1219–1230, 1992.
72. Koul, O., Microbial biopesticides: Opportunities and challenges CAB. *Rev. Persp. Agric. Vet. Sci. Nutr. Nat. Resour.*, 6, 056, 1–26, 2011.
73. Koul, O., Plant biodiversity as a resource for natural products for insect pest management, in: *Biodiversity and Insect Pests Key Issues for Sustainable Management*, G.M. Gurr, S.D. Wratten, W.E. Snyder, D.M.Y. Read (Eds.), pp. 85–105, Wiley, West Sussex, 2012.
74. Koul, O., *Insect antifeedants*, p. 1010, CRC Press, Boca Raton, FL, 2005.

75. Koul, O., Microbial biopesticides: Opportunities and challenges. *CAB Rev. Perspect. Agric. Vet. Sci. Nutr. Nat. Resour.*, 6, 056, 1–26, 2011.
76. Koul, O., *Naturally occurring insecticidal toxins*, p. 850, CAB International, Wallingford, 2016.
77. Kozempel, M., Goldberg, N., Craig, Jr, J.C., The vacuum/steam/vacuum process. *Food Technol.*, 57, 12, 30, 2003.
78. Krishnamurthy, S., Veerasamy, M., Karruppaya, G., A review on plant sources for nano biopesticide production. *Lett. Appl. NanoBioScience*, 9, 3, 1348–1358, 2020.
79. Kumar, S. and Singh, A., Biopesticides: Present status and the future prospects. *J. Biofertil. Biopestic.*, 6, 1–2, 2015.
80. Kyomuhimbo, H.D., Michira, I.N., Mwaura, F.B., Derese, S., Feleni, U., Iwuoha, E.I., *Silver-zinc oxide nanocomposite antiseptic from the extract of Bidens pilosa*, vol. 1, SN Applied Sciences, Switzerland, 2019, https://doi.org/10.1007/s42452-019-0722-y.
81. Lade, B.D., Gogle, D.P., Nandeshwar, S.B., Nano bio pesticide to constraint plant destructive pests. *J. Nanomed. Res.*, 6, 3, 1–9, 2017.
82. Lade, B.D. and Patil, A.S., Silver nano fabrication using leaf disc of Passiflora foetida Linn. *Appl. Nanosci.*, 7, 5, 181–119, 2017.
83. Larrouturou, L., *The nanosciences*, p. 42, Centre national De La Recherché Science, France, 2005.
84. Larson, E.L., APIC guidelines for handwashing and hand antisepsis in health care settings. *AJIC Am. J. Infect. Control*, 23, 251–269, 1995.
85. Le Chevallier, M.W., Cawthon, C.D., Lee, R.G., Factors promoting survival of bacteria in chlorinated water supplies. *Appl. Environ. Microbiol.*, 54, 649, 1988.
86. Lehr, P., *BCC Research global markets for bio pesticides, market research reports*. Code - CHM029E, BCC Publishing, United state, 2014.
87. Liu, S., Yuan, L., Yue, X., Zheng, Z., Tang, Z., Recent advances in nanosensors for organophosphate pesticide detection. *Adv. Powder Technol.*, 19, 419–441, 2008.
88. Lu, W., Senapati, D., Wang, S., Tovmachenko, O., Singh, A.K., Yu, H., Ray, P.C., Effect of surface coating on the toxicity of silver nanomaterials on human skin keratinocytes. *Chem. Phys. Lett.*, 487, 92–96, 2010.
89. Lyons, K. and Scrinis, G., Under the regulatory radar? Nanotechnologies and their impacts for rural Australia, in: *Tracking Rural Change: Community, Policy and Technology in Australia, New Zealand and Europe*, F. Merlan and D. Raftery (Eds.), pp. 151–171, Australian National University Press, Canberra, 2009.
90. Manjunatha, S.B., Biradar, D.P., Aladakatti, Y.R., Nanotechnology and its applications in agriculture: A review. *J. Agric. Sci.*, 29, 1, 1–13, 2016.
91. Mishra, A., Kumari, M., Pandey, S., Chaudhry, V., Gupta, K.C., Nautiyal, C.S., Biocatalytic and antimicrobial activities of gold nanoparticles synthesized by Trichoderma sp. *Bioresour. Technol.*, 166, 235–242, 2014.

92. Mishra, S., Singh, B.R., Singh, A., Keswani, C., Naqvi, A.H., Singh, H.B., Biofabricated silver nanoparticles act as a strong fungicide against Bipolaris sorokiniana causing spot blotch disease in wheat. *PLoS One*, 9, 5, e97881, 2014.
93. Misra, A.N., Misral, M., Singh, R., Nanotechnology in agriculture and food industry. *Int. J. Pure. Appl. Sci. Technol.*, 16, 1–9, 2013.
94. Monica, R.C. and Cremonini, R., Nanoparticles and higher plants. *Caryologia*, 62, 2, 161–165, 2009.
95. Nag, S.K. and Raikwar, M.K., Persistent organochlorine pesticide residues in animal feed. *Environ. Monit. Assess.*, 174, 327–335, 2011.
96. Nair, R., Varghese, S.H., Nair, B.G., Maekawa, T., Yoshida, Y., Kumar, D.S., Nanoparticulate material delivery to plants. *Plant Sci.*, 179, 154–163, 2010.
97. Narayanan, K.B. and Sakthivel, N., Biological synthesis of metal nanoparticles by microbes. *Adv. Colloid Interf. Sci.*, 156, 1–13, 2010.
98. Naturland, *Naturland standards for organic aquaculture*, Association of Organic Agriculture, Germany, 2011, http://www.naturland.de/fileadmin/MDB/documents/Richtlinien_englisch/Naturland Standards_Aquaculture.pdf. Accessed 1 Mar 2012.
99. Nazir, A., Itrat, N., Shahid, A., Mushtaq, Z., Abdulrahman, S.A., Egbuna, C., Adetuyi, B.O., Khan, J., Uche, C.Z., Toloyai, P.Y., Orange peel as a source of nutraceuticals, in: *Food and Agricultural Byproducts as Important Source of Valuable Nutraceuticals*, 1st ed, p. 400, Springer, Berlin, 2022.
100. Nicolopoulou-Stamati, P., Maipas, S., Kotampasi, C., Stamatis, P., Hens, L., Chemical pesticides and human health: The urgent need for a new concept in agriculture. *Front. Public Health*, 4, 148, 1–8, 2016. https://doi.org/10.3389/fpubh.2016.00148.
101. Noveriza, R., Trisno, J., Rahma, H., Yuliani, S., Effectiveness of several dosage formula of oil and nano emulsion of citronella against vascular streak dieback (VSD) disease on cocoa, in: *IOP conference series: Earth and Environmental Science*, vol. 122 (1), p. 012028, 2018.
102. Nuruzzaman, M., Rahman, M.M., Liu, Y., Naidu, R., Nanoencapsulation, nano-guard for pesticides: A new window for safe application. *J. Agric. Food Chem.*, 64, 7, 1447–1483, 2016.
103. Oancea, S., Padureanu, S., Oancea, A.V., Growth dynamics of corn plants during anionic clays action. *Lucrari Stiintifice Seria Agronomie*, 51, 212–217, 2009.
104. Ogunlana, O.O., Adetuyi, B.O., Adekunbi, T.S., Adegboye, B.E., Iheagwam, F.N., Ogunlana, O.E., Ruzu bitters ameliorates high–fat diet induced non-alcoholic fatty liver disease in male Wistar rats. *J. Pharm. Pharmacogn. Res.*, 9, 3, 251–260, 2021.
105. Ogunlana, O.O., Adetuyi, B.O., Rotimi, M., Esalomi, I., Adeyemi, A., Akinyemi, J., Ogunlana, O., Adetuyi, O., Adebisi, O., Okpata, E., Baty, R., Batiha, G., Hypoglycemic activities of ethanol seed extract of *Hunteria*

umbellate (Hallier F.) on Streptozotocin-induced diabetic rats. *Clin. Phytoscience*, 7, 1, 1–9, 2021.

106. Ogunlana, O.O., Ogunlana, O.E., Adekunbi, T.S., Adetuyi, B.O., Adegboye, B.E., Iheagwam, F.N., Anti-inflammatory mechanism of Ruzu bitters on diet-induced nonalcoholic fatty liver disease in male wistar rats. *J. Evid. Based Complementary Altern. Med.*, 23, 1–8, 2020. https://doi.org/10.1155/2020/5246725.
107. Ogunlana, O.O., Ogunlana, O.E., Popoola, J.O., Adetuyi, B.O., Adekunbi, T.S., David, O.L., Adeleye, O.J., Udeogu, S.A., Adeyemi, A.O., Twigs of Andrographis paniculata (Burn. F) Nees attenuates carbon tetrachloride (CCl4) induced liver damage in Wistar albino rats. *RASAYAN J. Chem.*, 14, 4, 2598–2603, 2021.
108. Ogunlana, O.O., Adetuyi, B.O., Esalomi, E.F., Rotimi, M.I., Popoola, J.O., Ogunlana, O.E., Adetuyi, O.A., Antidiabetic and antioxidant activities of the twigs of Andrograhis paniculata on Streptozotocin-induced diabetic male rats. *BioChem.*, 1, 3, 238–249, 2021.
109. Olajide, P.A., Adetuyi, O.A., Omowumi, O.S., Adetuyi, B.O., Anticancer and antioxidant phytochemicals as speed breakers in inflammatory signaling. *World News Nat. Sci.*, 44, 231–259, 2022.
110. Olajide, P.A., Omowumi, O.S., Okunlola, D.D., Adetuyi, B.O., Deadly Pandemia: Monkeypox disease, A case study. *Afr. J. Adv. Pure Appl. Sci. (AJAPAS)*, 1, 34–37, 2022.
111. Olajide, P.A., Omowumi, S.O., Odine, G.O., Pathogenesis of reactive oxygen species: A review. *World News Nat. Sci.*, 44, 150–164, 2022.
112. Owolade, O.F., Ogunleti, D.O., Adenekan, M.O., Titanium dioxide affects disease development and yield of edible cowpea. *Agric. Food Chem.*, 7, 2942–2947, 2008.
113. Popp, J., Pető, K., Nagy, J., Pesticide productivity and food security. A review. *Agron. Sustain. Dev.*, 33, 243–255, 2013.
114. Prasad, K.G., *Advances in application of nanotechnology training module chapter 5, electrospinning for production of nano materials*, pp. 48–53, ICAR-CIRCOT, MS, India, Sep 24–28 2018.
115. Prasad, R., Pandey, R., Barman, I., Engineering tailored nanoparticles with microbes: Quo vadis? *Wiley Interdiscip. Rev. Nanomed. Nanobiotechnol.*, 8, 2, 316–330, 2016.
116. Ragaei, M. and Sabry, A.H., Nanotechnology for insect pest control. *Int. J. Sci. Environ. Technol.*, 3, 2, 528–545, 2014.
117. Rajakumar, G. and Rahuman, A.A., Larvicidal activity of synthesized silver nanoparticles using Eclipta prostrata leaf extract against filariasis and malaria vectors. *Acta Trop.*, 118, 196–203, 2011.
118. Randall, C., Hock, W., Crow, E., Hudak-Wise, C., Kasai, J., *National Pesticide Applicator Certification Core Manual*, pp. 1–17, National Association of State Departments of Agriculture Research Foundation, Washington, DC, 2013.

119. Rebeca, Balancing the Benefits and Costs of Regulating Biological Plant Protection Products, in: *Regulation of Biological Control Agents (REBECA)*, WS 6 Synthesis, Deliverable No 25, 2007.
120. Sasson, Y., Levy-Ruso, G., Toledano, O., Ishaaya, I., Nanosuspensions: Emerging novel agrochemical formulations, in: *Insecticides Design Using Advanced Technologies*, I. Ishaaya, R. Nauen, A.R. Horowitz (Eds.), pp. 1–32, Springer-Verlag, Netherlands, Dordrecht, 2007.
121. Schmidt, H., Jonschker, G., Goedicke, S., Mennig, M., The sol-gel process as a basic technology for nanoparticle-dispersed inorganic-organic composites. *J. Solgel Sci. Technol.*, 19, 39–51, 2000.
122. Schwerdtfeger, P., Gold goes nano-from small clusters to low-dimensional assemblies. *Angew. Chem. Int. Ed.*, 42, 1892–1895, 2003.
123. Scrinis, G. and Lyons, K., Nanotechnology and the techno-corporate agri-food paradigm, in: *Food Security Nutrition and Sustainability*, G. Lawrence, K. Lyons, T. Wallington (Eds.), pp. 252–270, Earthscan, London, 2010.
124. Sekhar, K., Prasad, R.D., Venkatesham, M., Synthesis, characterization and *in vitro* antifungal activity of chitosan. *Nanoparticles*, 2, 767, 2013.
125. Shah, V. and Belozerova, I., Influence of metal nanoparticles on the soil microbial community and germination of lettuce seeds. *Water Air Soil Pollut.*, 197, 143–148, 2009.
126. Sharma, A., Sharma, N.K., Srivastava, A., Kataria, A., Dubey, S., Sharma, S., Kundu, B., Clove and lemongrass oil based non-ionic nanoemulsion for suppressing the growth of plant pathogenic Fusarium oxysporum f. sp. lycopersici. *Ind. Crop Prod.*, 123, 353–362, 2018.
127. Shi, W.J., Shi, W.W., Gao, S.Y., Lu, Y.T., Cao, Y.S., Zhou, P., Effects of nanopesticide chlorfenapyr on mice. *Environ. Toxicol. Chem.*, 92, 1901–1907, 2010.
128. Singh, P., Pandit, S., Garnæs, J., Tunjic, S., Mokkapati, V.R., Sultan, A., Thygesen, A., Mackevica, A., Mateiu, R.V., Daugaard, A.E., Baun, A., Mijakovic, I., Green synthesis of gold and silver nanoparticles from Cannabis sativa (industrial hemp) and their capacity for biofilm inhibition. *Int. J. Nanomed.*, 13, 3571–3591, 2018. https://doi.org/10.2147/IJN.S157958.
129. Stadler, T., Buteler, M., Weaver, D.K., Novel use of nanostructured alumina as an insecticide. *Pest Manag. Sci.*, 66, 577–579, 2010. https://doi.org/10.1002/ps.1915.
130. Taniguchi, N., On the basic concept of nanotechnology, in: *Proceedings of the International Congress on Production Engineering*, JSPE, Tokyo, 1974.
131. Teodorescu, M. and Bercea, M., Poly(vinylpyrrolidone)–A versatile polymer for biomedical and beyond medical applications. *Polym. Plast. Technol. Eng.*, 54, 923–943, 2015.
132. Tewari, S., Leskey, T.C., Nielsen, A.L., Pinero, J.C., Rodriguez-Saona, C.R., Use of pheromones in insect pest management, with special attention to weevil pheromones, in: *Integrated Pest Management: Current Concepts and Ecological Perspectives*, D.P. Abrol (Ed.), pp. 141–168, Academic Press, New York, 2014.

133. The Royal Society & The Royal Academy of Engineering, *Nanoscience and nanotechnologies: Opportunities and uncertainties*, pp. 1–127, Nanosci Nanotechnol, London, 2004.
134. USA National Nanotechnology Initiative, *Nanogov national nanotechnology initiative*, National Nanotechnology Initiative, United State, 2018, https://www.nano.gov/nanotech-101/what/definition.
135. Waage, J., *Global developments in biological control and the implications for Europe*, vol. 27, pp. 5–13, Bull. OEPP, France, 1997.
136. Wen, L.X., Ding, H.M., Wang, J.X., Chen, J.F., Porous hollow silica nanoparticles as carriers for controlled delivery of ibuprofen to small intestine. *J. Nanosci. Nanotechnol.*, 6, 3139–3144, 2006. https://doi.org/10.1166/jnn.2006.410.
137. World Health Organisation, WHO Guidelines on hand hygiene in health care, in: *First Global Patient Safety Challenge: Clean Care Is Safer Care*, World Health Organisation, Geneva, Switzerland, 2009.
138. Zammer, C., Food irradiation: Is it a matter of taste? *Food Qual.*, 11, 3, 44, 2004.
139. Adetunji, C.O., Titilayo, O., Mathew, J.T., Inobeme, A., Singh, K.R.B., Olaniyan, O.T., Singh, J., Singh, R.P., *Carbon Dots: Next-generation materials for biomedical applications*, pp. 8–12, IOP Publishing, Bristol, England, 2022.
140. Farid, A., Muzammal, M., Ghazanfar, S., Ismail, S., Ullah, A., Dauda, W.P., Adetunji, C.O., Shah, M., Dadrasnia, A., Maryam Idrees, S. *et al.*, Green cleaning activity of *Bacillus salmalaya* 139Sl: A novel strain for removing common household stains. *Biomass Conv. Bioref.*, 2022, 1, 1–8, 2022. https://doi.org/10.1007/s13399-022-03147-z.
141. Wali, S., Zahra, M., Okla, M.K., Wahidah, H.A., Tauseef, I., Haleem, K.S., Farid, A., Maryam, A., AbdElgawad, H., Adetunji, C.O., Akhtar, N., Akbar, S., Rehman, W., Yasir, H., Shakira, G., Brassica oleracea L. (Acephala Group) based zinc oxide nanoparticles and their efficacy as antibacterial agent. *Braz. J. Biol. Sci.*, 84, 1–7, 2022. https://doi.org/10.1590/1519-6984.259351.
142. Adetunji, C.O., Mathew, J.T., Singh, K.R.B., Inobeme, A., Nayak, V., Olaniyan, O.T., Singh, J., Singh, R.P., Zero-dimensional carbon nanomaterials for targeted drug delivery and photothermal therapy applications, in: *Zero-Dimensional Carbon Nanomaterials: Fundamentals and Applications*, pp. 1–10, IOP Publishing, Bristol, England, 2022.
143. Adetunji, C.O., Mathew, J.T., Singh, K.R.B., Inobeme, A., Nayak, V., Olaniyan, O.T., Singh, J., Singh, R.P., Zero-dimensional carbon nanomaterials for targeted drug delivery and photothermal therapy applications, in: *Zero-Dimensional Carbon Nanomaterials: Fundamentals and Applications*, pp. 1–12, IOP Publishing, Bristol, England, 2022.
144. Adetunji, C.O., Michael, O.S., Nwankwo, W., Anani, O.A., Adetunji, J.B., Olayinka, A.S., Akram, M., Biogenic nanoparticles based drugs derived from medicinal plants: A sustainable panacea for the treatment of malaria, in: *Green Synthesis in Nanomedicine and Human Health*, Boca Raton, Florida, United state, pp. 103–122, CRC Press, 2021.

145. Adetunji, O., Michael, O.S., Akram, M., Oseni, K., Anani, O.A., Nwankwo, W., Anwar, H., Adetunji, J.B., Olayinka, A.S., Recent advances in the utilization of bioengineered plant-based nanoparticles: A sustainable nanobiotechnology for the management of extensively drug-resistant tuberculosis, in: *Green Synthesis in Nanomedicine and Human Health*, Boca Raton, Florida, United state, pp. 149–166, CRC Press, 2021.
146. Adetunji, C.O., Michael, O.S., Akram, M., Oseni, K., Temidayo, A.K., Anani, O.A., Olayinka, A.S., Nwankwo, W., Ghaffar, I., Adetunji, J.B., *Green Synthesis in Nanomedicine and Human Health*, pp. 81–100, CRC Press, Boca Raton, Florida, United State, 2021.
147. Islam, S., Thangadurai, D., Adetunji, C.O., Michael, O.S., Nwankwo, W., Kadiri, O., Anani, O.A., Makinde, S., Adetunji, J.B., Nanomaterials: Applications in biomedicine and biotechnology, in: *Handbook of Nanomaterials and Nanocomposites for Energy and Environmental Applications*, O.V. Kharissova, L.M. Torres-Martínez, B.I. Kharisov (Eds.), Springer, Cham, 2021, https://doi.org/10.1007/978-3-030-36268-3_4.
148. Thangadurai, D., Prakash, L., Sangeetha, J., Mohammad Said Al-Tawaha, A.R., David, M., Islam, S., Adetunji, C.O., Adetunji, J.B., Sustainable synthesis of greener nanomaterials: Principles, processes, and products, in: *Handbook of Nanomaterials and Nanocomposites for Energy and Environmental Applications*, O.V. Kharissova, L.M.T. Martínez, B.I. Kharisov (Eds.), Springer, Cham, 2021, https://doi.org/10.1007/978-3-030-11155-7_30-1.
149. Adetunji, C.O. and Varma, A., Biotechnological application of trichoderma: A powerful fungal isolate with diverse potentials for the attainment of food safety, management of pest and diseases, healthy planet, and sustainable agriculture, in: *Trichoderma: Agricultural Applications and Beyond. Soil Biology*, vol. 61, C. Manoharachary, H.B. Singh, A. Varma (Eds.), Springer, Cham, 2020, https://doi.org/10.1007/978-3-030-54758-5_12.

6

Application of Plant-Based Nanobiopesticides Slimicides Against Slime-Producing Microorganisms

Ojo, S.K.S.[1*], Ayo, I.O.[2], Otugboyega, J.O.[2], Oluwole, B.R.[3] and Ojo, A.M.[4]

[1]Department of Microbiology, Federal University Oye-Ekiti, Ekiti, Nigeria
[2]Department of Environmental Management and Toxicology, Federal University Oye-Ekiti, Ekiti, Nigeria
[3]Department of Plant Science and Biotechnology, Federal University Oye-Ekiti, Ekiti State, Nigeria
[4]Chemistry Department, Faculty of Science, Federal University Oye Ekiti, Ekiti State, Nigeria

Abstract

Conventional pesticides are hazardous to humans and the environment. Synthetic agrochemicals continue to provide effective protection for agricultural crops; however, their continuous application has undesirable and deleterious repercussions, including carcinogenic effects, environmental instability, as well as other issues. As a result, new pesticides containing unique naturally derived constituents that are less hazardous, unique to the targeted pest, efficient in little amounts, and also disintegrate faster than traditional pesticides, are being created. This chapter appraised the application of plant-based nanopesticides slimicides that are capable of killing slime-producing microorganisms like algae, bacteria and fungi. It takes a cursory look at biopesticides, examines the application of nanoparticles in water treatment, biofouling and paper production, and further considers the applications of plant-based nanobiopesticide slimicides capable of killing slime-producing organisms in plant protection and preservation while placing emphasis over the control of varying pests and slime-producing microorganisms and an assessment of the benefits and demerits of plant-based nanobiopesticide slimicides. The study finds, among other things, that the combined effect of urbanization, rising food demands and persisting constraints on cultivation areas have resulted in the

**Corresponding author*: stephen.ojo@fuoye.edu.ng

Charles Oluwaseun Adetunji and Julius Kola Oloke (eds.) Handbook of Agricultural Biotechnology: Volume I Nanopesticides, (171–192) © 2024 Scrivener Publishing LLC

excessive usage of pesticide and the wrong application of nanobiopesticides slimicides. The study relies mostly on secondary data. The secondary sources include textbooks, journals, articles, monographs, international, and policy documents on agriculture, botany, and microbial science.

Keywords: Nanobiopesticides, biopesticides, slimicides, slime-producing microorganisms, ecofriendliness

6.1 Introduction

Plant-based nanobiopesticides is an emerging research field in environmental, botany science and microbiology that offers new frontiers of wide knowledge, encompassing novel methods of active ingredients within the purview of nanoscale boundaries. It is contended that Plant-based nanobiopesticide formulations possess the potential and bright future for the development of an efficient, all purposive, utterly effective and more environmentally friendly and safe biopesticides. Biopesticides have achieved great prominence in comparison with chemical pesticides as a result of their eco-friendliness, great efficiency and efficacy in these past years [34]. Much study has been carried out on nanobiopesticides as an alternative as a result of target pests developing resistance against chemical pesticides, causing harsh environmental consequences, which are of human health concerns [46].

Chemical pesticides and slimicides have over the years caused havoc in several sectors including the health and environmental sectors [6]. Their use has brought about disruption in the ecosystem, critical ill health in humans to mention a few [48]. Their application also results in the occurrence of resistant plant pathogen strains, leading to the reappearance of diseases in the ecosystem [34]. Thus, new and innovative botanical biopesticides on the nanometer range are required to develop nanobiopesticides for the control of pests. The toxicity of biopesticides is under debate as few tests have been done on the metabolites to ascertain the consequences of its long-term use. Thus, estimating a biopesticide's safety is complicated [34].

This study appraises the frontiers of biopesticides, microbial biopesticides, biochemical biopesticides, botanical biopesticides, examines the application of nanoparticles in water treatment, future projections and further considers the applications of plant-based nanobiopesticide slimicides capable of killing slime-producing organisms like Algae, bacteria and fungi; the application of nanoparticles in plant protection and preservation; while placing emphasis over the control of varying pests and slime-producing microorganisms; categorising nano-based technology being mostly used in different sectors of agro-ecosystems, an assessment of the benefits and demerits of plant-based nanobiopesticide slimicides.

6.2 Biopesticides

Pesticides remain a severe hazard to humanity and the environment, as stated by the World Health Organization (WHO) [83]. As a result, new pesticides containing unique biological constituents are being created with reduced toxicity, increased target specificity, efficient in little amounts, and disintegrates faster compared to regular pesticides [17]. It is on this note that synthetic pesticides continue to provide effective protection for agricultural crops, however, their continuous application have undesirable and deleterious repercussions to include cancer formation in humans and environmental instability [3]. Thus, it is against this background that the development of new insecticides became necessary to solve these challenges [39]. Biopesticides have been researched as they are both harmless and environmentally safe.

It is noted that any pesticide (bactericide, fungicide, algicide or slimicide) that impedes microbial growth or cause their death while causing negligible damage to the environment is considered a "green biocide" [9]. The important prerequisites for an eco-friendly biocide are thus: it should be active and efficient in dealing with offending microbes, cost-effective and harmonious with the natural environment [9].

A slimicide is generally a biocidal antimicrobial substance used in killing microorganisms that produce slime for example, bacteria, fungi, algae and slime molds [10]. There are several important usages slimicides are put to. Slimicides are used in the paper mill industry, water treatment redistribution systems to include cooling systems, petroleum reservoirs, and in conjunction with oil extraction fluids. Similarly, slime-forming bacteria are important in biofilm formation because they produce extracellular complex carbohydrates that ensure microbial cell adhesion to surfaces, offer protection from antimicrobial agents, acts as a nutrient reservoir and create unique structure to promote additional microbial adhesion [24, 53]. It is also noted that the efficient functioning of paper mills, marine structures and pest control would not be possible without the use of slimicides used in preventing the formation of slime from the offending microorganisms [10].

Biopesticides, also known as biological pesticides, are substances of biological origin used to inhibit, control and destroy pests [46]. They are acquired naturally from animals, plants, microbial and minerals sources [3, 46]. Examples include *Beauveria* sp. (fungi) and *Bacillus* sp. (Bacteria). The earliest use of biopesticides was in the late 1800s, when spores of fungi were utilized to manage insect pests. Agostine Bassi also demonstrated in 1835 that *Beauveria bassiana* spores might protect silkworms from illness [3].

Since then, biopesticides have been used continuously throughout modern agricultural history, though at a far lower level than conventional crop protection. The mode of action is the major distinction between synthetic pesticides and biopesticides. The former act by neurotoxicity to pests, while the latter have different mechanisms of action, such as dehydration, anti-feeding, disruption in mating patterns and oxygen deprivation [48]. Biological pesticides have gained global attention as they are less toxic and safe for the ecosystem. They target only specific pests, are biodegradable and have the ability to biodegrade quickly. This indicates lower exposure and helps in controlling pollution. Also, they are economical and have a rather high efficacy [46].

Biopesticides are grouped into three with respect to their sources. They are as follows: (i) microbial biopesticides, (ii) botanical biopesticides, (iii) plant-incorporated protectants [48]. These groups are therefore considered one after the other.

6.2.1 Microbial Biopesticides

Microbial biopesticides are microbes that are effective against diseases and field pests [55]. These biopesticides are made from bacteria, viruses, fungi and protozoa [48]. For example, *Metarhizium anisopliae* and *Beauveria bassiana* are fungi that infect insect pests, eventually leading to their death. Also, microorganisms such as *Trichoderma* and *Pseudomonas* have been used as biopesticides to extensively combat soil-borne diseases [51].

Trichoderma, a filamentous fungus found growing on organic materials like rotting wood, is effective against an array of disease-causing fungi, for example, *Candida albicans*, *Sclerotia*, *Fusarium*, and *Phytophagthora*. Many *Trichoderma* species (*Trichoderma virens, Trichodermaviride*, and *Trichoderma harzianum*) possess biocontrol abilities, including mycoparasitism. They release enzymes that degrade cell wall, such as chitinases, proteases, and glucanases. They can also cause antibiosis using antibiotics and their compounds, such as heptelidic acid, almethicins, glisoprenins, and viridian [50].

The ability of microbial biopesticides to cause disease in the intended pest differs between species [59]. Disease-causing ability is an essential role of pathogens' insecticidal toxins [3]. Most of the toxins generated by pathogenic microbes are classified as peptides although their configuration and deadliness differ significantly. It is thus posited that microbial pesticides outperform chemical insecticides in terms of efficacy. They are environmentally benign and eliminate risks posed by other pests. This gives rise to proliferation of diverse species in the ecosystem. Consequently, microbial biopesticides are particularly precise against the target pests, thus permitting useful species to flourish [25].

6.2.2 Biochemical Biopesticides

Natural substances sourced from animal origin, botanical origin or insects that are neither detrimental to the producing organisms, nor cause them harm are termed biochemical biopesticides. They occur organically and curb pests in an innocuous manner without killing them [3]. Biochemical pesticides thus have a distinct mechanism of action than conventional pesticides. Conventional pesticides have a direct effect on their target and destroy it, whereas biochemical pesticides have an indirect effect. They, for example, interfere with their target's sexual functions. These biopesticides act by repelling pest via pheromones and work as Plant Growth Regulators (PGR) [33].

6.2.3 Botanical Biopesticides/Plant-Based Biopesticides

Plants with pesticidal characteristics have been an old technique of pest control that has been used for thousands of years. Botanical pesticides, are natural pesticides derived from plants for the eradication of pests [46]. These plants have evolved as an innate defence against pests years ago. Several species of plants release compounds that have an impact on pests [46]. They utilize a range of secondary metabolites including alkaloids, organic acids, and glycoalkaloids, essential oils, phenolics and terpenes to combat pests and inhibit microbial growth [37, 73]. It is thus noted that these plant-based biopesticides are non-persistent and non-toxic in the environment [3].

Furthermore, the essential oils' antioxidant nonoxidative characteristics play a protective role in plants from pro-oxidants observed in proteins and Nucleic acids, which end up causing cytotoxic effects, the emergence of oxygen radicals, and the destruction of the cell membranes in the microbes that afflict plants [63]. It is further noted that several factors affect the effectiveness of a biological pesticide: they include the method of extraction used, the mist of the substance harvested, plant age, where the essential oil is obtained, among others [34].

It is restated that botanical pesticides were extensively used in commercial farming before the creation of pesticides of chemical origin in the 1940s. Although showing good activity against pests, the use of chemical pesticides led to negative impacts environmentally and on human well-being [37, 40]. The effects of chemical pesticides include destruction of the environment and health difficulties such as cancer, detrimental effects on the body's immunity, neurodegenerative problems as well as metabolic conditions like diabetes and sterility [46].

Table 6.1 Botanical pesticides and their target pests.

Botanical name	Common name	Pest	Reference
Allium sativum	Garlic	*Candida albicans, Pseudomonas aeruginosa, Staphylococcus* spp., *Bipolaris sorokiniana, Drechslera tritici-repentis Trichophyton rubrum*	[1, 43, 67, 78]
Azadirachta indica	Neem	*Vibrio cholerae, Bemisia tabaci, B. subtilis*	[14, 18, 70]
Calendula officinalis	Pot Marigold	*Klebsiella pneumoniae, Escherichia coli,*	[13]
Thymus vulgaris	Thyme	*Xanthomonas vesicatoria, Escherichia coli, Salmonella typhimurium*	[5, 42, 82]
Cinnamomum zeylanicum	Ceylon cinnamon	*Fusarium solani, Penicillium expansum, Botrytis cinerea, Escherichia coli, Staphylococcus aureus, Colletotrichum musae*	[21, 62, 85]
Zingiber officinale	Ginger	*Fusarium oxysporum, Pythium aphanidermatum, Rhizoctonia solani*	[8]

Pursuant to the World Health Organization (WHO), there are 25 million cases of exposure to pesticides resulting in poisoning and 20,000 fatalities in underdeveloped nations each year [77]. It is stated that the World Wildlife Fund called a ban globally on the manufacture and use of Dichlorodiphenyltrichloroethane (DDT) by the year 2007 [81]. Methyl bromide, a pesticide that was efficient against soil-borne pathogens, invasive pests, and worms in just about all plants, was equally banned in the year 2015 following the Montreal Protocol due to its ozone depletion negativity [36]. Likewise, some compounds, such as chloropicrin and dazomet, were similarly regulated in some locations due to fears concerning the safety of agricultural food and health impacts [36].

Although botanical biopesticides have a lot of merits, their use is not widespread because of various reasons: first, they do not maintain stability over time, the type of extraction techniques being used, purification process and the complexity of the extracted fractions [11]. Consequently, a number of plants have been researched and found to have pesticidal activities as shown in Table 6.1.

Ginger, turmeric, pepper, lemon, and garlic ethanolic extracts have been demonstrated to strongly limit the growth of *Fusarium oxysporum, Alternaria solani, Pythium ultimum,* and *Lycopersicium* sp. [60].

6.3 Nanobiopesticides

It should be noted that nanobiotechnology is an emerging field in modern research that deals with substances (nanoparticles) of 1 to 100 nm sizes [46, 48]. Metals including silver, copper, gold or platinum are commonly used to create nanoparticles [61]. Nanotechnology does have countless applications in various fields like agriculture, pest management, medicine, pharmacology, and parasitology [46, 72]. Because of their attributes of being environmentally benign, cost-efficient, compact size (1–100 nm), and high surface area, nanobiopesticides have an advantage over ordinary biological pesticides and other conventional pesticides and proffer solutions to several environmental concerns [34]. They also produce the desired results in a matter of hours and are biodegradable [72]. Upon combination with agrochemicals that can freely penetrate the plants, their minute size allows them to be an efficient vehicle. Furthermore, nanobiopesticides do not harm soil microbes and their activity. Ag-based nanomaterials were found to be phototoxic and this was reduced by nanocoating them with biologically compatible biomolecules of polyvinyl pyrrole [54]. Nanobiopesticides are appealing because of their small dimensions, enhanced solubility,

Table 6.2 Plant-derived nanobiopesticides.

Botanical name	Common name	NPs present	Plant part utilized	Function	References
Artemisia nilagirica	Wormwood	Ag	Leaves	Prevents moths and repels mosquito	[79]
Bidens pilosa	Black jack	Ag and ZnO	Leaves, seeds and root	Inhibits *S. aureus* and *C. albicans*	[52]
Carica papaya	Pawpaw	Ag and Zn	Leaves and seeds	Numerous	[69]
Cymbopogan nardus	Citronella grass	Ag	Entire plant	Repels mosquito	[41]

increased surface area-to-volume ratio, resilience, mobility, high potency, and reduced toxicological effects [34]. Table 6.2 provides examples of some plants that have been studied for their potential as nanobiopesticides.

6.3.1 Synthesis of Nanobiopesticides

Generally, nanobiopesticides can be produced in two ways:

1. Extraction of the natural Active Pesticidal Compound (APC) derived from the plants, blended together with nanoparticles, and inserting into an appropriate polymer which functions as a supportive agent.
2. Secretion of the metallic salt from the Active Pesticidal Compound (APC), which combines with nanoparticles, and almagamate into a suitable polymer that includes micelles, liposomes, nanospheres, polymers, and nanofibers [34]. These are utilized in a spray to exterminate pests in order to protect food crops [68].

In a related vein, designed nanoparticles are now being utilized as new pesticide delivery mediums. A variety of different formulations have been proposed, such as nanoemulsions, nanoencapsulations, nanovesicles, nanogels, and nanofibers. These can be used to enhance the eco-friendly status and potency of active ingredients in already known biological pesticides [45]. A number of these formulation types are therefore considered hereunder:

(a) *Nanoemulsion*

A nanoemulsion is dispersion mechanism composed of the active ingredients solubilized in two immiscible phases: both the oil and water phases, along with surfactants resulting in a single phase [76]. The nano-sized droplets (20–200 nm) are said to give this system proper kinetic stability and a transparent or translucent appearance [57]. Pesticide nanoemulsion is therefore an oil-in-water (O/W) type dispersion capable of dissolving water-insoluble pesticides into comparatively tiny oil droplets, thereby significantly improving potency and making it more bioavailable [28].

Furthermore, when compared to traditional pesticide formulations, organic solvent and emulsifier use are vastly reduced in nanoemulsion, which has piqued the interest of scholars recently [66]. Because of the smaller size of the particle of 200 nm and the reduced surface tension of this nanoemulsion, droplets were delivered homogeneously on the leaf surfaces a reduced angle of contact and improved absorption, distribution, and permeability [3].

(b) *Nanoencapsulation*

Pesticide nanoencapsulation is a pesticide delivery method in which active ingredients are encased in different nanomaterials that are then released gradually [64]. Nanomaterial encapsulation protects active ingredients from untimely degradation breakdown from photolysis, hydrolysis, and biodegradation. It also prevents unnecessary extraneous losses due to leachate and thermal decomposition. Nanoencapsulated pesticides outperform conventional pesticide treatments in real-world practical uses [36, 80]. Furthermore, nanomaterials can also allow the encapsulated pesticides demonstrate a continued release behavior, thereby extending the response time span, or provide sensory-responsive discharge characteristics, resulting in accurate control [16, 49, 56].

Organic materials, such as polymers of natural and synthetic sources, lipids, nanoparticles procured from plants, synthetic and natural polymers, lipids, and inorganic materials, such as silica, carbon, calcium and also clay-based nanoparticles, have already been utilized for pesticide nanoencapsulation [48]. Active ingredients are frequently encased into nanocarriers through incorporation, electrostatic interactions, or covalent bonding. This improves the active ingredient's absorption rate, dispersibility, movement, cohesion, as well as mediated or direct release, resulting in greater bioavailability and viable longevity. Nanoencapsulated pesticides made from polymers, such as nanomicelles, nanospheres, nanofibers, and nanogels, have received considered research owing to its remarkable biodegradability and biocompatibility [76].

Nanoencapsulated pesticides typically exhibit two release behaviors: steady (slow) release and release in response to stimuli. In the former, the active ingredient concentration levels stay within an effectual margin of control, thus resisting pests' species for a longer duration and reducing the need for recurrent application of pesticides. It is critical to design nanoencapsulated agrochemicals that are stimuli-responsive and can accurately transport active ingredients to the target in order to achieve maximal control efficacy. Thus far, several stimuli (such as temperature, pH, redox, light, enzyme, ionic strength, and moisture) have been identified. It has been observed that nanoencapsulated pesticides enhance the intelligent delivery of active chemicals in reaction to living (botanical diseases, bugs, and vegetation) or nonliving (temperature, salinity, sunlight, water shortages, flooding, and soil properties) stimuli [3]. Thus, the controlled release capabilities of nanoencapsulated pesticides suggest that they have the ability to improve the overall effectiveness of pesticides.

(c) *Nanoparticles*

The nanoparticle-based pesticides covered in this chapter fall into two categories: nanoparticles employed directly simply as pesticides in their own right and nanoparticles infused pesticides. Organic/inorganic nanoparticles are being widely explored as pesticides as well as bactericidal and fungicidal agents for plant microbial pathogens. Due to their antifungal or antibacterial action, several metalloids (such as gold, silver, iron, copper, platinum and zinc) and metal oxides (such as Ag_2O, FeO, CeO_2, ZnO, CuO, and TiO_2) are employed directly as nanopesticides Several metalloids (such as gold, silver, copper, and platinum) and metal oxides (such as silver oxide, zinc oxide, copper oxide, and titanium dioxide) can be used in their own right as nanopesticides due to their antifungal or antibacterial action [27, 31, 75]. Pesticides are nanosized to increase their solubility, penetration, permeability, bioactivity, and dispersion [22]. Silver nanoparticles results showed prospective antimicrobial activity against disease-causing pathogens viz a viz *B. subtilis, S. aureus, P. aeruginosa,* and *L. Plantarum* [47].

(d) *Nano-gels*

Scientists are interested in nanogels as nanosized drug delivery vehicles, primarily for distributing biologically active mediators in a location or period-specific way [3]. Water-repelling polymers can soak up large amounts of water because they are cross-linked using van der Waals forces or covalent bonding. Considerable research has been done on essential oil and pheromone carriers as nanogels [4, 12]. In fact, Pheromones are, in fact, ecologically benign and extremely specialized biocontrol substances that must be kept from breakdown in ambient settings and delivered slowly. Pheromone evaporation is greatly reduced in those conveyed by nanogel compared to the active components alone. It thus increases their effectiveness by up to 33 weeks in comparison with the efficacy of the plain active components, which is only 3 weeks [12]. Brunel *et al.* [15] claimed that antifungal treatment, which is copper-based, can be enhanced by nanogels obtained from pure chitosan.

(e) ***Electrospun nanofibers***

It is worth noting that electrospinning has popped up as a viable nanotech system for the creation of nanofiber that can be used in a variety of applications [15]. This is especially true given the ability of electrospinning technology to scale up and fulfil the needs of industrial manufacturing. Despite the fact that they are still in the beginning stages of its application in agriculture, electrospun nanofibers offer the advantage of avoiding indiscriminate release, making the application of essential oils and pheromones in the field much easier [3]. Pheromones can indeed be easily integrated in to the nanofibers consisting of cellulose acetate or polyamide as discovered and reported by Hellman *et al.* [35], and a virtually linear release over many weeks can be obtained.

6.4 Application of Nanobiopesticides

6.4.1 Application of Nanobiopesticide Slimicides in Agriculture

Biopesticides containing nanotreatments like micelles and nanocomposite materials decrease the incidence of ecological concerns that would have been posed by the application of chemical pesticides. Equally, clay-based nanotubes deliver biopesticides to control pests [71]. This technology has been used by farmers to stimulate plant growth, diagnose diseases in plants, prevent pests, and improve crop varieties [19, 58]. Plants as well as microbes have been used in the production of nanoparticles. For example, nanoparticles such as silver, nickel, copper and zinc are successfully prepared from plants and used to control offending organisms [23]. Mesoporous Silica Nanoparticles (MSNs) have been earlier utilized as a water-solvable pesticide against validamycin, while nanoparticles of

Alumina were used against *Sitophilus oryzae* and *Rhyzopertha dominica*, pests found in storage food [26].

6.4.2 Application of Nanoparticles in the Environment and Agro-Waste Water Treatment

Nanoparticles are among the most appealing and low-cost nanomaterials with multiple ecological uses. Protection of the environment, surveillance, water treatment and treatment of agrarian waste water are among its uses [44]. It should be noted that pristine and high-quality clean water is one of each and every country's top concerns. Silver and gold nanoparticles biosynthesized from plants can be used for the treatment of water. Water and sewage treatment, for instance, can be accomplished through sorption process, photodegradation, and nanofiltration methods utilizing nanoparticles [38].

Multiple research findings, however, have considered the technique of using gold and silver nanoparticles for the removal of agrochemical in water in the instances of chlorpyrifos, malathion, and atrazine. Furthermore, agrochemicals and other contaminants can be removed by adsorption onto nanoparticles. This holds the pollutants on the surface of the nanoparticles and engage them for an extended stretch of time till the composite precipitates. As a result, many such nanoparticles posit an appropriate, soothing, efficient, and less-expensive method of pesticide removal for water intake as well as inundation systems. Furthermore, the contamination of water with disease-causing microbes poses a significant risk for water-borne ailments. As a result, the antimicrobial attributes of metal nanoparticles have been confirmed for being successful water detoxification [65].

6.4.3 Application in the Paper Mill Industry

It is posited that the paper mill industry uses mainly recirculating water. The condition of the water used ticks all the boxes of the environmental factors that support microbial growth including temperature and pH, thus making it ideal for many slime-producing microorganisms to thrive. Also, the chemicals added in paper production helps to boost the finished product quality and aid their growth [10].

It is thus instructive to note that the growth of these microorganisms in these machines pose a number of challenges viz-a-viz the increased incidences of clogged pipes, blockage of spray nozzles, offensive odours, reduced thermal efficiency and flow and corrosion of equipment. All these cause operational troubles leading to increased energy loss and productivity [10].

Slimicides prevent bacteria, slime mold, fungi, and algae growth, resulting in a number of benefits for the industry such as:

1. Reduction in paper web breaks.
2. Improved machine running ability.
3. Enhancement of drainage and thermal transfer efficiency.
4. Stoppage of machine corrosion, including clogged drains and blockages.
5. Removal of unwanted smell due to the action of slimicides being strong oxidizing biocides.
6. Improvement of overall productivity.

Apart from the above, slimicides are used in the production of paper as well as cardboard packaging as antimicrobial agents to inhibit the production of slime. They may also be used safely in the manufacture of paper and paper boards used for packaging food. For all the aforementioned to happen however, they must follow the conditions outlined below.

i. Slimicides are incorporated toward the process water used in the production of paper or paperboard, and the amount of slimicides incorporated into the water used in paper manufacture shall not surpass the quantity required to achieve the desired result.
ii. The label of slimicides must entail with adequate usage instructions to guarantee proper usage.
iii. Slime-control compounds authorised to be used in the formation of slimicides must include materials that have undergone extensive approbation for that use.
iv. Adjuvant substances allowed to be used for the production of slimicides must comprise various substances regarded as safe in food, paper and paperboard production, under appropriate legislations.

6.4.4 Application in the Marine Industry

An undesirable accumulation of microorganisms on a wetted structure is termed biofouling [9]. Thus, it is posited that this has been a crucial challenge to the shipping and aquatic processing industry. In the marine environment, biofouling involves the undesirable growth of microorganisms on immersed equipment and machinery such as fishery net enclosures, navigation system equipment, ship hulls, and pipelines [9].

Organisms' adherence to surfaces in the marine environment is the main step in biofouling process [20]. Proteins, carbohydrates, and lipids are examples of organic substances which are adsorbed and form a conditioning film on the surface [2]. This process of adsorption is reversible, and is caused by electrostatic interactions, van der Waals forces, and other factors like Brownian motion [84].

The next step in the process is the settling of bacteria species. Bacteria settlement starts off with the attachment of bacteria species to the film by physical forces (electrostatic interactions, van der Waals forces), followed by an irreversible adherence of more bacteria population due to the extracellular binding polymers they secrete on the surface [2]. With increasing population of bacteria settlement, a biofilm is formed. It thus creates an ideal setting for the attachment of some of other microbes like protozoa, algae, and diatoms to the exposed surface of the marine environment [7]. This behavior of accumulation to form biofilm benefits the organisms as it protects them from toxins and predatory species and provides access to nutrients [29, 32].

6.4.5 Control of Microbial Biofilm

It is difficult to restrict the growth of biofilms using conventional biocides [9]. This difficulty in control is due to protective polymers that cover the biofilm formers and the production of extracellular enzymes that hinder the activity of biocides. Low concentration of biocides, which they are less susceptible to, reach the deep colonies in the biofilms resulting in their survival. This can result in the occurrence of many more immune species with efflux pumps altered, making biocides inept against the offending microorganisms [9]. In addition, the abuse of biocides is deleterious, causing the microorganisms to evolve and develop resistance to various antimicrobials.

The capability of citrate-capped silver nanoparticles (AgNPs) to check the forming of bacterial biofilms by *S. aureus* and *P. Aeruginosa* was examined by Flores *et al.* [30]. These microorganisms, both Gram-positive and Gram-negative, are clinically significant. The researchers went about it in two ways. The first method involved dispersing AgNPs to investigate their antibacterial activity on free-living bacteria, and the second one involved adsorbing AgNPs on a titanium medium to investigate their antibacterial properties on sessile bacteria. Because of their distinct cell wall architectures, *P. aeruginosa* which is planktonic was discovered being more vulnerable to nanoparticles than *S. aureus*. For sessile species, similar findings were obtained. From their report, it also showed that titanium-based nanoparticles increased performance, even in very small doses.

Similarly, Singh *et al.* [74] investigated the ability of *Cannabis sativa* gold and silver nanoparticles against slime-producing microorganisms in biofilm, primarily *S. epidermidis*, *P. aeruginosa*, and *E. coli*. Industrial hemp AgNPs showed strong activity against *E. coli* and *P. aeruginosa* but no activity against *S. epidermidis*, whereas AuNP were not inhibitory to all the tested organisms. AgNPs had significant inhibition on the formation of biofilm in all of the bacteria species examined. They proposed that the AgNPs' antimicrobial effect is due to their distinctive size and morphology, as well as biological substituents found on the surface of the nanoparticles. They proposed in their research that nanoparticles can hinder the formation of biofilm by preventing the microbes from adhering to the surface.

6.5 Conclusion

Nanobiotechnology is a comparatively novel biocide delivery system that is gaining traction, particularly in the domain of plant-based nanobiopesticides slimicides capable of destroying offending microbes and their product. It is instructive to state that nanotechnology and modern nanostructured materials are being progressively used in the marine, paper and agricultural industry. This is as a result of improved effectiveness, stability, efficiency, and potency of nano-based products due to improved changes in miscibility and attachment to substrate surface, ability to penetrate membranes, and target-specific release of the nanopesticides. Nanobiopesticides, which are biologically based compounds infused with nanoparticles, with strong impact on pests, are currently being developed for the benefit of the end users. It is worthy of note that since the introduction of marketed innocuous and target-specific nanobiopesticides, there is a growing demand for safer results with an efficient, especially as the active ingredients of conventional chemical pesticides are more toxic.

Nanobiopesticides, like most other natural chemicals, have limitations including working for a short period per time and not being perfectly stable. Beyond examining the application of plant-based nanobiopesticides slimicides capable of killing slime-producing microbes, it is recommended that more scientific education programs be carried out to bring the consumers and specialists to the table to acknowledge potential end-user preferences in the application and use of nanobiopesticides and slimicides, as there is insufficient research in this aspect.

Future research is required to enhance and optimize nanoparticle delivery to the target pests, particularly as natural pesticides have a number of advantages over traditional agrochemicals, the most crucial of which is they have negligible impact on public health and little to no damaging effect on the ecosystem. This is not to imply that the nanobiopesticides be used carelessly, indiscriminatory or with no regard for the consequences of their actions. Despite the fact that a substance is approved for use as a biopesticide, it is possible that it will cause health issues in the general population. Consequently, experts should estimate the impact of nanomaterials on public health and the environment, as well as perfect the methods for evaluating and preventing any dangers they could potentially cause in the short and long run.

Furthermore, researchers should embark on renewable approaches in the manufacture of nanoscale materials used in the application of nanobiopesticides.

References

1. Aala, F., Yusuf, U., Nulit, R., Rezaie, S., Inhibitory effect of allicin and garlic extracts on growth of cultured hyphae. *Iran. J. Basic Med. Sci.*, 17, 3, 150–161, 2014.
2. Abarzua, S. and Jakubowski, S., Biotechnological investigation for the prevention of biofouling: Biological and biochemical principles for the prevention of biofouling. *Mar. Ecol. Prog. Ser.*, 123, 301–312, 1995.
3. Abdollahdokht, D., Gao, Y., Faramarz, S., Poustforoosh, A., Abbasi, M., Asadikaram, G., Nematollahi, M., Conventional agrochemicals towards nano-biopesticides: An overview on recent advances. *Chem. Biol. Technol. Agric.*, 9, 1, 119, 2022.
4. Abreu, F.O., Oliveira, E.F., Paula, H.C., de Paula, R.C., Chitosan/cashew gum nanogels for essential oil encapsulation. *Carbohydr. Polym.*, 89, 4, 1277–1282, 2012.
5. Abtew, A., Subramanian, S., Cheseto, X., Kreiter, S., Garzia, G., Martin, T., Repellency of plant extracts against the legume flower thrips *Megalurothrips Sjostedti* (Thysanoptera: Thripidae). *Insect*, 6, 3, 608–625, 2015.
6. Alengebawy, A., Abdelkhalek, S.T., Qureshi, S.R., Wang, M.Q., Heavy metals and pesticides toxicity in agricultural soil and plants: Ecological risks and human health implications. *Toxics*, 9, 3, 42, 2021.
7. Almeida, E., Diamantino, T., Sousa, O., Marine paints: The particular case of antifouling paints. *Prog. Org. Coat.*, 59, 2–20, 2007.

8. Al-Rahmah, A.N., Mostafa, A.A., Abdel-Megeed, A., Yakout, S.M., Hussein, S.A., Fungicidal activities of certain methanolic plant extracts against tomato phytopathogenic fungi. *Afr. J. Microbiol. Res.*, 7, 6, 517–524, 2013.
9. Ashraf, M., Ullah, S., Ahmad, I., Qureshi, A., Balkhair, K., Abdur-Rehman, M., Green biocides, a promising technology: Current and future applications to industry and industrial processes. *J. Sci. Food Agric.*, 94, 3, 388–403, 2014.
10. Bajpai, P., The control of microbiological problems. *Pulp Pap. Sci.*, 103, 1–3, 2015.
11. Bardin, M., Ajouz, S., Comby, M., Lopez-Ferber, M., Graillot, B., Siegwart, M., Is the efficacy of biological control against plant diseases likely to be more durable than that of chemical pesticides? *Front. Plant Sci.*, 6, 566–576, 2015.
12. Bhagat, D., Samanta, S.K., Bhattacharya, S., Efficient management of fruit pests by pheromone nanogels. *Sci. Rep.*, 3, 1, 1–8, 2013.
13. Bissa, S. and Bohra, A., Antibacterial potential of pot marigold. *J. Microbiol. Antimicrob.*, 3, 3, 51–54, 2011.
14. Biswas, G.C., Comparative effectiveness of neem extracts and synthetic organic insecticide against mustard aphid. *Bangladesh J. Agric. Res.*, 38, 2, 181–187, 2013.
15. Brunel, F., El Gueddari, N.E., Moerschbacher, B.M., Complexation of copper (ii) with chitosan nanogels: Toward control of microbial growth. *Carbohydr. Polym.*, 92, 2, 1348–1356, 2013.
16. Camara, M.C., Campos, E.V.R., Monteiro, R.A., do Espirito Santo Pereira, A., de Freitas Proença, P.L., Fraceto, L.F., Development of stimuli-responsive nano-based pesticides: Emerging opportunities for agriculture. *J. Nanobiotechnology*, 17, 1, 1–19, 2019.
17. Cantrell, C.L., Dayan, F.E., Duke, S.O., Natural products as sources for new pesticides. *J. Nat. Prod.*, 75, 6, 1231–1242, 2012.
18. Castillo-Sánchez, L.E., Jiménez-Osornio, J.J., Delgado-Herrera, M.A., Candelaria-Martínez, B., Sandoval-Gío, J.J., Effects of the hexanic extract of neem *Azadirachta indica* against adult whitefly *Bemisia tabaci*. *J. Entomol. Zool. Stud.*, 5, 95–99, 2015.
19. Chaloupka, K., Malam, Y., Seifalian, A.M., Nanosilver as a new generation of nanoproduct in biomedical applications. *Trends Biotechnol.*, 28, 11, 580–588, 2010.
20. Clare, A.S., Marine natural product antifoulants: Status and potential. *Biofouling*, 9, 3, 211–229, 1996.
21. Cruz, M.E.S., Schwan-Estrada, K.R.F., Clemente, E., Itako, A.T., Stangarlin, J.R., Cruz, M.J.S., Plant extracts for controlling the post-harvest anthracnose of banana fruit. *Rev. Bras. Plantas Medicinais*, 15, 4, 727–733, 2013.
22. Cui, B., Lv, Y., Gao, F., Wang, C., Zeng, Z., Wang, Y., Cui, H., Improving abamectin bioavailability via nanosuspension constructed by wet milling technique. *Pest Manage. Sci.*, 75, 10, 2756–2764, 2019.

23. Devakumar, C. and Dureja, P., Global news on pesticides. *Pestic. Res. J.*, 14, 2, 365–370, 2002.
24. Di Martino, P., Bacterial adherence: Much more than a bond. *AIMS Microbiol.*, 4, 3, 563, 2018.
25. Dillon, R.J. and Dillon, V.M., The gut bacteria of insects: Nonpathogenic interactions. *Annu. Rev. Entomol.*, 49, 1, 71–92, 2004.
26. Dureja, P. and Johnson, S., Photodegradation of Azadirachtin-A: A neem-based pesticide. *Curr. Sci.*, 79, 12, 1700–1703, 2000.
27. Elmer, W. and White, J.C., The future of nanotechnology in plant pathology. *Annu. Rev. Phytopathol.*, 56, 111–133, 2018.
28. Feng, J., Zhang, Q.I., Liu, Q.I., Zhu, Z., McClements, D.J., Jafari, S.M., Application of nanoemulsions in formulation of pesticides, in: *Nanoemulsions*, pp. 379–413, Academic Press, Massachusetts, 2018.
29. Flemming, H.C., Griebe, T., Schaule, G., Antifouling strategies in technical systems—A short review. *Water Sci. Technol.*, 34, 5–6, 517–524, 1996.
30. Flores, C.Y., Miñán, A.G., Grillo, C.A., Salvarezza, R.C., Vericat, C., Schilardi, P.L., Citrate-capped silver nanoparticles showing good bactericidal effect against both planktonic and sessile bacteria and a low cytotoxicity to osteoblastic cells. *ACS Appl. Mater. Interfaces*, 5, 8, 3149–3159, 2013.
31. Garg, D., Sarkar, A., Chand, P., Bansal, P., Gola, D., Sharma, S., Bharti, R.K., Synthesis of silver nanoparticles utilizing various biological systems: Mechanisms and applications—A review. *Prog. Biomater.*, 9, 3, 81–95, 2020.
32. Gilbert, P., Allison, D.G., McBain, A.J., Biofilms *in vitro* and *in vivo*: Do singular mechanisms imply cross-resistance? *J. Appl. Microbiol.*, 92, 98S–110S, 2002.
33. Gupta, S. and Dikshit, A.K., Biopesticides: An ecofriendly approach for pest control. *J. Biopestic.*, 3, Special Issue, 186, 2010.
34. Hazafa, A., Murad, M., Masood, M.U., Bilal, S., Khan, M.N., Farooq, Q., Iqbal, M., Shakir, M., Naeem, M., Nano-biopesticides as an emerging technology for pest management, in: *Insecticides*, IntechOpen, York, UK, intechopen.com, 2021, http://dx.doi.org/10.5772/intechopen.101285.
35. Hellmann, C., Greiner, A., Wendorff, J.H., Design of pheromone releasing nanofibers for plant protection. *Polym. Adv. Technol.*, 22, 4, 407–413, 2011.
36. Huang, X., Zhao, J., Zhou, X., Han, Y., Zhang, J., Cai, Z., How green alternatives to chemical pesticides are environmentally friendly and more efficient. *Eur. J. Soil Sci.*, 70, 3, 518–529, 2019.
37. Ibanez, S., Gallet, C., Després, L., Plant insecticidal toxins in ecological networks. *Toxins*, 4, 4, 228–243, 2012.
38. Jain, K., Patel, A.S., Pardhi, V.P., Flora, S.J.S., Nanotechnology in wastewater management: A new paradigm towards wastewater treatment. *Molecules*, 26, 6, 1797, 2021.

39. Jampílek, J. and Kráľová, K., Nanobiopesticides in agriculture: State of the art and future opportunities, in: *Nano-Biopesticides Today and Future Perspectives*, pp. 397–447, 2019.
40. Jayaraj, R., Megha, P., Sreedev, P., Organochlorine pesticides, their toxic effects on living organisms and their fate in the environment. *Interdiscip. Toxicol.*, 9, 3-4, 90–100, 2016.
41. Kamarudin, N.S., Jusoh, R., Setiabudi, H.D., Jusoh, N.W.C., Jaafar, N.F., Sukor, N.F., Cymbopogon nardus mediated synthesis of ag nanoparticles for the photocatalytic degradation of 2, 4-dicholorophenoxyacetic acid. *Bull. Chem. React. Eng. Catal.*, 14, 1, 173, 2019.
42. Karaca, G., Bilginturan, M., Olgunsoy, P., Effects of some plant essential oils against fungi on wheat seeds. *Indian J. Pharm. Educ. Res.*, 3, S385–S388, 2017.
43. Khalegi, M., Salahi Eshlaghi, B., Ghotaslou, R., Antibacterial effects of Azerbaijan garlic (Allium sativum). *J. Zoonotic Dis.*, 2, 1, 30–34, 2017.
44. Khan, I., Saeed, K., Khan, I., Nanoparticles: Properties, applications and toxicities. *Arab. J. Chem.*, 12, 7, 908–931, 2019.
45. Kookana, R.S., Boxall, A.B., Reeves, P.T., Ashauer, R., Beulke, S., Chaudhry, Q., Van den Brink, P.J., Nanopesticides: Guiding principles for regulatory evaluation of environmental risks. *J. Agric. Food. Chem.*, 62, 19, 4227–4240, 2014.
46. Krishnamurthy, S., Veerasamy, M., Karruppaya, G., A review on plant sources for nanobiopesticide production. *Lett. Appl. NanoBioScience*, 9, 3, 1348–1358, 2020.
47. Kumar, G. and Badoni, P.P., Arisaema tortuosum leaf extract mediated synthesis of silver nanoparticles, characterization and their antibacterial activity. *Asian J. Chem.*, 11, 2, 419–422, 2018.
48. Kumar, J., Ramlal, A., Mallick, D., Mishra, V., An overview of some biopesticides and their importance in plant protection for commercial acceptance. *Plants*, 10, 6, 1185, 2021.
49. Kumar, S., Nehra, M., Dilbaghi, N., Marrazza, G., Hassan, A.A., Kim, K.H., Nano-based smart pesticide formulations: Emerging opportunities for agriculture. *J. Controlled Release*, 294, 131–153, 2019.
50. Kumar, V.V., Plant growth-promoting microorganisms: Interaction with plants and soil. *Plant, Soil Microbes*, 1, 1–16, 2016.
51. Kumar, V.V., Biofertilizers and biopesticides in sustainable agriculture, in: *Role of Rhizospheric Microbes in Soil*, Meena, V.S., (ed.), pp. 377–398. Springer: Singapore, 2018.
52. Kyomuhimbo, H.D., Michira, I.N., Mwaura, F.B., Derese, S., Feleni, U., Iwuoha, E., II, Silver–zinc oxide nanocomposite antiseptic from the extract of bidens pilosa. *SN Appl. Sci.*, 1, 7, 1–17, 2019.
53. Limoli, D.H., Jones, C.J., Wozniak, D.J., Bacterial extracellular polysaccharides in biofilm formation and function. *Microbiol. Spectr.*, 3, 3, 3, 2015.

54. Lu, W., Senapati, D., Wang, S., Tovmachenko, O., Singh, A.K., Yu, H., Ray, P.C., Effect of surface coating on the toxicity of silver nanomaterials on human skin keratinocytes. *Chem. Phys. Lett.*, 487, 1–3, 92–96, 2010.
55. Mathew, I.L., Singh, D., Singh, R.P., Tripathi, C.P.M., *Bacillus thuringiensis*: The biocontrol agent in a food web perspective. *Biolife*, 2, 1, 348–362, 2014.
56. Mattos, B.D., Tardy, B.L., Magalhães, W.L., Rojas, O.J., Controlled release for crop and wood protection: Recent progress toward sustainable and safe nanostructured biocidal systems. *J. Controlled Release*, 262, 139–150, 2017.
57. McClements, D.J., Nanoemulsions versus microemulsions: Terminology, differences, and similarities. *Soft Matter*, 8, 6, 1719–1729, 2012.
58. Misra, A.N., Misra, M., Singh, R., Nanotechnology in agriculture and food industry. *Int. J. Pure Appl. Sci. Technol.*, 16, 2, 1, 2013.
59. Montesinos, E., Development, registration and commercialization of microbial pesticides for plant protection. *Int. Microbiol.*, 6, 4, 245–252, 2003.
60. Muthomi, J.W., Lengai, G.M., Wagacha, M.J., Narla, R.D., *In vitro* activity of plant extracts against some important plant pathogenic fungi of tomato. *Aust. J. Crop Sci.*, 11, 6, 683–689, 2017.
61. Nag, S.K. and Raikwar, M.K., Persistent organochlorine pesticide residues in animal feed. *Environ. Monit. Assess.*, 174, 1, 327–335, 2011.
62. Nikkhah, M., Hashemi, M., Habibi Najafi, M.B., Farhoosh, R., Synergistic effects of some essential oils against fungal spoilage on pear fruit. *Int. J. Food Microbiol.*, 257, 285–294, 2017.
63. Ntalli, N.G. and Menkissoglu-Spiroudi, U., Pesticides of botanical origin: A promising tool in plant protection, in: *Pesticides-Formulations, Effects, Fate*, pp. 1–23, 2011.
64. Nuruzzaman, M.D., Rahman, M.M., Liu, Y., Naidu, R., Nanoencapsulation, nano-guard for pesticides: A new window for safe application. *J. Agric. Food Chem.*, 64, 7, 1447–1483, 2016.
65. Olvera, R.C., Silva, S.L., Robles-Belmont, E., Lau, E.Z., Review of nanotechnology value chain for water treatment applications in mexico. *Resource-Efficient Technol.*, 3, 1, 1–11, 2017.
66. Pavoni, L., Pavela, R., Cespi, M., Bonacucina, G., Maggi, F., Zeni, V., Benelli, G., Green micro-and nanoemulsions for managing parasites, vectors and pests. *Nanomaterials*, 9, 9, 1285, 2019.
67. Perello, A.E., Noll, U., Slusarenko, A.J., *In vitro* efficacy of garlic extract to control fungal pathogens of wheat. *J. Med. Plant Res.*, 7, 1809–1817, 2013.
68. Ragaei, M. and Sabry, A.K.H., Nanotechnology for insect pest control. *Int. J. Sci. Environ. Technol.*, 3, 2, 528–545, 2014.
69. Rathnasamy, R., Thangasamy, P., Thangamuthu, R., Sampath, S., Alagan, V., Green synthesis of ZnO nanoparticles using carica papaya leaf extracts for photocatalytic and photovoltaic applications. *J. Mater. Sci. Mater. Electron.*, 28, 14, 10374–10381, 2017.

70. Raut, R.R., Sawant, A.R., Jamge, B.B., Antimicrobial activity of *Azadirachta indica* (Neem) against pathogenic microorganisms. *J. Acad. Ind. Res.*, 3, 7, 327–329, 2014.
71. Sasson, Y., Levy-Ruso, G., Toledano, O., Ishaaya, I., Nanosuspensions: Emerging novel agrochemical formulations, in: *Insecticides Design Using Advanced Technologies*, pp. 1–39, Springer, Berlin, Heidelberg, 2007.
72. Sayed, A.M., Kim, S., Behle, R.W., Characterisation of silver nanoparticles synthesised by bacillus thuringiensis as a nanobiopesticide for insect pest control. *Biocontrol Sci. Technol.*, 27, 11, 1308–1326, 2017.
73. Schuster, C., Konstantinidou-Doltsinis, S., Schmitt, A., *Glycyrrhiza glabra* extract protects plants against important phytopathogenic fungi. *Commun. Agric. Appl. Biol. Sci.*, 75, 4, 531–540, 2010.
74. Singh, P., Pandit, S., Garnæs, J., Tunjic, S., Mokkapati, V.R., Sultan, A., Mijakovic, I., Green synthesis of gold and silver nanoparticles from Cannabis sativa (industrial hemp) and their capacity for biofilm inhibition. *Int. J. Nanomed.*, 13, 3571, 2018.
75. Smitha, S.L. and Gopchandran, K.G., Surface enhanced raman scattering, antibacterial and antifungal active triangular gold nanoparticles. *Spectrochim. Acta A Mol. Biomol. Spectrosc.*, 102, 114–119, 2013.
76. Solans, C., Izquierdo, P., Nolla, J., Azemar, N., Garcia-Celma, M.J., Nano-emulsions. *Curr. Opin. Colloid Interface Sci.*, 10, 3-4, 102–110, 2005.
77. Stadler, T., Buteler, M., Weaver, D.K., Novel use of nanostructured alumina as an insecticide. *Pest Manag. Sci.*, 66, 6, 577–579, 2010.
78. Suleiman, E.A. and Abdallah, W.B., *In vitro* activity of garlic (Allium Sativum) on some pathogenic fungi. *Eur. J. Med. Plants*, 4, 10, 1240–1250, 2014.
79. Vijayakumar, M., Priya, K., Nancy, F.T., Noorlidah, A., Ahmed, A.B.A., Biosynthesis, characterisation and anti-bacterial effect of plant-mediated silver nanoparticles using *Artemisia nilagirica. Ind. Crops Products*, 41, 235–240, 2013.
80. Vurro, M., Miguel-Rojas, C., Pérez-de-Luque, A., Safe nanotechnologies for increasing the effectiveness of environmentally friendly natural agrochemicals. *Pest Manag. Sci.*, 75, 9, 2403–2412, 2019.
81. Wen, L.X., Ding, H.M., Wang, J.X., Chen, J.F., Porous hollow silica nanoparticles as carriers for controlled delivery of ibuprofen to small intestine. *J. Nanosci. Nanotechnol.*, 6, 9–10, 3139–3144, 2006.
82. Witkowska, D., Sowińska, J., Żebrowska, J.P., Mituniewicz, E., The antifungal properties of peppermint and thyme essential oils misted in broiler houses. *Braz. J. Poult. Sci.*, 18, 629–638, 2016.
83. World Health Organization (WHO), *Preventing disease through healthy environments: exposure to highly hazardous pesticides: a major public health concern,* World Health Organization, 2019. https://apps.who.int/iris/handle/10665/329501. (accessed 15th June, 2023)

84. Yebra, D.M., Kiil, S., Dam-Johansen, K., Antifouling technology—Past, present and future steps towards efficient and environmentally friendly antifouling coatings. *Prog. Org. Coat.*, 50, 2, 75–104, 2004.
85. Zhang, Y., Liu, X., Wang, Y., Jiang, P., Quek, S., Antibacterial activity and mechanism of cinnamon essential oil against *Escherichia coli* and *Staphylococcus aureus*. *Food Control*, 59, 282–289, 2016.

7

Application of Plant-Based Nanobiopesticides That Could Be Applied for the Rejuvenation of Heavily Contaminated Environments

Babatunde Oluwafemi Adetuyi[1]*, Peace Abiodun Olajide[1] and Charles Oluwaseun Adetunji[2]

[1]Department of Natural Sciences, Faculty of Pure and Applied Sciences, Precious Cornerstone University, Ibadan, Nigeria
[2]Applied Microbiology, Biotechnology and Nanotechnology Laboratory, Department of Microbiology, Edo University Iyamho, Auchi, Edo State, Nigeria

Abstract
Insect pests limit the production of vegetable crops. The use of pesticides derived from plants is one alternative insect pest management strategy that is becoming more and more relevant. According to reports, plant-derived pesticides with biodegradable organic components, a variety of bioactivities, and low toxicity to nontarget organisms are efficient at controlling a variety of insect pests by natural methods. Many researchers have thoroughly investigated an antifeedant strategy for pest control in crop management; however, it was only done at the preliminary level or using plant-based compounds. Biopesticide formulations using nanodelivery systems have a number of advantages, such as improved performance and productivity, as well as improved antifeedant properties (precise conveyance, hydrophilicity, and well dispersion). To control agricultural pest species without endangering the local biota and promote a lusher environment, nanotechnology has provided great replacements. Due to the practical implementation of nanotechnology in a variety of agricultural contexts, peculiarities have developed in novel nanoagrochemicals. Novel goods have been deemed necessary due to concerns about the need for them, and experts believe they have a great deal of propensity to enable the necessary increase in food production worldwide in a

**Corresponding author*: badetuyi@gmail.com

Charles Oluwaseun Adetunji and Julius Kola Oloke (eds.) Handbook of Agricultural Biotechnology: Volume I Nanopesticides, (193–228) © 2024 Scrivener Publishing LLC

sustainable manner. Synthetic pesticides used in farming have negative impacts on both people and the environment, and conventional solutions like integrated pest management are insufficient. In order to adopt a regulatory paradigm for nanopesticides, it is imperative to create a repository of secure and advantageous formulations. It is common knowledge in agriculture that nanopesticides can lower costs, boost agricultural product quality, increase shelf life, and improve nutrition. The creation of nanopesticides for aquatic usage, nanocapsules for fungicide administration, and nanosensing devices for insect identification, and other innovations have been the subject of much research.

Keywords: Plant, nanobiopesticides, environments, crop management, fungicide administration

7.1 Introduction

Food production must increase to meet the needs of the expanding human population. Agriculture must be intensified because resources are scarce [4–10, 62, 73, 148, 153]. However, as this too has limitations, it is necessary to move on to another potential tactic—better crop pest protection [22, 61]. To preserve the quantity, quality, and variety of food and feed that farmers produce, plant diseases must be managed. Plant diseases can be prevented, reduced, and controlled using a variety of strategies. In addition to standard "good farming" practices and soil care, various chemical and artificial agrochemicals are also discarded by farmers. Over the past 100 years, the use of these primarily chemical substances in agriculture has made a substantial contribution to notable advancements in crop quality and productivity. On the other side, agrochemical abuse and overuse contribute significantly to environmental damage. Additionally, the improper application of some agrochemicals has contributed to several health issues, which has significantly altered peoples' perceptions of the use of chemicals in agriculture. As a result, each nation has its own laws governing the use of chemical pesticides, also, a lot of pressure on the government to remove the most dangerous substances from circulation. Additionally, the scale at which chemical applications might need to be made may make it impossible to successfully combat plant diseases in natural environments. As a result, several pest-management researchers have concentrated their efforts on creating synthetic chemical-alternative inputs for pest and disease control. One of these alternate methods could be the application of biopesticides [2, 22, 122]. Rather than being chemical pesticides, chemicals used to control insect populations through particular biological effects are referred to as "biopesticides" or "biological pesticides" [3, 99, 100, 125, 137]. However, regulatory agencies in the United States and the European Union have distinct definitions of "biopesticides" [11, 137]. The Environmental Protection Agency (EPA) in the United States stated that "Biopesticides encompass naturally existing compounds which manage insects (metabolic insecticides), microbes which kill insects (microbial agrochemicals), and insecticidal compounds derived from plants that are genetically pests (plant-incorporated insecticides). Biopesticides are "a type of insecticide dependent on microbes or natural products," according to the European Commission (2008) [152]. Although the two definitions/concepts differ, agricultural insects ought to be effectively controlled by biopesticides without harming the ecosystem. As a result, they are utilized as a component of an IPM program for integrated pest management. Investments in the development of biopesticides may be hampered by the EU's regulation being governed by a system that was initially intended for chemical pesticides [11, 45, 60]. Since biopesticides often pose lower risks than traditional pesticides, the EPA normally requests significantly little documentation to certify a biopesticide than it is to license a traditional insecticide.

In contrast to conventional pesticides, which take an average of more than 3 years to get registered, innovative biopesticides are frequently approved in under a year [152]. A new, forward-thinking plan for the creation and use of biopesticides was adopted, not just in the US but also, for instance, in Canada [1]. Vegetation, microbes, fungus, protozoan, and other microorganisms are some of the sources of biopesticides. The following are the key benefits of using biopesticides as natural substances instead of conventional pesticides: Less poisonous, highly potent in very little quantities, only affect the target pest and closely related creatures, less toxic, quick decomposing, and low exposure with nearly no pollution issues.

Thus, biopesticides can significantly minimize the application of conventional pesticides when employed as a part of IPM programs while maintaining excellent crop yields. The biggest drawback is that users must have extensive knowledge of pest management techniques and strictly adhere to all label instructions for safe and effective application [22, 137, 152]. High sensitivity to targeted pests and low stability are two further drawbacks [100, 125]. Biopesticides can be categorized in three ways, according to Md and Polosky [100, 125]: (i) Bug sex fragrances and different fragrant phytoconstituents that attract insects to snares are examples of non-toxic natural substances that monitor and manage pests and microscopic diseases (e.g., pyrethrum, rote none, neem oil, and different essential oils) [13, 121]; (ii) Protozoa, microbes (such as viruses, bacteria, or entomopathogenic spores, including the most prevalent taxa and serotypes of Bacillus thuringiensis (Bt) and its toxic metabolites), and sometimes byproducts made by microbial pathogens make up microbiological insecticides [12, 41, 75] (iii) Pesticides known as "plant-incorporated protectants" (PIPs) are made by cultivars when various genetic materials are integrated into their genes (i.e., Genetically Modified crops) [85, 108]. It is really crucial to keep in mind that biological control agents have no effect on plant development, photosynthesis, or other fundamental functions. They work well against biological pests instead. There are numerous chemicals that plants make to defend themselves against pests. These materials can be used in practical applications because they are renewable and biodegradable [22, 100, 121, 125]. The use of nanotechnology, which has given ancient materials new qualities, is another topic of discussion when it comes to biopesticides [22, 78]. The food and agrochemical industries have embraced this technology, leading to the widespread production and sale of innovative "smart" nanostructured materials ideal for preserving diverse foods, novel nanocompositions of conventional insecticides, intelligent dynamic or effective storage containers, and edible coatings [46, 63, 77]. According to Garcia *et al.* [66], the steady improvement of crop yields is greatly aided by nanotechnology, and vice versa; the food industry and agricultural production uses nanotechnology in significant fields [49]. The accessibility and sustainability (protection from disintegration) of compounded chemical agents are often improved by nanoformulations, and controlled release of active substances is made possible. It has been proposed to create nanobiopesticides using similar design principles to those used in the creation and marketing of nanopesticides [46, 57].

7.2 Nanopesticides: State-of-the-Art

Researchers in the private, public, and scientific sectors have been working hard to understand nanopesticides. There is a direct link between agricultural systems and climate variability. In a downward spiral, while increasing external forces and the influence of rheological circumstances have a negative impact on both the amount and quality of food supply, existing farming systems intensify variables that cause global warming, increasing food instability and obstructing adequate nutrition and fair distribution of food. A similar revolution is currently being applied in technologies that dispense agrochemicals articulately known as "nanopesticides," which leverage nanotechnology to create efficient drug delivery mechanisms and diagnostic imaging equipment. A new technological development known as nanopesticides makes use of nanotechnology to create smart formulations of pesticides for use in agriculture. Ingenious compositions for lipid nanoparticles depend on a range of substances, including metals, ceramics, carbon, copolymers, polymers, lipids, silica, etc., are developed to study pesticide activities through the shape and distinctive properties of nanoscale components. These involve using nanomaterials to conjugate, encapsulate, and adsorb pesticides in order to improve the active ingredients' dispersion, permeability, stability, and biodegradability. The direct release of the active chemicals into the human body and its hazards are prevented by the encapsulation of pesticides in nanoparticles. Contrary to conventional pesticides, nanopesticides have a huge surface area, it raises their chance of being exposed to the plant pathogens at reduced concentrations. Conventional pesticides and nanopesticides clearly behave differently in the environment, therefore, extensive research on the future of nanopesticides is required to ensure that they follow applicable laws and standards [81, 82]. By replacing conventional pesticides with nanopesticides, the environmental impact is significantly reduced. There are two main subcategories of nanopesticides: (a) pesticides with coherent nanoscale constituents, often an emulsion of active pesticides in a nanodispersion, and (b) insecticides that have nanoparticles infused, encased, or coating them. Insecticides in the previously stated category often incorporate the potential benefits of conventional pesticides, with the protection, transportation, and control of pesticide exposure [23, 47, 80]. Contrary to nanopesticides, conventional pesticides have a lower employment rate, poorer stability, reduced dispersibility, coarse drug carrier particles, and lesser biological activity [148, 150]. Furthermore, nanopesticide carriers can boost insecticides' dissolvability while also expediting the distribution of the pesticides' advantageous components to the targeted organisms to raise overall accessibility [141]. Due to these advantages, nanopesticides have been widely used to address the drawbacks of conventional pesticides' high doses and low efficacy. It has been established that nanopesticides may have adverse effects on health. The US Environmental Protection Agency has linked these effects to: (a) nanopesticides' ability to penetrate cellular components and their incredibly small dimension both contribute to their cutaneous absorption; (b) their ability to cross the blood-brain barrier to enter the lungs; (c) environmental problems stem from some nanostructures' catalytic activity and endurance; and (d) their ability to enter the lungs [129]. Well-known active compounds that have been reformulated to have fungicidal, insecticidal, and herbicidal properties make up the variety of nanopesticides currently being produced [154]. Aside from this, some active compounds, like copper, which has been used for its fungicidal effects, are inorganic. Organic molecules make up the majority of active ingredients. Graphene oxide, silica, and carbon nanotube nanoparticles are a few examples of hard nanoparticles. Typically, lipid nanocomposites are soft nanostructures such solid lipids, composites, etc. [21, 44]. High pesticide dosages are ineffective at achieving their goals, and regular pesticide use contaminates both aquatic and terrestrial habitats. Reducing the number of agrochemicals required to defend plants is one of the key reasons for utilizing nanotechnology. This can be done in a variety of ways, for instance, by enhancing the environmental leaf adhesion, persistence, focused distribution, perceived dissolution, and accessibility of the active substances. Less insecticide use, the emergence of insect species that are impervious to it due to contamination of the soil and groundwater, and overexposure to chemicals are both prevented by the controlled administration of pesticides. Contrary to conventional formulations, controlled administration makes sure that the active ingredient is released at a set rate to provide the nanopesticides in the field the appropriate efficacy lifespan. To properly investigate the perceived benefits of the industrialized or licensed nanoformulation, it must be evaluated to a traditional composition that serves as a benchmark item. In a perfect world, this comparison would occur in a field setting while following the instructions on the packaging for a farming pattern that is approved by the Australian Licensing System. The market is currently dominated by a historic study on "diyarex gold" nanopesticides developed utilizing nanotechnology. In certain plants, it has been employed as an antibacterial for powdery mildew and rust [38].

7.3 Nanobiopesticide

The creation of nanobiopesticides can be accomplished in one of two ways. In the first category, nanobiopesticides are active pesticide chemicals or compounds (APCs) generated biologically that have been combined with nanoparticles and a suitable polymer. In the second category, a metallic salt is reduced by an active pesticide compound (APC) to produce metallic nanoparticles (NPs), which are then homogenized into an appropriate composite. In order to exhibit a number of benefits, including regulated and gradual delivery, increased activity, and renewability of the insecticide ingredient and composite, APC was conjugated with nanostructures and mixed with appropriate vehicles such as polymers, micelles, nanofibers, liposomes, or nanospheres. Additionally, different composites could be employed to create different nanocomposites. To defend plants, effective insecticide components are delivered precisely to control bugs, extensive study into the usage of nanobiopesticide will reduce the amount of chemicals sprayed [19].

7.3.1 Benefits of Nanobiopesticides

Nanobiopesticides are considered due to the nanosize, large specific surface area ratio, sustainability, improved effectiveness, superior dissolution, portability, and reduced side effects [134]. Although a harmful agent or agrochemicals used in nanoformulations is modified in some ways to avoid nanopesticide discharge in the atmosphere, groundwater, and land, the toxicants could buildup in the food supply chain, posing major threats to the environment. To maintain and increase the pesticidal efficacy of delivery materials, nanopesticides compounds are included as biodegradable polymers, microemulsions, or polymer blends while also minimizing environmental buildup to prevent such occurrences. A trigger from the surroundings, such as a change in temperature, moisture, or sunshine, causes the release of these nanobiopesticides. Clay nanostructures (halloysite), which are intended to carry pesticides, are generated by environmental stressors to interact more effectively with insects, similar to how biopolymers do. Similar to nanopesticides, nanofertilizers are encased in a particular carrier to allow controlled release for specific effects [36, 98]. The biopolymers are enhanced with nanobiopesticides to produce superior properties such as stiffness, high porosity, crystalline nature, heat resistance, dissolution, and renewability [24, 40, 79]. According to Shah and Belozerova [138], nanomaterials had no discernible effects on soil microorganisms, and the phototoxicity of powdered Ag nanoparticles was suppressed by covering them with biocompatible poly vinyl pyrrole, a nanocoating [95]. The likelihood of using nanomaterials in plant germination and growth increases as a result of research into such biocompatible coverings that counteract the toxicity of nanomaterials. Another possibility for organic farming, according to Oancea *et al.* [111], is the controlled release of active plant growth stimulators and other compounds enclosed in nanocomposites formed of layered double hydroxides (anionic clays). Nanoemulsions, nanoencapsulates, nanocontainers, and nanocages have recently been described as several nanopesticide delivery techniques with various functionalities for plant protection [24, 36, 96].

Because of their advantages over conventional pesticides, such as their inexpensive synthesis, effectiveness, and eco-compatibility, nanobiopesticides have taken the place of chemical or phytochemical pest control. However, the synthesis of size-dependent nanoparticles or nanostructures requires precise setup, equipment, and methodology.

7.3.2 Formulation of Nanobiopesticides

There are numerous chemical processes that can be used to create nanoparticles, but they are sometimes quite expensive and include dangerous substances. Because they are safer than chemical approaches, researchers are interested in using biologically produced nanostructures that contain microbial and botanical concentrates [105]. Additionally, processes of natural source happen at ambient temperature and use very little energy. They are also economical and recyclable [14, 35]. An intriguing field of nanotechnology is the production of nanoparticles from naturally occurring plant materials for the creation of nanobiopesticides (NBP). A polymer made of protein or carbohydrates contains the nanoparticles and the pesticide plus nanoparticles or (APC) reduced nanoparticles [33]. With the benefit of cross-linking, poliglusam, polyoxyethylene (POE), and gelling agent are examples of biodegradable polymers that are used to coerce NBP and help with their systemic delivery. The essential oils (EO) and polymers are also complexed in the multifunctional nanoparticles to enhance the anti-inflammatory, antimicrobial, and odor-masking properties. This triptych structure helps with the controlled release of the active ingredient for a long-lasting action of insecticidal activity. It is made up of a cross-linking composite material, active biological pest control, and nanostructures [19]. Predicting the potential of nanobiopesticides or nanoparticles is aided by comparison of crude extract, metals made from plant extract, nanostructures, biodegradable polymers, and artificial nanobiopesticides. This potential can be determined by deducting significant efficacy of the nanoscale biopesticides' crude extract [89].

7.3.3 Approaches for Synthesizing Nanobiopesticides

The simplest method for making nanobiopesticides is when APC or APC-reduced nanostructures are mixed with nanomaterials; this can be done in a magnetic mixer, an RT, a sterilizer, a microwave oven, or via blending using high intensity or exceptionally high intensity, among other methods. More complex methods include electro-spraying, crushing, spherical grinding, and electrospinning. This article only discusses the electrospinning method of NBP production and cross-linking in polymers.

7.3.3.1 Electrospinning

Recently, the use of electrospinning processes in a variety of industries, including medicine, wound care, agriculture, and pharmaceuticals, has been investigated. The basic idea behind electrospinning is that it turns conductor polymers that contain nanoparticles (NPs) into nanofibers. Under high voltage, globules are created by forcing liquified polymer via needles. The melting fluid generated at the receiver vaporizes when an increased electric current is applied due to electrostatic force till it approaches the collecting chamber and creates nanostructures. The fundamental elements of electrospinning are (1) The needle pump, this enables a steady and regulated flow of the fluid through the spinneret, is one of the fundamental parts of electro-spinning. It joins the spinneret's metallic needle to the injector that receives the current-carrying composite material. (2) The high-voltage setup: The principal force drawing nanofibers is high voltage DC power. The negative end is attached to an electrically conductive receiver wrapped in insulating material, and the positive charge is connected to the high-voltage DC power source together with the tip of the injector (which contains the composite material). Typically, 30-50 kV is used. (3) The charged fibers are collected by the collector, which is aluminum or another conducting material, and are then deposited on the metal making a nanocomposites network. Receivers come in a variety of shapes and sizes, including metal mesh, pin arrays, linear or matrix ded rods, spinning bars, and spinning fluid troughs. They can also be stationary or revolving.

7.3.3.2 Method

The produced NBPs are mixed with the current-carrying polymers and solubilized in a definite diluent. The solvent is injected, the liquid polymer is mixed with the NBP, composites are gathered in the collecting chamber as a result of electricity flowing across the system. NBP + polymer nanofibers are created. The NBP + NP entrapped in the conductive filler is electrospun resulting in a high surface/volume ratio, reducing the need for chemicals [127].

7.3.4 Possible Nanostructures

Different nations formulate, research, and even patent the structures of nanobiopesticides. Here, we explore these nanoparticles that can hold nanostructures and APC. (1) Nanogels are hydrophilic (typically cross-linked) polymeric materials with a high water absorption capacity. (2) Nanospheres are granules with a uniform distribution of the main ingredient all across the polymer network. (3) Nanocapsules are aggregates, it has a high concentration of the main ingredient close to the center core.

(4) Micelles: aggregation made up of molecules with both hydrophilic and hydrophobic moieties that forms in aqueous liquids [19]; (5) Liposomes: tiny water droplet-enclosed sphere of phospholipid molecules, often created artificially to deliver medications to organs; (6) Dendrimer: macromolecules with a high branching number; (7) Hydrogel: a complex molecular polymeric gel composed of a network of molecular chains that have been cross-linked, which include hydrophilic groups, with the aid of an effective cross-linker to promote directional pattern; (8) Biodegradable polymers with layers; (9) Mesoporous silica: MCM-41 and SBA-15, two mesoporous forms of silica employed in imaging, distribution of drugs, and catalysis; and (10) Nanofibers: fibers made of various conducting polymers with sizes in the nanoscale range.

7.4 *In Vitro* Nanobiopesticides Bioassay

A formulated nanomaterial or NBP is tested against a host insect in the lab to determine its effectiveness. Ag, alum, titania, or calamine are used to create nanobiopesticides (NBPs), which are then purified and used against pests at defined quantities. APC is an active pesticidal chemical [22, 24, 26]. Additionally, applying agarose plate dilution technique and permeability, the minimum inhibitory concentration (MIC) can be calculated to get the lethality dosage [8, 10, 12, 84]. This technique specifically feeds the insect through oral feeding at a specified composition; similarly, filter paper is coated with NBP. Both dead and living insects are identified in both cases 20–40 days after, and the deadliest doses are estimated statistically using Polo Plus (version 2.0). (LC50 and LC90); the Abbott formula may be employed to account for the morbidity factor [25, 90].

7.4.1 Antifeedent Activity

One third-instar larva is put onto a plant platter of insect chow, 4 cm across, after the NBP has been applied to it. Every 24 hours, comparison of the amounts of leaves chewed by experimental and reference larvae. The finest illustration is a 92.4% feeding deterrence utilizing the antifeedant assay discovered by Ag nanocomposites in cotton bollworm [7, 9, 144]; similarly, the NBP might be examined [3–6, 25].

7.4.2 Larvicidal Activity

On a plant's leaves that have been immersed in the NBP, insect naiad larvae are released for testing using the leaf dip method used by the NBP. After 96 hours, larvicidal activity is measured after the introduction of the one-third naiad larvae with chow. *Ambrosia artemisiifolia* leaf extract produces the most thoroughly studied reference of NP, with LC50 = 0.28 ppm and LC90 = 0.43 ppm [14, 18, 21, 88, 102].

7.4.3 Pupicidal Activity

Similar to larvicidal and pupicidal experiments, the target insect's pupae are exposed to known doses of NBP, and the mortality rate is determined every 24 hours. According to Sivapriyajothi *et al.* [144] research, The mosquito vectors, yellow fever mosquito (*Aedes aegypti*) and primary mosquito vector of malaria (*Anopheles stephensi*) first to fourth naiad larvae and pupa are highly susceptible to NPs made by a crude leaf extract of Thumbai (LC50 and LC90).

7.5 *In Vivo* Nanobiopesticide Application

Nanobiopesticides (NBPs) have the capacity and ability to remove bugs and many other insects, as shown by the concentration level (LC50 and LC90) utilized in *In vitro* studies. Such dosages are increased to incorporate per acre in real-world settings. Before the onset of pathological conditions or during certain periods, NBPs are used to protect plants. When particular environmental factors, such as heat, moisture, or metabolic deficiency, occur, the NBPs are compelled into progressive delivery on insects. These NBPs are sprayed directly on pest-infested plants in the field and act as active NPs, which directly kill bugs [27, 29, 32, 35].

7.5.1 Mechanism of Action of Biopesticides

Numerous habitats can benefit from the use of biopesticides, including agriculture, they exhibit a multitude of different, one-of-a-kind patterns of operation. Microorganisms produce pesticides that are dangerous for both people and animals by a variety of techniques, such as scavenging, antibiosis, parasitism, and many more [36, 39, 40]. Since they kill insects and thwart the spread of diseases, botanical insecticides have been proven to be quite effective [45–48]. By being poisoned or parasitized, the invasion results in the death of the target, which causes their death. Because of the administration of the treatment, pests are drawn to the treatment area, where they are killed or sterilized. Plant pathogenic fungi's mycelia have been found to change structurally and experience hyphal growth inhibition when extracts from plants in the Asteraceae family are used [41, 44, 110]. The terpenoids, coumarin alkaloids, and flavonoids found in asteraceae plants make them completely toxic to fungi. Some compounds cause modifications to the shape of cellular organelles together with the cell membrane [15]. As a result, bioactive compounds may cause partitioning and penetration when they come into contact with the membranes of fungal cells. This would cause a breach in the segmentation that will enable the elements of the cell to exit [50]. Additionally, it has been demonstrated that plant bioactive compounds can cause the cytoplasmic membrane to separate, which causes the breakdown of intracellular components, cell proliferation, and finally cell death [58]. Sabadilla, pyrethrum, azadirachtin, and fluoroacetate are a few examples of biopesticides that exhibit various modes of action against pests. For instance, by damaging the nerve cell membrane of insects, the sabadilla alkaloid toxin considerably contributed to the loss of the mechanism for nerve cell membrane. Most insects were found to be killed by sabadilla soon after application, but some were shown to be able to survive for a few days in a paralyzed state before passing away [112]. Additionally, new research showed that pyrethrins considerably promote the instantaneous death of insects at modest doses. Pyrethrins are not poisonous to people or warm-blooded animals. However, allergic reactions to people happen frequently. Inhaling the dust can result in headaches and illness, and it may result in a rash. Pyrethrins kill insects by altering the process of Na+ and K+ ion fluxes in the sensory neurons, which prevents normal motor neuron delivery. Pyrethrin-based insecticides operate very swiftly and cause paralysis in insects very quickly. Despite their acute toxicity, pyrethrins can nonetheless be quickly metabolized (broken down) by a variety of insects. However, a combination therapy using pyrethrin and piperonyl butoxide (PBO) could be utilized to treat these insects [17, 54–57]. According to a recently published study, feeding third instar larvae azadirachtin dramatically lowers food intake than the control group [37, 51, 52]. But the type of bug and dosage amount clearly affect its antifeedant effectiveness [139]. According to reports, the suppression of feeding behaviors caused by the stimulation of deterrent receptors and sucrose receptors following an azadirachtin dose resulted to food restriction, hunger, and poor nutrition [67, 70, 101]. Numerous research conducted recently show that azadirachtin causes weight loss in a variety of insects, among them are *Phalaena eridania*, American cockroach, fruit fly, and cotton bollworm [28, 32, 60–62]. Fluoroacetate's pesticide method of action is well established; it has been observed that after being consumed by insects, fluoroacetate transforms into fluoroacetyl-CoA and then fluorocitric acid. The energy generation approach was abandoned because the structural analogue of citric acid (fluorocitric acid) reduced the ability of such an enzyme to transform citric acid into cis-aconitic acid. Because the nervous system of insects is extremely sensitive to these acids, especially glutamic acid, an essential neurotransmitter, the concentration of α-ketoglutaric acid, calcium, and glutamic acid decreased as a result of the accumulation of citric acid inside the cell [71–74, 85, 112]. Insects that come into contact with pesticides like allicin, which is present in garlic bulbs (Allium sativum), choke and die. Allicin affects insects through interfering with their nervous systems' neurotransmitter receptors. Allicin is one of the chemicals that causes suffocation. Through ionic and hydrophobic interactions, phytotoxins and terpenoids affect insects biochemically. Additionally, a significant number of proteins are targeted for destruction, leading to physiological degeneration. A variety of chemicals found in plant extracts and essential oils may interfere with an insect's nervous system and coordination, disrupting it to the point where the insect dies as a result [131].

7.6 Fate of Nanopesticides

Nanopesticides potentially interfere with the photosynthetic pathway in the metabolic or photon emission stages, which in turn influences the development of plants, based on its physical and chemical properties or concentration of the nanostructures. The mechanical impact of nanostructures on plants is associated with their size and shape, which limits the distribution of nanomaterials to particular locations on the outside of structures or perhaps in vegetation. However, the effects are perceptible across the herbaceous body, which can either impede or promote plant growth. Nanopesticides interact with chloroplasts in the cytoplasm of the cell, which, by adhering to the photosystems as well as inhibiting its performance, influences specific photorespiration processes at specific locations in plants. Studies indicate that metal oxide nanocomposites may impair photosynthesis by posing structural and physiological issues [75–78, 149]. Therefore, appreciation of how nanopesticides interact with the mechanism of photosynthesis could lead to a better knowledge of how nanopesticides affect the system of antioxidant defense in plants, photoinduced peroxidation, and electron carrier's suppression of nanostructures.

For instance, tomato vegetative growth can be inhibited by 3 nm of nanostructures (*Solanum lycopersicum* L.) because of the effect of the moisture content of seedlings on the entry [79–81, 86]. Conversely, carbon-based nanostructures may prevent development by mechanically harming organelles or cellulose membranes as a result of their perforating impact, which also incorporates the main cytotoxic mechanism [26, 49, 71, 82, 83]. Additionally, clay-based colloidal suspensions or titanium-based nanoparticles may alter the corn's (*Zea mays* L.) mechanical conductance, that prevents plant development and water loss [30]. Consequently, examining overall role on nanostructures in insecticide nanoformulation and the anticipated action as a result is crucial. When examining how nanoparticles function, the following circumstances can be taken into consideration:

(a) The fungicide or bactericide capabilities of nanoparticles, plant growth regulators generated from silica, and nanometals including Al, Cu, and Ag, with Ti-based oxides serving as key examples [32, 83, 87].

(b) The essential component of a nanoformulation is the active ingredient, which is attached to a nanocarrier that, even after spraying, maintains a complex structure. Next, a controlled release of the active component occurs. Here, the persistence of the nanopesticide—that is, how long the active component has been associated with the lipid nanoparticles—is linked to nanoformulation, which in turn affects the pesticide's fate. There are many different types of nanoformulations, including lipid-based nanomaterials (like liposomes and liposomes vehicles with nanostructures), nanoparticles composed of polymers (like nanospheres, nanogels, micelles, and nanocapsules), and built on clay nanostructures [88–91, 109].

(c) To improve the active ingredient's apparent dispersion or solubility or to protect it from deterioration, the characteristics of the nanoformulation are regulated. Both nanodispersions and nanoemulsions fall within this group. In this case, the product is referred to in the risk assessment as a conventional pesticide [20, 94]. As a result, the approaches used to study the fate of chemicals should be appropriate for nanopesticides [106–109].

For example, the existence of virtually similar natural colloids to nanostructures makes it difficult to characterize and detect nanomaterials in complex matrices like soil [68, 110, 111]. As a result, the quantity of the active components determines how long nanopesticides survive in soils. For instance, one can determine the persistence of the nanoformulation through pore spaces in water or soil solution by analyzing the soil release kinetics, it is next conditioned to techniques for differentiating between encapsulated and unbound insecticides. Depending on the dosage and type of the material, nanoparticles may or may not interact well with plants [33, 70, 142]. A change in the particle properties may also be caused by the active interactions between nanostructures and their surroundings, as well as the absorption of various molecules onto surface of the particles. Nanoparticles' activity, and environmental persistence can also be significantly influenced by the biotransformation that occurs as a result of their interactions with biofilms, living things, and plants. For diverse environmental compartments, a variety of transport and fate models, for industrial substances, FOCUS designs like TOXSWA, MACRO, PEARL, PRZM, PEARL, and others are currently accessible to assist in decision-making [23, 54, 92]. Furthermore, it is well known also that majority of physical, chemical and biological parameters that affect both outcome and functionality of nanostructures generally apply to a wide range of substances.

Regarding the soil compartment, consider the following factors: the type and amount of naturally occurring organic material, particularly humic acids, and electrolytes in the soil, as well as its porosity, water movement, nutrient content, acidity, bacterial population, and heat (especially divalent cations) [96–99].

pH, salinity, ionic strength, dissolved (and suspended) organic matter, and microorganisms for the aqueous compartment [23, 100, 102–104].

7.7 Current Methods of Reducing Soil Pollution Through Biomimicry

Physical, biological, and chemical processes have been used to synthesize agricultural composites based on nanotechnology, which have improved environmental toxicity, cost, cytotoxicity, etc. [113–117]. As a result, questions have been raised about the requirement to create such formulations without the use of toxic substances. Green chemistry, a panacea in the current situation, has produced several innovations and been used to the agriculture industry [118, 119]. Sonochemistry and microwave irradiation have recently attracted interest as hygienic, sustainable, and eco-friendly ways to produce better environmental nanocomposites [121–123]. In contrast to other methods highlighted as increasing energy efficiency, Anastas *et al.* developed a Green ChemisTREE that lists the sustainable chemistry fundamentals, including atomic efficiency, the use of safe agents, practicability, resource conservation, time-saving techniques, and straightforward analytical techniques [59]. Due to their environmentally beneficial qualities, the manufacturing of nanomaterials utilizing microbes and plants has gained the attention of researchers all over the world. Different nanoparticles have been created for crop protection using a variety of plants and bacteria. A number of naturally occurring substances are widely recognized to possess pesticidal properties, however the majority of them are typically unstable and need to be protected against early deterioration [125, 128]. Table 7.1 already discusses several polymer-based formulations with greener methods in a variety of

Table 7.1 Typical examples of nanomaterial that serve as carrier for biopesticides.

Polymer	Material	Characteristics	Outcome	Ref.
Chitosan	Zinc nanoparticles	Inhibiting activity against harmful bacteria.	Silver nitrate bio-reduction by *Pleurotus ostreatus.*	[53]
Chitosan	Copper nanoparticles	Improved growth metrics, increased antioxidant activity, and enzyme defense.	The Curvularia leaf spots on maize have been combated with a biodegradable nanocomposite.	[16]
Chitosan	Copper (II) ion nanogel	Nanogels are being developed to solubilize Cu(II) ions, and it was shown that they were significantly resistant.	A good medium to use in inhibiting *Gibberella zeae* development, as well as a pH-sensitive delivery mechanism and applications in biopesticides.	[42]
Chitosan/Cashew gum	*Alecrim pimenta*	Oil from *Alecrim pimenta* added to nanogel.	Slower, systemic absorption can be seen in *in vitro* pharmacokinetic profiles.	[2]

(*Continued*)

Table 7.1 Typical examples of nanomaterial that serve as carrier for biopesticides. (*Continued*)

Polymer	Material	Characteristics	Outcome	Ref.
Chitosan	Copper and zinc nanoparticles	Food immobilization enzymes like amylase and protease have been evaluated for resistance to nanoparticles.	Plant defense enzymes shield them against illnesses.	[11]
Polyamide 6 cellulose acetate	Nanofibers	Electrospinning was used to integrate pheromones in high concentration.	Pheromone release from polymer carriers.	[69]
Chitosan-copoly (lactic acid)	Imidacloprid	Particles with imidacloprid put on them displayed a prolonged release process.	Lipophilic pesticide delivery for systems utilizing imidacloprid.	[34]
Chitosan	Spinosad and permethrin	Chitosan nanoparticles with agrochemicals loaded on them were studied on Fruit fly under various circumstances.	Outstanding after effect in controlling insect infestations.	[140]
Chitosan	Gum Arabic nanoparticles	It was produced and its biological effectiveness towards *Bemisia tabaci* was examined in encapsulated geraniol. Protection from UV deterioration was noticed.	Potential technique for use in managing insect pests.	[120, 150, 151]
Polyethylene glycol	Essential oil (peppermint and palmarosa)	Essential oil effects on croton bug that are both enhanced deadly and sublethal.	Created to eradicate German cockroaches.	[148, 155]

(*Continued*)

Table 7.1 Typical examples of nanomaterial that serve as carrier for biopesticides. (*Continued*)

Polymer	Material	Characteristics	Outcome	Ref.
Poly(N-isopropyl acrylamide-come-thacrylic acid) methacrylic acid	Hollow mesoporous silica nanoparticles	Persistent defense against brown planthopper in rice plants, showed strong adhesion and successfully protected the model pesticide thiamethoxam from UV light.	Pesticide temperature coefficient.	[65]
	Azadirachta indica oil with zein nanostructures	French bean is unharmed by phytotoxicity. Nanoencapsulated margosa oil proved effective at combating red spider mite and silverleaf whitefly.	The nanobiopesticide's capacity to manage insect populations.	[124]
Poly(lactic-coglycolic acid)	Nanoparticles	The encapsulated nanoparticles had properties that were thermodynamically advantageous for *Phytophthora infestans* disease management using cyzofamid.	Tomato leaf infection suppression and utilization in novel load reduction approaches.	[64, 147]
Bridged disulfide bond in dextrose	Mesoporous silica nanoparticles	Avermectin's early release and the deterioration of active components could both be avoided with the use of manufactured nanovehicles.	Utilizing a non-toxic organic solvent and superior insecticidal properties, stimuli-responsive nanocarriers.	[93]

(*Continued*)

Table 7.1 Typical examples of nanomaterial that serve as carrier for biopesticides. (*Continued*)

Polymer	Material	Characteristics	Outcome	Ref.
Poly ε-caprolactone	Nanostructures of chitosan	The solvent removal technique is used to encapsulate thiamethoxam in nanocapsules. Preservation against early deterioration. To assess the toxicity, the microalgae and microcrustaceans were studied.	Low ecotoxicity and ability to protect citrus trees from huanglongbing illness.	[31]
Chitosan	b-Cyclodextrin containing linalool and carvacrol	Nanoencapsulation results in a decrease in toxicity and shows insecticidal effectiveness towards cotton bollworm and red spider mite.	For mites, repellent action and a reduction in oviposition were seen.	[43]

forms, including nanoparticles, nanofibers, nanogels, nanoemulsions, nanospheres, etc. During a research on the production of Silver Nanocomposite, cell-free isolates of Sphaeria rufa Pers. were employed for *in vivo* and *in vitro* testing with a primary vector, i.e., early blight, to treat infestation with blight disease in tomato plants. After three days of therapy, the spore count was reduced by 100% according to the antifungal results. In addition, after seven days of therapy, there was a 74% decrease in the fungal biomass [129–132]. The influence of the generated nanostructures on the foliage allowed for a 47.57% decrease in the amount of mold spores in comparison to unmanaged infected plants, whereas the spotless mass and composition of the seedlings' overall photosynthetic pigment increased by 33.58% and 22.52%, correspondingly [126]. Although microorganisms have been extensively explored for their being able to create biogenic nanostructures, there are still some gaps that need to be filled. It is challenging to control the processes as it is unknown exactly how microbes reduce substances to form nanostructures. It is challenging to scale up these procedures to the industrial level since the metabolic activities inside the agar medium can be impacted by saline, acid concentration, and heat. As a result, maintaining overall consistency of the agar medium is essential. Additionally, there are difficulties in employing plants to produce biologically active nanostructures. For example, changes in form and size and lower recovery lead to a decreased rate of nanoparticle synthesis [135–137].

As a result, a grasp of the reduction mechanisms as well as a greater awareness of the right choice of phytoconstituents to be used as reductants are required. In conclusion, in order to establish commercially feasible methods for the development of industrially manufactured nanoparticles with the correct physical and chemical qualities, nontoxicity, and substantial microbial properties, it is essential to fully understand the mechanisms governing the fabrication of nanoparticles through living organisms [142, 143, 145]. The previously stated nanocomposites have the underlying capacity for agricultural operations and may be employed to gauge fertilizer and pesticide application rates and fine-tune them, detect soil moisture to modify supply of water, which reduces water waste, and evaluate plant development and viability.

7.8 Toxicology, a Barrier for Nanopesticides

There is concern about the potential entry of produced nanoparticles the introduction of those substances' ensuing environmental contamination into the food supply chain as nanotechnology develops quickly. But little is understood about the negative consequences that pesticides and nanomaterial toxicity have on living things. We must thus look at these issues because they will have profound effects on both humans and biota. The effectiveness of pesticides depends heavily on their toxicity, which is linked to how well they work against the intended targets. Toxic substances such as stabilizing agents, reductases, and alcohols are utilized to stop the aggregation of undesirable colloidal particles. Nanocomposites application is indeed not necessarily harmful; for instance, numerous conventional meals, such as milk proteins, mayonnaise fat globules, carbohydrates, DNA, etc., contain nanoscale elements. However, certain intended nanoscale materials may be detrimental towards the wellness of people and nature when used in agriculture, water, and food [63]. Utilization of nanoscale compounds for agribusiness should be accelerated, it is therefore important to innovate and create methods for their non-hazardous, effective, and commercially feasible manufacturing [153, 156].

Although assessments across different insects or hosts could not be linked, it has been found that perhaps the lethality of nanopesticides proved significantly greater than that of conventional compositions almost an absolute scale [56, 132]. According to Sasson *et al.'s* [134] paper, nanopesticides are more hazardous to target organisms than commercial treatments are. This highlights how important it is to compare nanopesticides to their actual, current analytical counterparts. The variance in lethality amongst standard preparations is not rare, but because the formulations and component techniques are secret and secretive, it frequently goes unrecognized. The effectiveness of nanopesticides was compared to that of their pure bioactive components in formulations used in industry in a different study. These studies are necessary to compare the effects of nanoformulations to those of conventional formulations while taking into account the toxicity of fungicides and insecticides. Because the length of the research period was demonstrated to affect the goods' activity, the items designed to regulate the adsorption capability of the bioactive constituents must be examined [133].

7.9 Environmental Repercussions

The action and outcome of nanopesticides when used in the ecosystem are the primary determinants of their health implications as well as the biosphere. The active ingredients must be released gradually in nanoformulations in order to stop them from moving around and becoming overly unsustainable mostly in the field of agriculture. In the aforementioned situations, the procedure that controls how nanopesticides behave in the environment is bioactive substances' release via lipid nanoparticles models. Another Gordian knot is the ineffectiveness of pesticide application and delivery. It has been noted that 10 to 75 percent of pesticides miss their intended target. The drawbacks of using pesticides include their toxicity to creatures other than the ones they are intended to kill, inherent tenacity, the development of tolerance, and the accumulation of them in the environment.

The lessened environmental pressure brought on by agricultural production might result in enormous linked advantages if agronomic health, fertilizer application, and watering effectiveness may be enhanced by nanostructures. To properly compare the impacts of nanopesticides to those of conventional treatments, it is required to quantify the unbound dissolved ions of latent toxic micronutrients. These days, it is believed that such investigations are essential for the evaluation of nanostructures' potential ecological risks. Agribusiness needs to undergo a transformation immediately to supply the growing need for food yet eradicate any detrimental impacts on the microbiota. Important research must be conducted in order to find new products that are both extremely competitive and thus improve the efficacy of agricultural output in the coming decades. Positively, the successful implementation of nanoscience can significantly contribute to reaching the said objective.

7.10 Second-Generation Nanobiopesticides

Specific species of plants may evolve defense mechanisms (pest taxa enzyme) to fend off subsequent attacks from such an insect as a result of excessive insect exposure. Such a pest taxa enzyme can be accessed and employed to create a nanobiopesticide formulation that is highly effective against pests. Second-generation nanopesticides are the name given to these protease-specific nanoparticles. In addition, a recombinant protease that produces a poly-protease inhibitor for insects might be created from the gene library. This nanoformulation based on poly-inhibitors will help protect plants while making insects more harmful. In this approach, one inhibitor may have two targets or two distinct targets may be shared by two inhibitors. Bifunctional inhibitors include those that inhibit both alpha-amylase and trypsin as well as trypsin and carboxypeptidase A. They also work to control the larvae of the red flour beetle and the codling moth. In comparison to transgenic tobacco plants that alone expressed the Bt-toxin, these were more resistant to the effects of cotton bollworm when they additionally expressed the cowpea trypsin inhibitor (CpTI). Additionally, cotton bollworm and rice cutworm larvae died substantially more quickly when they ate bioengineered tobacco leaflets encoding avidin (from chicken egg white) which was coated with Cry1Ba protein [146].

7.11 Conclusion

It is getting much harder to provide enough food for the population due to the growing human population and climatic changes. Due to lack of resources, agriculture must be intensified through the use of watering and additional fertilizers, which will result in higher agricultural yields. On the other side, excessive fertilizer use degrades the soil and denies it the opportunity to regain its "intrinsic" fertility, which, despite intensification, causes yields to stagnate or even fall. Climate change and ongoing global warming make it possible to see the development of every type of pest on animals, fruits, vegetables, and agricultural products. Farmers and plants are now more vulnerable as a result of pests expanding their range into previously uninhabited areas. Thus, to tackle such bugs, producers often employ chemical-based insecticides, but when these chemicals are used excessively, they harm numerous non-pathogenic creatures and further degrade the quality of soil and water. Pathogens found in already prepared foods have also increased. Additionally, numerous very hazardous persistent synthetic pesticides that were previously misused and persistently present in the soil may be unduly emitted into the atmosphere amid severe flooding or periods of severe rain, posing a long-term significant risk for living things.

The use of novel nanomaterials and nanotechnology is rapidly advancing in the agriculture and food sectors. There are more commercial nanoformulated pesticide products than there are articles about them in the world's marketplaces. Because of modified diffusion coefficient via biomembranes, targeted, and increased bioavailability, as well as changed dissolution and adhesion to membranes, these products should have the advantages of greater stability and increased efficiency/potency. These altered qualities enable the use of lower pesticide doses. Currently, foods are also contained in packaging materials made of "smart" nanoparticles.

The creation of biopesticides has started using substances of biological source that are effective towards certain insects, in response to the extensive public opposition the application of several insecticides. The majority of these natural chemicals have short half-lives and poor stability, which are their main drawbacks. In order to formulate sustainable biopesticides with prolonged or long-lasting impacts or to "green" synthesize active inorganic nanomaterials that have antibacterial, antifungal, and antiparasitic properties, nanotechnologies have been applied in this field.

These unique natural-source nanomaterials, including different biocidal plant oils or metabolites that have been nanoencapsulated, but rather the use of reducing characteristics of phytoconstituents to synthesize or produce metal or metallic oxide NPs, do have a wide range of possibilities. They are well-liked by the general public due to their natural active ingredients, and farmers are interested in them due to their great efficacy even in little doses. It is possible to see the regulatory agencies' proactive attitude to their approval process. However, when applied, these nanobiopesticides call for extra care and knowledge. Additionally, the so-called third world's agricultural areas cannot use them due to their high cost.

Genetically engineered plants and products are a sensitive topic. Due to the fact that NPs encoding distinctive genetic code or gene products are incorporated in plant genomic sequences, this is also true when it comes to using nanobiopesticides, allowing the plant to produce natural biocides that specifically kill plant pests on its own. Although the universe is split into two competing factions over these engineered eras of safeguarding vegetation, this is certainly the goal of both produced plants and conserved produce. It must be highlighted that even if they have better qualities and are more effective than bulky biopesticides, yet still need to be used carefully given that they will still be naturally occurring. Minimal "potential" to contribute to global warming, lesser side effects to humans and other living organisms, and smooth manufacturing are characteristics of nanobiopesticides. Because of this, there needs to be a serious discussion about the impact of nanostructures mostly on humans and the environment. Contrarily, the development of genetically modified crops with the ability to protect themselves, so reducing the usage of any natural or synthetic substances and obviating the need for pesticides, appears to hold promise.

However, nanobiopesticides have demonstrated considerable effectiveness against mosquito larvae, which are serious pests. In actuality, adult mosquitoes transmit a variety of human infections responsible for serious tropical diseases including trypanosomiasis, malaria, etc. There is a significant risk that these illnesses will spread to other locations as a result of ongoing climate change. The best strategy to stop the spread of these protozoans is to eradicate their vectors—mosquitoes—since treating these infections is difficult, particularly because of either outdated medications or the multidrug resistance of these pathogens. This kind of combination of bioactive, naturally sourced ingredients has a bright future ahead of it given the strong larvicidal activity of several nanobiopesticides.

Nanobiopesticides were approved as well as supported by regulatory authorities and also the community as "ecofriendly agro-chemicals," much like how biological control agents are thought for being relatively safe and reliable substances because of their lower detrimental effect on the people and the environment. Drawing a concrete safety protocol would be necessary, though, for scientific study in medicine, pharmaceuticals, and agriculture. By comparability, it is assumed that the majority of these nanoformulated biocontrol agents, especially those belonging to the category of natural pesticides and herbal extracts, will be able to satisfy all strict requirements into becoming brand-new, reliable agro-products.

References

1. AAFC, *Biopesticides: Strategies for discovery, development, and adoption*, Agriculture and Agri-Food Canada, Canada, 2009, Cat. No. A52-120/2009E-PDF, https://www.gov.mb.ca/sd/envprograms/initiatives/pesticide_red/pdf/biopesticide.pdf.

2. Abreu, F.O., Oliveira, E.F., Paula, H.C., de Paula, R.C., Chitosan/cashew gum nanogels for essential oil encapsulation. *Carbohydr. Polym.*, *89*, 4, 1277–1282, 2012.
3. Adeleke, D.A., Olajide, P.A., Omowumi, O.S., Okunlola, D.D., Taiwo, A.M., Adetuyi, B.O., Effect of monosodium glutamate on the body system. *World News Nat. Sci.*, *44*, 1–23, 2022.
4. Adetunji, C.O. and Varma, A., Biotechnological application of Trichoderma: A powerful fungal isolate with diverse potentials for the attainment of food safety, management of pest and diseases, healthy planet, and sustainable agriculture, in: *Trichoderma: Agricultural Applications and Beyond. Soil Biology*, vol. 61, C. Manoharachary, H.B. Singh, A. Varma (Eds.), Springer, Cham, 2020, https://doi.org/10.1007/978-3-030-54758-5_12.
5. Adetunji, C.O., Mathew, J.T., Singh, K.R.B., Inobeme, A., Nayak, V., Olaniyan, O.T., Singh, J., Singh, R.P., Zero-dimensional carbon nanomaterials for targeted drug delivery and photothermal therapy applications, in: *Zero-Dimensional Carbon Nanomaterials: Fundamentals and Applications*, pp. 1–10, IOP Publishing, United Kingdom, 2022.
6. Adetunji, C.O., Mathew, J.T., Singh, K.R.B., Inobeme, A., Nayak, V., Olaniyan, O.T., Singh, J., Singh, R.P., Zero-dimensional carbon nanomaterials for targeted drug delivery and photothermal therapy applications, in: *Zero-Dimensional Carbon Nanomaterials: Fundamentals and Applications*, pp. 1–12, IOP Publishing, United Kingdom, 2022.
7. Adetunji, C.O., Michael, O.S., Akram, M., Oseni, K., Temidayo, A.K., Anani, O.A., Olayinka, A.S., Nwankwo, W., Ghaffar, I., Adetunji, J.B., *Green Synthesis in Nanomedicine and Human Health*, pp. 81–100, CRC Press, United Kingdom, 2021.
8. Adetunji, C.O., Michael, O.S., Nwankwo, W., Anani, O.A., Adetunji, J.B., Olayinka, A.S., Akram, M., Biogenic nanoparticles based drugs derived from medicinal plants: A sustainable panacea for the treatment of malaria, in: *Green Synthesis in Nanomedicine and Human Health*, pp. 103–122, CRC Press, United Kingdom, 2021.
9. Adetunji, C.O., Titilayo, O., Mathew, J.T., Inobeme, A., Singh, K.R.B., Olaniyan, O.T., Singh, J., Singh, R.P., *Carbon dots: Next-generation materials for biomedical applications*, pp. 8–12, IOP Publishing, United Kingdom, 2022.
10. Adetunji, O., Michael, O.S., Akram, M., Oseni, K., Anani, O.A., Nwankwo, W., Anwar, H., Adetunji, J.B., Olayinka, A.S., Bottom of formrecent advances in the utilization of bioengineered plant-based nanoparticles: A sustainable nanobiotechnology for the management of extensively drug-resistant tuberculosis, in: *Green Synthesis in Nanomedicine and Human Health*, pp. 149–166, CRC Press, United Kingdom, 2021.
11. Adetuyi, B.O. and Oluwole E.O Dairo, J.O., Chemoprotective potential of ethanol extract of *Ganoderma Lucidum on* liver and kidney parameters in

Plasmodium Beghei-Induced Mice. *Int. J. Chem. Chem. Processes (IJCCP)*, 1, 8, 29–36, 2015.

12. Adetuyi, B.O., Oluwole, E.O., Dairo, J.O., Biochemical effects of shea butter and groundnut oils on white albino rats. *Int. J. Chem. Chem. Processes (IJCCP)*, 1, 8, 1–17, 2015.
13. Adetuyi, B.O., Dairo, J.O., Didunyemi, M.O., Anti-hyperglycemic potency of *Jatropha Gossypiifolia in* alloxan induced diabetes. *Biochem. Pharmacol. (Los Angel)*, 4, 5, 193, 2015.
14. Adetuyi, B.O., Omolabi, F.K., Olajide, P.A., Oloke, J.K., Pharmacological, biochemical and therapeutic potential of milk thistle (Silymarin): A review. *World News Nat. Sci.*, 37, 75–91, 2021.
15. Adetuyi, B.O. and Farombi, E.O., 6-Gingerol, an active constituent of ginger, attenuates lipopolysaccharide-induced oxidation, inflammation, cognitive deficits, neuroplasticity, and amyloidogenesis in rat. *J. Food Biochem.*, *45*, 4, e13660, 2021.
16. Adetuyi, B.O., Adebayo, P.F., Olajide, P.A., Atanda, O.O., Oloke, J.K., Involvement of free radicals in the ageing of cutaneous membrane. *World News Nat. Sci.*, 43, 11–37, 2022.
17. Adetuyi, B.O., Adebisi, O.A., Adetuyi, O.A., Ogunlana, O.O., Toloyai, P.-B., Egbuna, C., Uche, C.Z., Khan, J., Adumanya, O.C., Patrick-Iwuanyanwu, K.C., Ficus exasperata attenuates acetaminophen-induced hepatic damage via NF-κB signaling mechanism in experimental rat model. *BioMed. Res. Int.*, 2022, 6032511, 2022.
18. Adetuyi, B.O., Adebisi, O.A., Awoyelu, E.H., Adetuyi, O.A., Ogunlana, O.O., Phytochemical and toxicological effect of ethanol extract of *Heliotropium indicum* on liver of male albino rats. *Lett. Appl. NanoBioscience (LIANB)*, 10, 2, 2085–2095, 2020, https://doi.org/10.33263/LIANBS102.20852095.
19. Adetuyi, B.O., Odine, G.O., Olajide, P.A., Adetuyi, O.A., Atanda, O.O., Oloke, J.K., Nutraceuticals: Role in metabolic disease, prevention and treatment. *World News Nat. Sci.*, 42, 1–27, 2022.
20. Adetuyi, B.O., Ogundipe, A.E., Ogunlana, O.O., Egbuna, C., Estella, O.U., Mishra, A.P., Akram, M., Achar, R.R., Banana peel as a source of nutraceuticals, in: *Food and Agricultural Byproducts as Important Source of Valuable Nutraceuticals*, 1st ed, p. 400, Springer, Berlin, 2022.
21. Adetuyi, B.O., Okeowo, T.O., Adetuyi, O.A., Adebisi, O.A., Ogunlana, O., Oretade, O.J., Marraiki, N., Beshbishy, A.M., Welson, N.N., Batiha, G.E., Ganoderma lucidum from red mushroom attenuates formaldehyde-induced liver damage in experimental male rat model. *Biology*, *9*, 10, 313, 2020, https://doi.org/10.3390/biology9100313.
22. Adetuyi, B.O., Olajide, P.A., Awoyelu, E.H., Adetuyi, O.A., Adebisi, O.A., Oloke, J.K., Epidemiology and therapeutic measure for COVID-19; A review. *Afr. J. Reprod. Health* (Special Edition COVID-19), 24, 2, 142, June 2020, https://www.ajrh.info/index.php/ajrh/article/view/2300.

23. Adetuyi, B.O., Olajide, P.A., Omowumi, O.S., Odine, G.O., Okunlola, D.D., Taiwo, A.M., Opayinka, O.D., Blockage of Alzheimer's gene: Breakthrough effect of Apolipoprotein E4. *Afr. J. Adv. Pure Appl. Sci. (AJAPAS)*, 1, 3, 26–33, 2022.
24. Adetuyi, B.O., Toloyai, P.Y., Ojugbeli, E.T., Oyebanjo, O.T., Adetuyi, O.A., Uche, C.Z., Olisah, M.C., Adumanya, O.C., Jude, C., Chikwendu, J.K., Akram, M., Awuchi, C.G., Egbuna, C., Neurorestorative roles of microgliosis and astrogliosis in neuroinflammation and neurodegeneration. *Scicom J. Med. Appl. Med. Sci.*, 1, 1, 1–5, 2021.
25. Adetuyi, B.O., Olajide, P.A., Oluwatosin, A., Oloke, J.K., Preventive phytochemicals of cancer as speed breakers in inflammatory signaling. *Res. J. Life Sci. Bioinform. Pharm. Chem. Sci.*, 8, 1, 30–61, 2022.
26. Adewale, G.G., Olajide, P.A., Omowumi, O.S., Okunlola, D.D., Taiwo, A.M., Adetuyi, B.O., Toxicological significance of the occurrence of selenium in foods. *World News Nat. Sci.*, *44*, 63–88, 2022.
27. Agrow, *Global Biopesticide Regulations 2017*, Informa – Agribusiness Intelligence, London, UK, 2017.
28. Ahmad, S., Ansari, M.S., Muslim, M., Toxic effects of neem based insecticides on the fitness of Helicoverpa armigera (Hübner). *Crop Prot.*, *68*, 72–78, 2015.
29. Aslani, F., Bagheri, S., Muhd Julkapli, N., Juraimi, A.S., Hashemi, F.S.G., Baghdadi, A., Effects of engineered nanomaterials on plants growth: An overview. *Sci. World J.*, PMID: 25202734; PMCID: PMC4150468. 2014.
30. Asli, S. and Neumann, P.M., Colloidal suspensions of clay or titanium dioxide nanoparticles can inhibit leaf growth and transpiration via physical effects on root water transport. *Plant Cell Environ.*, *32*, 5, 577–584, 2009.
31. Assalin, M.R., dos Santos, L.D.L., Souza, D.R.C., Rosa, M.A., Duarte, R.C.R.M., Castanha, R.F., Donaire, P.P.R., Duran, N., Ecotoxicity evaluation: Preparation of poly-ecaprolactone and chitosan nanoparticles as carriers of thiamethoxam pesticide. *J. Physics: Conf. Ser.*, IOP Publishing, 1323, 1, 012017, 2019.
32. Athanassiou, C.G., Kavallieratos, N.G., Benelli, G., Losic, D., Usha Rani, P., Desneux, N., Nanoparticles for pest control: Current status and future perspectives. *J. Pest Sci.*, *91*, 1, 1–15, 2018.
33. Awoyelu, E.H., Oladipo, E.K., Adetuyi, B.O., Senbadejo, T.Y., Oyawoye, O.M., Oloke, J.K., Phyloevolutionary analysis of SARS-CoV-2 in Nigeria. *New Microbes New Infect.*, 36, 100717, 2020, https://doi.org/10.1016/j.nmni.2020.100717.
34. Batiha, G.B., Awad, D.A., Algamma, A.M., Nyamota, R., Wahed, M., II, Shah, M.A., Amin, M.N., Adetuyi, B.O., Hetta, H.F., Cruz-Marins, N., Koirala, N., Ghosh, A., Sabatier, J., Diary-derived and egg white proteins in enhancing immune system against COVID-19. *Front. Nutr. (Nutritional Immunology)*, 8, 629440, 2021.

35. Benelli, G., Plant-mediated biosynthesis of nanoparticles as an emerging tool against mosquitoes of medical and veterinary importance: A review. *Parasitol. Res.*, *115*, 1, 23–34, 2016.
36. Bergeson, L.L., Nanosilver pesticide products: What does the future hold? *Environ. Qual. Manag.*, 19, 4, 73–82, 2010b.
37. Bezzar-Bendjazia, R., Kilani-Morakchi, S., Maroua, F., Aribi, N., Azadirachtin induced larval avoidance and antifeeding by disruption of food intake and digestive enzymes in Drosophila melanogaster (Diptera: Drosophilidae). *Pestic. Biochem. Physiol.*, *143*, 135–140, 2017.
38. Bhattacharyya, A., Duraisamy, P., Govindarajan, M., Buhroo, A.A., Prasad, R., Nano-biofungicides: Emerging trend in insect pest control. *Adv. Appl. Fungal Nanobiotechnol.*, 1, 307–319, 2016.
39. Bioinicia, Innovative material application, in: *Electrospinning Process Bioinicia*, 2017, https://bioinicia.com/electrospinning-equipment-fiber-fabrication/electrospinning-process-bioinicia-2/.
40. Bordes, P., Pollet, E., Avérous, L., Nano-biocomposites: Biodegradable polyester/nanoclay systems. *Prog. Polym. Sci.*, 34, 125–155, 2009.
41. Borgio, J.F., Sahayaraj, K., Susurluk, I.A., *Microbial insecticides: Principles and applications*, Nova Publishers, USA, 2011.
42. Brunel, F., El Gueddari, N.E., Moerschbacher, B.M., Complexation of copper (II) with chitosan nanogels: Toward control of microbial growth. *Carbohydr. Polym.*, *92*, 2, 1348–1356, 2013.
43. Campos, E.V.R., Proença, P.L.F., Oliveira, J.L., Pereira, A.E.S., Nunes de Morais Ribeiro, L., Fernandes, O., Gonçalves, K.C., Carvacrol and linalool co-loaded in b-cyclodextrin-grafted chitosan nanoparticles as sustainable biopesticide aiming pest control. *Sci. Rep.*, 8, 1, 1–14, 2018.
44. Cao, L., Zhou, Z., Niu, S., Cao, C., Li, X., Shan, Y., Huang, Q., Positive-charge functionalized mesoporous silica nanoparticles as nanocarriers for controlled 2, 4-dichlorophenoxy acetic acid sodium salt release. *J. Agric. Food Chem.*, *66*, 26, 6594–6603, 2017.
45. Chandler, D., Davidson, G., Grant, W.P., Greaves, J., Tatchell, G.M., Microbial biopesticides for integrated crop management: An assessment of environmental and regulatory sustainability. *Trends Food Sci. Technol.*, *19*, 5, 275–283, 2008.
46. Chaudhary, S., Kanwar, R.K., Sehgal, A., Cahill, D.M., Barrow, C.J., Sehgal, R., Progress on Azadirachta indica based biopesticides in replacing synthetic toxic pesticides. *Front. Plant Sci.*, 8, 610, 2017, https://doi.org/10.3389/fpls.2017.00610.
47. Chaudhry, N., Dwivedi, S., Chaudhry, V., Singh, A., Saquib, Q., Azam, A., Musarrat, J., Bio-inspired nanomaterials in agriculture and food: Current status, foreseen applications and challenges. *Microb. Pathog*, *123*, 196–200, 2018.

48. Chen, H., Seiber, J.N., Hotze, M., ACS select on nanotechnology in food and agriculture: A perspective on implications and applications. *J. Agric. Food Chem.*, 62, 1209–1212, 2014.
49. Chen, K.L. and Bothun, G.D., Nanoparticles meet cell membranes: Probing nonspecific interactions using model membranes. *Environ. Sci. Technol.*, 48, 2, 873–880 2014.
50. Choudhary, R.C., Kumaraswamy, R.V., Kumari, S., Pal, A., Raliya, R., Biswas, P., Saharan, V., Synthesis, characterization, and application of chitosan nanomaterials loaded with zinc and copper for plant growth and protection, in: *Nanotechnology*, pp. 227–247, Springer, Singapore, 2017.
51. Choudhary, R.C., Kumaraswamy, R.V., Kumari, S., Sharma, S.S., Pal, A., Raliya, R., Saharan, V., Cu-chitosan nanoparticle boost defense responses and plant growth in maize (Zea mays L.). *Sci. Rep.*, *7*, 1, 1–11, 2017.
52. De Oliveira, J.L., Campos, E.V.R., Pereira, A.E., Nunes, L.E., Da Silva, C.C., Pasquoto, T., Fraceto, L.F., Geraniol encapsulated in chitosan/gum arabic nanoparticles: A promising system for pest management in sustainable agriculture. *J. Agric. Food Chem.*, *66*, 21, 5325–5334, 2018.
53. Deshpande, P., Dapkekar, A., Oak, M.D., Paknikar, K.M., Rajwade, J.M., Zinc complexed chitosan/TPP nanoparticles: A promising micronutrient nanocarrier suited for foliar application. *Carbohydr. Polym.*, *165*, 394–401, 2017.
54. Di Guardo, A., Gouin, T., MacLeod, M., Scheringer, M., Environmental fate and exposure models: Advances and challenges in 21st century chemical risk assessment. *Environ. Sci.: Processes Impacts*, *20*, 1, 58–71, 2018.
55. Didunyemi, M.O., Adetuyi, B.O., Oyewale, I.A., Inhibition of lipid peroxidation and *in-vitro* antioxidant capacity of aqueous, acetone and methanol leaf extracts of green and red Acalypha wilkesiana Muell Arg. *Int. J. Biol. Med. Res.*, 11, 3, 7089–7094, 2020.
56. Didunyemi, M.O., Adetuyi, B.O., Oyebanjo, O.O., Morinda lucida attenuates acetaminophen-induced oxidative damage and hepatotoxicity in rats. *J. Biomed. Sci.*, *8*, 1–7, 2019, https://www.jbiomeds.com/biomedical-sciences/morinda-lucida-attenuates-acetaminopheninduced-oxidative-damage-and-hepatotoxicity-in-rats.php?aid=24482.
57. Duhan, J.S., Kumar, R., Kumar, N., Kaur, P., Nehra, K., Duhan, S., Nanotechnology: The new perspective in precision agriculture. *Biotechnol. Rep.*, 15, 11–23, 2017.
58. El-Wakeil, N.E., Botanical pesticides and their mode of action. *Gesunde Pflanzen*, 65, 4, 125–149, 2013.
59. Erythropel, H.C., Zimmerman, J.B., de Winter, T.M., Petitjean, L., Melnikov, F., Lam, C.H., Anastas, P.T., The Green ChemisTREE: 20 years after taking root with the 12 principles. *Green Chem.*, *20*, 9, 1929–1961, 2018.
60. European Commission, *Encouraging innovation in biopesticide development*, EU research and innovation magazine, Germany, 2008, http://ec.europa.eu/environment/integration/research/newsalert/pdf/134na5.pdf.

61. FAO, Global agriculture toward 2050, in: *High Level Expert Forum - How to Feed the World in 2050*, 2009, http://www.fao.org/fileadmin/templates/wsfs/docs/Issues_papers/ HLEF2050_ Global_Agriculture.pdf.
62. Farid, A., Muzammal, M., Ghazanfar, S., Ismail, S., Ullah, A., Dauda, W.P., Adetunji, C.O., Shah, M., Dadrasnia, A., Maryam Idrees, S. *et al.*, Green cleaning activity of Bacillus salmalaya 139Sl: A novel strain for removing common household stains, in: *Biomass Conversion and Biorefinery*, 2022, https://doi.org/10.1007/s13399-022-03147-z.
63. Farombi, E.O., Abolaji, A.O., Adetuyi, B.O., Awosanya, O., Fabusoro, M., Neuroprotective role of 6-gingerol-rich fraction of Zingiber officinale (Ginger) against acrylonitrile-induced neurotoxicity in male Wistar rats. *J. Basic Clin. Physiol. Pharmacol.*, *30*, 3, 1–11, 2019.
64. Fukamachi, K., Konishi, Y., Nomura, T., Disease control of Phytophthora infestans using cyazofamid encapsulated in poly lactic-co-glycolic acid (PLGA) nanoparticles. *Colloids Surf. A: Physicochem. Eng. Aspects*, *577*, 315–322, 2019.
65. Gao, Y., Xiao, Y., Mao, K., Qin, X., Zhang, Y., Li, D., He, S., Thermoresponsive polymer-encapsulated hollow mesoporous silica nanoparticles and their application in insecticide delivery. *Chem. Eng. J.*, *383*, 123169, 2020.
66. García, M., Forbe, T., Gonzalez, E., Potential applications of nanotechnology in the agro-food sector. *Food Sci. Technol.*, *30*, 573–581, 2010.
67. Grillo, R., Fraceto, L.F., Amorim, M.J., Scott-Fordsmand, J.J., Schoonjans, R., Chaudhry, Q., Ecotoxicological and regulatory aspects of environmental sustainability of nanopesticides. *J. Hazardous Mater.*, *404*, 124148, 2021.
68. Hassellöv, M., Readman, J.W., Ranville, J.F., Tiede, K., Nanoparticle analysis and characterization methodologies in environmental risk assessment of engineered nanoparticles. *Ecotoxicology*, *17*, 5, 344–361, 2008.
69. Hellmann, C., Greiner, A., Wendorff, J.H., Design of pheromone releasing nanofibers for plant protection. *Polym. Adv. Technol.*, *22*, 4, 407–413, 2011.
70. Hermes, P.H., Gabriela, M.P., Ileana, V.R., Fusaro, C., Fernando, L.V., Mariana, M.A., Fabián, F.L., Carbon nanotubes as plant growth regulators: Prospects, in: *Green Nanoparticles*, pp. 77–115, Springer, Cham, 2020.
71. Hu, X., Lu, K., Mu, L., Kang, J., Zhou, Q., Interactions between graphene oxide and plant cells: Regulation of cell morphology, uptake, organelle damage, oxidative effects and metabolic disorders. *Carbon*, *80*, 665–676, 2014.
72. Ingale, A.G. and Chaudhari, A.N., Biogenic synthesis of nanoparticles and potential applications: An eco-friendly approach. *J. Nanomed. Nanotechol.*, 4, 165, 1–7, 2013.
73. Islam, S., Thangadurai, D., Adetunji, C.O., Michael, O.S., Nwankwo, W., Kadiri, O., Anani, O.A., Makinde, S., Adetunji, J.B., Nanomaterials: Applications in biomedicine and biotechnology, in: *Handbook of Nanomaterials and Nanocomposites for Energy and Environmental Applications*, O.V. Kharissova, L.M. Torres-Martínez, B.I. Kharisov (Eds.), Springer, Cham, 2021, https://doi.org/10.1007/978-3-030-36268-3_4.

74. Isman, M.B., Botanical insecticides, deterrents, and repellents in modern agriculture and an increasingly regulated world. *Annu. Rev. Entomol.*, 51, 45–66, 2006.
75. Jackson, T.A., Entomopathogenic bacteria: Mass production, formulation, and quality control, in: *Microbial Control of Insect and Mite Pests: From Theory to Practice*, L.A. Lacey (Ed.), pp. 125–139, Elsevier, London, UK, 2017.
76. James-Okoro, P.O., Iheagwam, F.N., Sholeye, M., II, Umoren, I.A., Adetuyi, B.O., Ogundipe, A.E., Braimah, A.A., Adekunbi, T.S., Ogunlana, O.E., Ogunlana, O.O., Phytochemical and *in vitro* antioxidant assessment of Yoyo bitters. *World News Nat. Sci.*, 37, 1–17, 2021.
77. Jampílek, J. and Kráľová, K., Application of nanotechnology in agriculture and food industry, its prospects and risks. *Ecol. Chem. Eng.*, S 22, 321–361, 2015.
78. Jampílek, J., Záruba, K., Oravec, M., Kuneš, M., Babula, P., Ulbrich, P., Preparation of silica nanoparticles loaded with nootropics and their *in vivo* permeation through blood–brain barrier. *Biomed. Res. Int.*, 2015, 812673, 2015, https://doi.org/10.1155/2015/812673.
79. Yan, J., Huang, K., Wang, Y., Liu, S., Study on anti-pollution nanopreparation of dimethomorph and its performance. *Chin. Sci. Bull.*, 50, 2, 108–112, 2005.
80. Jindal, A.B., Bachhav, S.S., Devarajan, P.V., *In situ* hybrid nano drug delivery system (IHN-DDS) of antiretroviral drug for simultaneous targeting to multiple viral reservoirs: An *in vivo* proof of concept. *Int. J. Pharm.*, *521*, 1–2, 196–203, 2017.
81. Kah, M., Kookana, R.S., Gogos, A., Bucheli, T.D., A critical evaluation of nanopesticides and nanofertilizers against their conventional analogues. *Nat. Nanotechnol.*, *13*, 8, 677–684, 2018.
82. Kah, M., Walch, H., Hofmann, T., Environmental fate of nanopesticides: Durability, sorption and photodegradation of nanoformulated clothianidin. *Environ. Sci.: Nano*, 5, 4, 882–889, 2018.
83. Kah, M., Weniger, A.K., Hofmann, T., Impacts of (nano) formulations on the fate of an insecticide in soil and consequences for environmental exposure assessment. *Environ. Sci. Technol.*, *50*, 20, 10960–10967, 2016.
84. Kamaraj, C., Rajakumar, G., Rahuman, A.A., Velayutham, K., Bagavan, A., Zahir, A.A., Elango, G., Feeding deterrent activity of synthesized silver nanoparticles using Manilkara zapota leaf extract against the house fly, Musca domestica (Diptera: Muscidae). *Parasitol. Res.*, *111*, 6, 2439–2448, 2012.
85. Kamle, M., Kumar, P., Patra, J.K., Bajpai, V.K., Current perspectives on genetically modified crops and detection methods. *Biotech*, 7, 3, 219, 2017, https://doi.org/10.1007/ s13205-017-0809-3.
86. Khodakovskaya, M., Dervishi, E., Mahmood, M., Xu, Y., Li, Z., Watanabe, F., Biris, A.S., Carbon nanotubes are able to penetrate plant seed coat and

dramatically affect seed germination and plant growth. *ACS Nano*, *3*, 10, 3221–3227, 2009.
87. Koul, O., Microbial biopesticides: Opportunities and challenges *CAB Rev. Perspect. Agric. Vet Sci. Nutr. Nat. Resour.*, 6, 056, 1–26, 2011.
88. Koul, O., Plant biodiversity as a resource for natural products for insect pest management, in: *Biodiversity and Insect Pests Key Issues for Sustainable Management*, G.M. Gurr, S.D. Wratten, W.E. Snyder, D.M.Y. Read (Eds.), pp. 85–105, Wiley, West Sussex, 2012.
89. Lade, B.D. and Patil, A.S., Silver nano fabrication using leaf disc of Passiflora foetida Linn. *Appl. Nanosci.*, *7*, 5, 181–192, 2017.
90. Lade, B.D., Gogle, D.P., Nandeshwar, S.B., Nanobiopesticide to constraint plant destructive pests. *J. Nanomed. Res.*, *6*, 3, 1–9, 2017.
91. Lade, B.D., Gogle, D.P., Lade, D.B., Moon, G.M., Nandeshwar, S.B., Kumbhare, S.D., Nanobiopesticide formulations: Application strategies today and future perspectives, in: *Nano-Biopesticides Today and Future Perspectives*, pp. 179–206, Academic Press, United Kingdom, 2019.
92. Li, M., Huang, Q., Wu, Y., A novel chitosan-poly (lactide) copolymer and its submicron particles as imidacloprid carriers. *Pest Manag. Sci.*, *67*, 7, 831–836, 2011.
93. Liang, Y., Gao, Y., Wang, W., Dong, H., Tang, R., Yang, J., Cao, Y., Fabrication of smart stimuli-responsive mesoporous organosilica nano-vehicles for targeted pesticide delivery. *J. Hazardous Mater.*, *389*, 122075, 2020.
94. López-Cabeza, R., Methods for understanding the fate of nanopesticides in soil and water. *Nanopesticides*, 1, 111–136, 2020.
95. Lu, W., Senapati, D., Wang, S., Tovmachenko, O., Singh, A.K., Yu, H., Ray, P.C., Effect of surface coating on the toxicity of silver nanomaterials on human skin keratinocytes. *Chem. Phys. Lett.*, *487*, 1–3, 92–96, 2010.
96. Lyons, K. and Scrinis, G., Under the regulatory radar? nanotechnologies and their impacts for rural Australia, in: *Tracking Rural Change: Community, Policy and Technology in Australia, New Zealand and Europe*, pp. 151–171, Australian National University E Press, Canberra, 2009.
97. Malherbe, S., *Listing 17 microbes and their effects on soil, plant health and biopesticide functions*, ExploGrow™ International Ltd, London, UK, 2017, https://www.explogrow.com/farming-with-microbes/beneficial-microbialbio-fertilizer-plant-health-and-bio-control-functions.
98. Manjunatha, S.B., Biradar, D.P., Aladakatti, Y.R., Nanotechnology and its applications in agriculture: A review. *J. Farm Sci.*, *29*, 1, 1–13, 2016.
99. Mason, P.G. and Gillespie, D.R., *Biological control programmes in Canada 2001–2012*, CAB International, Wallingford, UK, 2013.
100. Md, A.A., *Biopesticides and bioagents: Novel tools for pest management*, CRC Press, Warentown, NJ, USA, 2017.
101. Mordue, A.J., Morgan, E.D., Nisbet, A.J., Gilbert, L., II, Gill, S.S., Azadirachtin, a natural product in insect control, in: *Insect Control: Biological and Synthetic Agents*, pp. 185–197, 2010.

102. Morejón, B., Pilaquinga, F., Domenech, F., Ganchala, D., Debut, A., Neira, M., Larvicidal activity of silver nanoparticles synthesized using extracts of Ambrosia arborescens (Asteraceae) to control Aedes aegypti L. (Diptera: Culicidae). *J. Nanotechnol.*, 2018, Article ID 6917938, 8, 2018.
103. Mossa, A.T.H., Green pesticides: Essential oils as biopesticides in insect-pest management. *J. Environ. Sci. Technol.*, *9*, 5, 354, 2016.
104. Mushtaq, W., Shakeel, A., Fazili, M.A., Chakrabartty, I., Sevindik, M., Pros and cons of nanotechnology, in: *Nanobiotechnology in Agriculture*, pp. 207–222, 2020.
105. Narayanan, K.B. and Sakthivel, N., Biological synthesis of metal nanoparticles by microbes. *Adv. Colloid Interface Sci.*, *156*, 1–2, 1–13, 2010.
106. Nazir, A., Itrat, N., Shahid, A., Mushtaq, Z., Abdulrahman, S.A., Egbuna, C., Adetuyi, B.O., Khan, J., Uche, C.Z., Toloyai, P.Y., Orange Peel as a Source of Nutraceuticals, in: *Food and Agricultural Byproducts as Important Source of Valuable Nutraceuticals*, 1st ed, p. 400, Springer, Berlin, Gebunden, 2022.
107. Nehra, M., Dilbaghi, N., Marrazza, G., Kaushik, A., Sonne, C., Kim, K.H., Kumar, S., Emerging nanobiotechnology in agriculture for the management of pesticide residues. *J. Hazardous Mater.*, *401*, 123369, 2021.
108. NPIC, National Pesticide Information Center, in: *Plant Incorporated Protectants (PIPs)/Genetically Modified Plants*, 2017, http://npic.orst.ed7u/reg/pip.html.
109. Nuruzzaman, M.D., Rahman, M.M., Liu, Y., Naidu, R., Nanoencapsulation, nano-guard for pesticides: A new window for safe application. *J. Agric. Food Chem.*, *64*, 7, 1447–1483, 2016.
110. Nuzhat, T. and Vidyasagar, G., Antifungal investigations on plant essential oils. A review. *Int. J. Pharm. Pharm. Sci.*, 2, 19–28, 2014.
111. Oancea, S., Padureanu, S., Oancea, A.V., Growth dynamics of corn plants during anionic clays action. *Lucrări Ştiinţifice*, *52*, 212–217, 2009.
112. Oguh, C.E., Okpaka, C.O., Ubani, C.S., Okekeaji, U., Joseph, P.S., Amadi, E.U., Natural pesticides (biopesticides) and uses in pest management-a critical review. *Asian J. Biotechnol. Genet. Eng.*, *2*, 3, 1–18, 2019.
113. Ogunlana, O.O., Adetuyi, B.O., Adekunbi, T.S., Adegboye, B.E., Iheagwam, F.N., Ogunlana, O.E., Ruzu bitters ameliorates high–fat diet induced non-alcoholic fatty liver disease in male Wistar rats. *J. Pharm. Pharmacogn. Res.*, 9, 3, 251–260, 2021.
114. Ogunlana, O.O., Adetuyi, B.O., Rotimi, M., Esalomi, I., Adeyemi, A., Akinyemi, J., Ogunlana, O., Adetuyi, O., Adebisi, O., Okpata, E., Baty, R., Batiha, G., Hypoglycemic activities of ethanol seed extract of *Hunteria umbellate* (Hallier F.) on Streptozotocin-induced diabetic rats. *Clin. Phytosci.*, 7, 1–9, 2021.
115. Ogunlana, O.O., Ogunlana, O.E., Adekunbi, T.S., Adetuyi, B.O., Adegboye, B.E., Iheagwam, F.N., Anti-inflammatory mechanism of Ruzu Bitters on diet-induced nonalcoholic fatty liver disease in male wistar rats. *Evid. Based Complement. Altern. Med.*, 2020, https://doi.org/10.1155/2020/5246725.

116. Ogunlana, O.O., Ogunlana, O.E., Popoola, J.O., Adetuyi, B.O., Adekunbi, T.S., David, O.L., Adeleye, O.J., Udeogu, S.A., Adeyemi, A.O., Twigs of Andrographis paniculata (Burn. F) Nees attenuates carbon tetrachloride (CCl4) induced liver damage in wistar albino rats. *RASAYAN J. Chem.*, 14, 4, 2598–2603, 2021.
117. Ogunlana, O.O., Adetuyi, B.O., Esalomi, E.F., Rotimi, M., II, Popoola, J.O., Ogunlana, O.E., Adetuyi, O.A., Antidiabetic and antioxidant activities of the twigs of Andrograhis paniculata on Streptozotocin-induced diabetic male rats. *BioChem*, 1, 3, 238–249, 2021.
118. Olajide, P.A., Adetuyi, O.A., Omowumi, O.S., Adetuyi, B.O., Anticancer and antioxidant phytochemicals as speed breakers in inflammatory signaling. *World News Nat. Sci.*, 44, 231–259, 2022.
119. Olajide, P.A., Omowumi, O.S., Okunlola, D.D., Adetuyi, B.O., Deadly Pandemia: Monkeypox disease, a case study. *Afr. J. Adv. Pure Appl. Sci. (AJAPAS)*, 34–37, 2022.
120. Olajide, P.A., Omowumi, S.O., Odine, G.O., Pathogenesis of reactive oxygen species: A review. *World News Nat. Sci.*, 44, 150–164, 2022.
121. Pal, G.K. and Kumar, B., Antifungal activity of some common weed extracts against wilt causing fungi, Fusarium oxysporum. *Curr. Discov.*, *2*, 1, 62–67, 2013.
122. Pal, K.K. and Gardener, B.M., Biological control of plant pathogens. *The Plant Health Instructor*, 1–25, 2006.
123. Pankaj, Shakil, N.A., Kumar, J., Singh, M.K., Singh, K., Bioefficacy evaluation of controlled release formulations based on amphiphilic nano-polymer of carbofuran against Meloidogyne incognita infecting tomato. *J. Environ. Sci. Health, Part B*, *47*, 6, 520–528, 2012.
124. Pascoli, M., de Albuquerque, F.P., Calzavara, A.K., Tinoco-Nunes, B., Oliveira, W.H.C., Gonçalves, K.C., Fraceto, L.F., The potential of nanobiopesticide based on zein nanoparticles and neem oil for enhanced control of agricultural pests. *J. Pest Sci.*, *93*, 2, 793–806, 2020.
125. Polosky, Z., *21st century homestead: Biological pest control*, Lulu Press, Inc, USA, 2015.
126. Ponmurugan, P., Biosynthesis of silver and gold nanoparticles using Trichoderma atroviride for the biological control of Phomopsis canker disease in tea plants. *IET Nanobiotechnol.*, *11*, 3, 261–267, 2017.
127. Prasad, K.G., *Advances in application of nanotechnology training module chapter 5, electrospinning for production of nano materials*, pp. 48–53, organised by ICAR–CIRCOT, MS, India, 2018.
128. Prasad, V. and Srivastava, S., Insect viruses, in: *Ecofriendly Pest Management for Food Security*, Omkar (Ed.), pp. 411–442, Academic Press, Oxford, UK, 2016.
129. Ragaei, M. and Sabry, A.H., Nanotechnology for insect pest control. *Int. J. Environ. Sci. Technol.*, 3, 2, 528–545, 2014.

130. Rattan, R.S., Mechanism of action of insecticidal secondary metabolites of plant origin. *Crop Prot.*, *29*, 9, 913–920, 2010.
131. Rodino, S., Butu, A., Butu, M., Cornea, C.P., *In vitro* efficacy of some plant extracts against damping off disease of tomatoes. *J. Int. Sci. Publications: Agric. Food*, *2*, 240–244, 2014.
132. Saini, P., Gopal, M., Kumar, R., Srivastava, C., Development of pyridalyl nanocapsule suspension for efficient management of tomato fruit and shoot borer (Helicoverpa armigera). *J. Environ. Sci. Health, Part B*, *49*, 5, 344–351, 2014.
133. Saini, P., Kumar, S., Sharma, V.K., Chobhe, K.A., *Nanopesticides: Manage food security and environmental pollution*, Biotech Articles, United Kingdom, 2018, https://www.biotecharticles.com/Agriculture-Article/Nanopesticides-ManageFood-Security-andEnvironmental-Pollution-4391.html#:B:text=Nanopesticides%20has%20been%20visualized%20as,materials%20to%20deliver%20pesticides.
134. Sasson, Y., Levy-Ruso, G., Toledano, O., Ishaaya, I., Nanosuspensions: Emerging novel agrochemical formulations, in: *Insecticides Design Using Advanced Technologies*, pp. 1–39, Springer, Berlin, Heidelberg, 2007.
135. Sasson, Y., Levy-Ruso, G., Toledano, O., Ishaaya, I., Nanosuspensions: Emerging novel agrochemical formulations, in: *Insecticides Design Using Advanced Technologies*, pp. 1–39, Springer, Berlin, Heidelberg, 2007.
136. Sayed, A.M. and Behle, R.W., Comparing formulations for a mixed-microbial biopesticide with Bacillus thuringiensis var. kurstaki and Beauveria bassiana blastospores. *Arch. Phytopathol. Plant Prot.*, *50*, 15–16, 745–760, 2017.
137. Seiber, J.N., Coats, J., Duke, S.O., Gross, A.D., Biopesticides: State of the art and future opportunities. *J. Agric. Food Chem.*, *62*, 48, 11613–11619, 2014.
138. Shah, V. and Belozerova, I., Influence of metal nanoparticles on the soil microbial community and germination of lettuce seeds. *Water, Air, Soil Pollut.*, *197*, 1, 143–148, 2009.
139. Shannag, H.K., Capinera, J.L., Freihat, N.M., Effects of neem-based insecticides on consumption and utilization of food in larvae of Spodoptera eridania (Lepidoptera: Noctuidae). *J. Insect Sci.*, *15*, 1, 152, 2015.
140. Sharma, A., Sood, K., Kaur, J., Khatri, M., Agrochemical loaded biocompatible chitosan nanoparticles for insect pest management. *Biocatal. Agric. Biotechnol.*, *18*, 101079, 2019.
141. Sharma, S., Singh, S., Ganguli, A.K., Shanmugam, V., Anti-drift nano-stickers made of graphene oxide for targeted pesticide delivery and crop pest control. *Carbon*, *115*, 781–790, 2017.
142. Siddiqui, M.H., Al-Whaibi, M.H., Firoz, M., Al-Khaishany, M.Y., Role of nanoparticles in plants. *Nanotechnol. Plant Sci.*, 19–35, 2015.
143. Singh, P., Kim, Y.J., Zhang, D., Yang, D.C., Biological synthesis of nanoparticles from plants and microorganisms. *Trends Biotechnol.*, *34*, 7, 588–599, 2016.
144. Sivapriyajothi, S., Kumar, P.M., Kovendan, K., Subramaniam, J., Murugan, K., Larvicidal and pupicidal activity of synthesized silver nanoparticles

using Leucas aspera leaf extract against mosquito vectors, Aedes aegypti and Anopheles stephensi. *J. Entomol. Acarological Res.*, *46*, 2, 77–84, 2014.
145. Sporleder, M. and Lacey, L.A., Biopesticides, in: *Insect Pests of Potato - Global Perspectives on Biology and Management*, A. Alyokhin, C. Vincent, P. Giordanengo (Eds.), pp. 463–497, Elsevier, Oxford, UK, 2013.
146. Stevens, J., Dunse, K., Fox, J., Evans, S., Anderson, M., Biotechnological approaches for the control of insect pests in crop plants, in: *Pesticides-Advances in Chemical and Botanical Pesticides*, R.P. Soundararajan (Ed.), pp. 269–308, InTech, London, United Kingdom, 2012.
147. Sun, Y., Liang, J., Tang, L., Li, H., Zhu, Y., Jiang, D., Zeng, G., Nanopesticides: A great challenge for biodiversity? *Nano Today*, *28*, 100757, 2019.
148. Thangadurai, D., Prakash, L., Sangeetha, J., Mohammad Said Al-Tawaha, A.R., David, M., Islam, S., Adetunji, C.O., Adetunji, J.B., Sustainable synthesis of greener nanomaterials: Principles, processes, and products, in: *Handbook of Nanomaterials and Nanocomposites for Energy and Environmental Applications*, O.V. Kharissova, L.M.T. Martínez, B.I. Kharisov (Eds.), Springer, Cham, 2021, https://doi.org/10.1007/978-3-030-11155-7_30-1.
149. Tighe-Neira, R., Carmora, E., Recio, G., Nunes-Nesi, A., Reyes-Diaz, M., Alberdi, M., Inostroza-Blancheteau, C., Metallic nanoparticles influence the structure and function of the photosynthetic apparatus in plants. *Plant Physiol. Biochem.*, *130*, 408–417, 2018.
150. Tong, Y., Shao, L., Li, X., Lu, J., Sun, H., Xiang, S., Wu, X., Adhesive and stimulus-responsive polydopamine-coated graphene oxide system for pesticide-loss control. *J. Agric. Food Chem.*, *66*, 11, 2616–2622, 2018.
151. US EPA, *Biopesticides*, United States Environmental Protection Agency, London, United Kingdom 2017, https:// www.epa.gov/pesticides/biopesticides.
152. US EPA, *What are Biopesticides?*, United States Environmental Protection Agency, United State of America, 2017, https://www.epa.gov/ingredients-used-pesticide-products/what-are-biopesticides.
153. Wali, S., Zahra, M., Okla, M.K., Wahidah, H.A., Tauseef, I., Haleem, K.S., Farid, A., Maryam, A., AbdElgawad, H., Adetunji, C.O., Akhtar, N., Akbar, S., Rehman, W., Yasir, H., Shakira, G., Brassica oleracea L. (Acephala Group) based zinc oxide nanoparticles and their efficacy as antibacterial agent. *Braz. J. Biol.*, 84, 1–9, 2022. https://doi.org/10.1590/1519-6984.259351.
154. Walker, G.W., Kookana, R.S., Smith, N.E., Kah, M., Doolette, C.L., Reeves, P.T., Navarro, D.A., Ecological risk assessment of nano-enabled pesticides: A perspective on problem formulation. *J. Agric. Food Chem.*, *66*, 26, 6480–6486, 2017.
155. Yeguerman, C., Jesser, E., Massiris, M., Delrieux, C., Murray, A.P., González, J.W., Insecticidal application of essential oils loaded polymeric nanoparticles to control German cockroach: Design, characterization and lethal/sublethal effects. *Ecotoxicol. Environ. Saf.*, *189*, 110047, 2020.
156. Zaytseva, O. and Neumann, G., Carbon nanomaterials: Production, impact on plant development, agricultural and environmental applications. *Chem. Biol. Technol. Agric.*, *3*, 1, 1–26, 2016.

8

Application of Plant-Based Nanobiopesticides for Effective Management of Pests and Diseases in Animal Husbandry

Etta, Hannah Edim

Animal and Environmental Biology, Faculty of Biological Sciences, Cross River University of Technology, Calabar, Cross River State, Nigeria

Abstract

Plant based nanobiopesticides are purely organic. They have been employed in different ways in livestock farms to keep off different kinds of pests. It is a well-known fact that chemical pesticides have serious after-effects on both animal and humans that consume them if not properly handled. There have been reports of triggering life-threatening ailments like Alzheimer's disease, cancer, ADHD, and certain infantile defects at birth. Certain systems in the animal body, like the reproductive system, endocrine system, and nervous system, have been harmed by Pesticides, including herbicides, fungicides, and insecticides. Hence, the need for biological alternatives like nanobiopesticides. Several plant sources have been tested and seen to be effective against animal pests. There are insecticides, repellents, attractants, rodenticides, etc. Their method of application and target pests in animal husbandry management will be highlighted.

Keywords: Nanobiopesticides, animal husbandry, animal pests, phytochemicals

Email: hannahetta76@gmail.com

Charles Oluwaseun Adetunji and Julius Kola Oloke (eds.) Handbook of Agricultural Biotechnology: Volume I Nanopesticides, (229–238) © 2024 Scrivener Publishing LLC

8.1 Introduction

Nanotechnology is a multidisciplinary and vast field of science that employs technology, novel approach, and available conventional advanced physics, chemistry, engineering and biological facilities. Nanobiotechnology on the other hand has to do with applying technological know-how to improve the field of biological sciences. This entails the use of nanoparticles of living organisms' origin to improve life. A nanoparticle is any particle ranging from 1 to 100 nanometers (nm) in size. These particles cannot be detected by the human eyes and actually display obviously various types of chemical and physical properties [1]. This definition, that seemingly is agreeable by most, admits plant extracts as nanoparticles.

These particles are biologically active against pathogens by either repelling them or outrightly destroying them. Metallic nanoparticles (NPs) have also been used to formulate nanobiopesticides. NPs of zinc, gold, nickel, and titanium are believed to be more effective than plant-based nanobiopesticides basically because of their increased solubilization and target-oriented delivery.

Plant extracts are a composition of different biological components. These biological components are microscopic and fall within the range of 1 to 100 nm. Hence, they can be termed *nanoparticles.* Some of these biological nanoparticles are saponnins, tannins, steroids, glycosides, alkaloids, flavonoids, carotenoids, etc. and their derivatives especially the essential oils.

Nanobiopesticides of plant origin, are plant based nanoproducts produced to keep off pests from our environments. Some of these pets that affects animals (livestock), are mice, rats, spiders, ants, lice, birds, termites, moths, geckos, and wildlife. These pests can be a real nuisance. Especially like rats. The gnaw on furniture, eat foodstuff, and drop disease-causing remains everywhere, contaminating food and exposing animals and humans to deadly diseases like Anthrax, cattle tick, Salmonella, vibrosis, Lassa fever, Tularemia, hantavirus, *Lymphocytic choriomeningitis* (LCMV), and leptospirosis. Nanobotanicals and nanopesticides are believed to be more advantageous than broad-spectrum conventional pesticides because of several reasons mostly that they are target-specific, attacking only the pests they are prepared for. Also, they decompose easily, and while attacking the target organism, they also affect organisms that are closely related to those ones. They are effective even in small quantities while bearing no adverse effects on the food and the entire environment [2].

Synthetic or chemical pesticides have been implicated in pest resurgence, pollution of the environment, resistance by pests to target pesticides, direct toxicity to users and deadly effects on other non-target organisms in the agro-ecosystem [3]. The residual effects of these chemicals on human beings and the safety of the environment should be the two top factors when applying synthetic pesticides in animal husbandry.

First record of nanotechnology in agriculture is silica nanomaterial used as an insecticide for pest control [4]. Due to the efficacy and safety of nanobiopesticides, it is certain that soon, plant-based nanopesticides either as nanopowders or nanoemulsions will replace chemical pesticides for household and farm pests control totally. These natural products serve as well as alternatives for controlling pests that cause diseases that affect plants and animals and yet not have any deleterious effects on the environment [5].

8.2 Common Pests in Animal Husbandry

Pests in animal husbandry can be very expensive and time-consuming to control. It is better to prevent their emergence on the farm. Some common pests that threaten the lives of our livestock include lice like the little blue cattle louse (*Solenopotes capillatus*), shortnosed cattle louse (*Haematopinus eurysternus*), cattle biting louse (*Bovicola bovis*), the cattle tail louse (*Haematopinus quadripertusus*), and the longnosed cattle louse (*Linognathus vituli*); mites including chicken mite (*Dermanyssus gallinae*), chorioptic scab mites, (*Chorioptes bovis*), cattle follicle mites (*Demodex bovis*), psoroptic ear mite, (*Psoroptes cuniculi*), goat follicle mite (*Demodex capra*), sheep itch mite, (*Psorobia ovisticks*), scabies mite, (*Sarcoptes scabiei*), cattle ear mite (*Raillietia auris*), and sheep scab mite, (*Psoroptes ovis*); houseflies—black fly, blow fly, fleas (one flea of importance is the cat flea), mosquitoes (different spp), and rats. Some of these pests are ectoparasites that live on the farm animals threatening their existence. Other ways that pests affect animal production is by helping to spread disease agents and parasitic worms. They cause loss of blood, and anemia, damage animals and animal products physically, reduce weight gain within the herd, reduce egg and milk production and negatively affect the animals' resistance to other diseases. These in turn affect productivity and yield. This is where pesticides come in. This chapter will discuss nanobiopesticides that have been discovered and used as pesticides to rid poultry and livestock farms of animal pests.

8.3 Plant-Based Nanobiopesticides

There is this age-long belief that some plants parts on exposure to pests, repel the pests. Many of such plant-based nanocompounds have been reported to include aromatic phenols, sapogenins, limonoids, naphtoquinones, alcohols, saponins ketones, polyolesthers aldehydes, terpenoids, alkaloids, saccharides, and flavonoids. Some reported plants with active nanoparticles include *Capsicum annuum*, *Origanum vulgarize*, and *Laurus nobilis.* From several laboratory experimental reports, the efficacy of plants like *Ocimum basilicum, Azadiracta indica, Eucalyptus camaldulensis, Melia azaderach, Capsicum frutescents, Zingiber officinale,* and wood ash of *Khaya senegalensis* have been established to be effective against farm pests [6]. Some of these nanobiopesticides act as repellents. They do not kill the pests but repel them, keeping them away from the animals' habitation. Some known and well-researched repellents are essential oils of rosemary (*Rosmarinus officinalis* L), Eucalyptus (*Eucalyptus globulus*), clove (*Eugenia caryophyllus* (Spreng.)), lemongrass (*Cymbopogon flexuosus*), vetiver (*Vetiveria zizanioides* L.), and thyme (*Thymus vulgaris* L.). Catnip oil, extracted from *Nepeta cataria* L. (Lamiaceae), is a highly effective repellent against mosquitoes, bees, and other flying insects and there is a belief that it repels *Aedes aegypti* more effectively than the synthetic DEET. The most effective constituent, nepatalactone and lactone are implicated in this strong repellent action of the nanobiopesticide [7].

Plant-based nanobiopesticides can be made by simple methods like extraction using either water or alcohol to formulate a solution or suspension; or as dust or powder. The seeds, leaves, bark or roots of the plants can be used. The potency of some of these plants are stronger than synthetic or chemical pesticides while some are lesser. One major advantage that the biopesticides have over synthetic pesticides is the non-residual nature of these plant-based pesticides. They are also pest-specific and non-zoonotoxic and cheap to get.

Souto *et al.* [8] reported several plant-based nanobiopesticides and the efficacy of their active ingredients against farm pests. Daisy, *Tanacetum cinerariifolium* (Trevir.), roots of *Decalepis hamiltonii*, seeds of *Schoenocaulon officinale* (Schltdl. and Cham.). *Ryania speciosa*, *Lantana camara* L., *Cassia fistula* L. the chilli pepper (*Capsicum annuum*) is an effective repellent against house rats. There have been extensive reports on the phytochemical components of *C. anuum* L. especially due to its' pest-repellent property. The Phenolic compounds found in the Chilli pepper fruits are β catenin, flavonoids, glycerolipids, vitamins A, B, C, and E, capsaicinoids carotenoids (capsanthin, capsorubin, and cryptocapsin glycolipids and esters) [9]. The principal pungent component of the chilli pepper is the phenolic substance named capsaicin (trans-8-methyl-N-vanillyl-6-non-enamide). The capsaicin content of hot peppers varies from 0.1% to 1%. Chili pepper fruits have been reported to also contain phenolic compounds such as flavonoids, β-catenin, capsaicinoids carotenoids (capsanthin, capsorubin, and cryptocapsin), vitamins A, B, C, and E, glycolipids, glycerolipids, and esters [9]. Neem, (*Azadirachta indica*), has also been applied as a biopesticide in its different forms as extract, powder, dust or oil. It is effective against lice on farm animals.

Origanum vulgare also known as wild marjoram or sweet majoram has been proven to be a nanobio-repellent. It has a very complex chemical composition [8]. *Origanum vulgare* species are some of the most popular medicinal aromatic plants with enormous socioeconomic value and many applications in pharmaceutical industries [10]. It is believed that the pungent aroma also repels rats. Bay leaf (*Laurus nobilis*) in the coastal areas of Nigeria is very popular and has been employed medicinally for various ailments. Both insecticidal and repellent effects of *L. nobilis* have been demonstrated. Bay leaves were scattered in poultry houses to repel insects like cockroach, flies etc. and other creatures like meal moth and silverfish [11, 12].

8.4 Method of Application

The mode of application and time of application of a nanobiopesticide does affect its effectiveness. The choice of the method of application depends to a great extent on the type of nanopesticide, formulations, site of application and the pest to be managed.

It is believed that most nanobiopesticides are more effective in form of extracts or as powders than as isolated active ingredients as a result of synergistic effect of compound mixture [13, 14]. There are several ways of applying these pesticides in our livestock and poultry farms. One must be very careful to follow instructions on application because the effects of misapplication could be unpleasant to the animals.

To control flies in a cattle ranch, controlled or limited use of Nanobiopesticides in sprayers, dust bags, back rubs, and ear tags can be employed. Louse too can be sprayed with an appropriate nanobiopesticide. There is a tendency for poultry farms to attract lots of houseflies. To control house fly populations in a poultry farm, the manure must be managed properly. This will prevent creating breeding location for flies. Albeit, larvicidal nanoproducts can either be directly sprayed on the manure pile or administered as feed additive or bolus. For blackflies, it is best to target the larvae since they have specific habitats and habits. To treat mite infestation, which can be deadly to cattle, the most appropriate method of application will be to dip the animal or use a spray-dip machine to apply the effective nanobio-pesticide. These methods work well for beef cattle or non-lactating dairy cattle. For controlling lice, spot-treatments work best. Mosquitoes exist everywhere on earth except in the Antarctica. They are implicated in carrying different disease pathogens and viruses and infecting their vertebrate hosts with same. Pest management for mosquitoes is best achieved by spreading drops of plant-based nanobiopesticides into water bodies that could serve as breeding areas for the mosquito larvae and pupae. To rid livestock farms of adult mosquitos, aerial thermal, or ultra-low volume (ULV) sprays are used to apply the appropriate nanobiopesticide.

8.5 Effects of Nanobiopesticides

As varied as the targeted pests are, so also the action exerted by the nanopesticide on the animal pest. As earlier stated, the plant derived nanobiopesticides present different effects on pests. These effects also have different

Table 8.1 Mode of action of plant-based nanobiopesticides.

Plant source	Targeted pest	Mode of action
Capsicum annuum	Rats; Rodents	Repellent
Origanum vulgarize	Rats; Rodents	Repellent
Laurus nobilis	Rats; Rodents; ants; cockroaches	Repellent; insecticidal
Azadiracta indica	Lice; Insects	Insecticidal
Melia azaderach	Tick	Insecticidal
Zingiber officinale	Tick	Insecticidal
Eucalyptus camaldulensis	Mosquitoe	Repellent
Ocimum basilicum	Mosquitoe	Repellent
Cymbopogon flexuosus	Mosquitoe	Repellent
Khaya senegalensis	Mosquitoe	Repellent
Vetiveria zizanioides (L.)	Lice; insects	Kills; repellent
Rosmarinus officinalis L	Flea	Kills
Eugenia caryophyllus	Mite	Repellent
Tanacetum cinerariifolium	Lice; Mite	Insecticidal
Decalepis hamiltonii	Mosquitoe	Larvicide
Lantana camara L.	Weevil	Insecticidal
Cassia fistula	Mosquitoe	Insecticidal
Schoenocaulon officinale	Lice; Tick	Insecticide
Nepeta cataria	Mosquitoes, bees, and other flying insects	Repellent; insecticidal
Allium sativum	Beetles; mites	Larvicide
Thymus vulgaris	House flies	Repellent

Source: Research, 2022.

modes of action in the said pest being targeted. Table 8.1 summarizes these modes of action.

8.6 Conclusion

In this chapter, Plant-based nanobiopesticides have been seen to be active against a broad spectrum of animal pests. Albeit, the wide commercialization and acceptability of these products is still being discussed. The efficacy of the products against pests have been shown. They are also preferred because of their nonresidual nature of the products. The methods of application and mode of action in the pests were also discussed. However, factors, such as formulation, availability of raw materials, registration and adoption, still mitigate against the acceptability of plant-derived nanobiopesticides. It is hoped that in the near future, they will completely replace synthetic, unhealthy chemical pesticides.

References

1. TWI, 2022. https://www.twi-global.com/technical-knowledge/faqs/what-are-nanoparticles.
2. Isman, M., Neem and other botanical insecticides: Barriers to commercialization. *Phytoparasitica*, 25, 339–344, 1997.
3. Prakash, A., Rao, J., Nandagopal, V., Future of botanical pesticides in rice, wheat, pulses and vegetables pest management. *J. Biopestic.*, 1, 154–169, 2008.
4. El-Bendary, H.M. and El-Helaly, A.A., First record nanotechnology in agriculture: Silica nano-particles potential new insecticides for pest control. *Appl. Sci. Reports.*, 4, 3, 241–246, 2013.
5. Isman, M.B., Neem and other botanical insecticides: Barriers to commercialization. *Phytoparasitica*, 25, 4, 339–344, 1997.
6. Yusuf, S., Flather, M., Pogue, J., Hunt, D., Varigos, J., Piegas, L., Avezum, A., Anderson, J., Keltai, M., Budaj, A., Keith, F., Ceremuzynski, L., Variations between countries in invasive cardiac procedures and outcomes in patients with suspected unstable angina or myocardial infarction without initial ST elevation. *Lancet*, 352, 9127, 507–514, 1998. https://doi.org/10.1016/S0140-6736(97)11162-X.
7. Tripathi, A.K., Prajapati, V., Aggarwal, K.K., Kumar, S., Prajapti, V., Kumar, S., Kukreja, A.K., Dwivedi, S., Singh, A.K., Effect of volatile oil constituents of mentha species against stored grain pests, Callosobruchus maculatus and Tribolium castaneum. *J. Med. Arom. Plant Sci.*, 22, 549–556, 2000.

8. Souto, A.L., Sylvestre, M., Tölke, E.D., Tavares, J.F., Barbosa-Filho, J.M., Cebrián-Torrejón, G., Plant-derived pesticides as an alternative to pest management and sustainable agricultural production: Prospects, applications and challenges. *Molecules*, 10, 26, 16, 4835, 2021. https:/doi.org/10.3390/molecules26164835.
9. Chamikara, M.D.M., Dissanayake, D.R.R.P., Ishan, M., Sooriyapathirana, S.D.S.S., Dietary, anticancer and medicinal properties of the phytochemicals in chili pepper (Capsicum spp.). *Ceylon J. Sci.*, 45, 3, 5–20, 2016. http://doi.org/10.4038/cjs.v45i3.7396.
10. Karch, A.M., *Lippincott's Nursing Drug Guide*, 1486 pp., Lippincott Williams & Wilkin, Philadelphia, 2010.
11. Smart, J., *British Museum (Natural History) Instructions for Collectors No. 4A. Insects*, Trustees of the British Museum, London, 1963.
12. Palacios, S., Bertoni, A., Rossi, Y., Santander, R., Urzua, A., Efficacy of essential oils from edible plants as insecticides against the house fly, musca domestica L. *Molecules*, 14, 5, 1938–1947, 2009.
13. Mitscher, L.A., Drake, S., Gollapudi, S.R., Okwute, S.K., A modern look at folkloric use of anti-infective agents. *J. Nat. Prod.*, 50, 1025–1040, 1987.
14. MacKinnon, S., Durst, T., Arnason, J.T., Angerhofer, C., Pezzuto, J., Sanchez-Vindas, P.E., Poveda, L.J., Gbeassor, M., Antimalarial activity of tropical meliaceae extracts and gedunin derivatives. *J. Nat. Prod.*, 60, 336–341, 1997.

9

Application of Plant-Based Nanobiopesticides for Mitigation of Several Biotic Stress

Babatunde Oluwafemi Adetuyi[1]* and Charles Oluwaseun Adetunji[2]

[1]Department of Natural Sciences, Faculty of Pure and Applied Sciences, Precious Cornerstone University, Ibadan, Nigeria
[2]Applied Microbiology, Biotechnology and Nanotechnology Laboratory, Department of Microbiology, Edo University Iyamho, Auchi, Edo State, Nigeria

Abstract
Numerous biotic and abiotic stressors have substantial effects on plant growth and development. Seed priming is an effective technique for increasing productivity under adverse situations by hastening plant growth and germination. Plants become established when their roots grow deep and their seeds germinate. Nanopriming's small size and one-of-a-kind physicochemical features make it more effective than other seed priming procedures. When it comes to the uptake of nanoparticles during nanopriming, different plant species have different uptake rates due to physiological differences. Previous studies have mostly concentrated on how nanoparticles affect the growth of an individual plant. However, few research have investigated the impact of nanopriming, the use of several particles at simultaneously, on the development of forage and medicinal plant seeds. This literature review focuses on studies that examine how nanopriming with different particles affects the germination and development of seedlings of food and medicinal plants. There is also an evaluation of how various nanoparticles affect the most important morphological, physiological, and germination parameters in the establishment, growth, and output of these plants. In respect of seed germination, seedling growth and development, strength, and pace of growth, nearly all varieties of medicinal and forage plants benefited from nanopriming. Nanopriming negatively impacted seed germination, seedling development, and plant growth characteristics, despite the usage of nanoparticles making these plants more

**Corresponding author*: badetuyi@pcu.edu.ng

Charles Oluwaseun Adetunji and Julius Kola Oloke (eds.) Handbook of Agricultural Biotechnology: Volume I Nanopesticides, (239–278) resistant to environmental stress. There is also a cursory discussion of research priorities for the future.

Keywords: Nanobiopesticides, biotic stress, medicinal plants, nanopriming

9.1 Introduction

Because of agricultural product deterioration due to biotic stress, world hunger is a growing concern [1, 2]. A plant's life is fraught with biotic stresses like competition, fluctuating temperatures, lack of nutrients, bad weather, and interactions with phytopathogens [3]. The term "biologic stress" describes the damage that organisms including fungal, viral, bacterial, pathogenic nematodes, insects, and weeds do to plants [1]. To respond to environmental challenges, plants employ a number of molecular, phenotypic, and biochemical methods. Leaf spots, vascular wilts, and cankers are all symptoms of a fungal infection, which frequently happens in conjunction with bacterial diseases [4]. The existence of biological stress can have an impact on both farm productivity and output. Developing tolerance in host plants requires a thorough understanding of how viruses interact with their hosts. Among the biotic agents that plants confront include insects, fungi, bacteria, and viruses. The regular metabolic processes of plants are disrupted by these creatures, leading to biotic stress that ultimately leads to the plants' demise. On the other hand, some biotic organisms have mutually beneficial relationships with the plants that play host to them.

Some bacteria are beneficial to plants because they operate as biocontrol agents and/or biofertilizers, which mimic the actions of synthetic fertilizers and pesticides. Numerous species, including viruses, bacteria, nematodes, weeds, and arachnids, can cause biotic stress on plants. The plants that are meant to be protected by toxins experience the opposite impact, withering or killing them. They also cause field crops fatalities before and after collection. Plants are negatively affected by biotic and abiotic stresses in terms of their proliferation, maturity, yield, and biomass output.

The rhizosphere, the area of soil immediately around a plant's roots and the principal location of nutrient uptake, is the site of numerous physiological, chemical, and biological activities. The vast majority of the rhizosphere's microorganisms are bacteria. Some species of Bacillus are able to create growth-promoting compounds and disease-resistant, stress-resistant spores [5, 41, 48]. The addition of beneficial bacteria to the rhizosphere can help plants better withstand environmental stresses, such as those brought on by climate change. In addition, Bacillus subtilis dramatically improves resistance to biotic stress. For *B. subtilis*, chemotaxis is an absolute must when searching for and colonizing new root systems [6, 26, 30]. No matter how small or large their genomes may be, all bacterial species have evolved their own distinct chemotaxis mechanism. There are many genes in the bacterial genome that are involved in chemoreception, cell proliferation, and interspecies communication [7, 21–24].

To aid in the development of a healthy symbiotic interaction between bacteria and plants is the major role of a bacterial chemoreceptor. Microorganisms, such as *Pseudomonas, Azotobacter chroococcum, Rhizobium,* and *Sinorhizobium meliloti* thrive in root exudates [8]. Chemoattractant studies focused mostly on Gram-negative bacteria, such as *Bacillus subtilis, Escherichia coli,* and a serovar of *Salmonella enterica.* The *B. subtilis* genome encodes ten chemoreceptors (or ligands) made of amino acids, carbon, and oxygen (*G. Bacillus subtilis* uses chemoreceptors to detect its environment, specifically plant roots [112]. *B. subtilis*, in particular, is an important rhizosphere bacterium for plant development [56].

9.2 Systemic Resistance Can Be Caused by Physiological Stress

Chester [46] claimed that plants had developed a wide range of biological responses against biotic stress. Over the years, developed systemic resistance has gone by many various names. Examples include induced resistance, translocated resistance, and plant vaccination. Plants may efficiently defend themselves from a wide variety of phytopathogens and insects by inducing resistance if given the right signals. Systemic acquired resistance (SAR) and induced systemic resistance (ISR) are complicated defense mechanisms that some organisms might generate in response to biotic stresses [9–11, 153]. It is suggested that SAR and ISR share a phenotypic similarity; both involve a plant's own defense mechanisms working to lessen the severity of damage produced by a pathogen-induced sickness. Different signaling molecules distinguish each phytopathogen [12–15, 102]. ISRs are activated as a defense mechanism in plants with commensal rhizobacteria in their roots, boosting the host plant's disease resistance [90]. Auxin, cytokinin, ACC decarboxylase, gibberellic acid, phosphate solubilization, and nutrient absorption are all boosted by rhizobacteria, rhizobacterial fungi, and plant growth-promoting bacteria living in the rhizosphere [17–20, 44].

These microorganisms increase or produce pathogen resistance in place of directly combating phytopathogens [62, 68, 140]. One of the most studied biocontrol agents for triggering ISR against pathogenic bacteria is the *Pseudomonas* species [129].

Bacteria such as *Trichoderma* spp., *Piriformospora indica, Penicillium simplicissimum, Phoma, Bacillus, Pseudomonas,* and *Serratia* are only a few examples of the many elicitors for ISR induction. These bacteria are therefore thought to be potential candidates for the purpose of biological control agents in the future [141]. In order to create ISR during initiation, pathogens such as viruses and insects are utilized [131]. When the pathogen is successfully immunized against, a variety of latent resistance mechanisms are concurrently activated, leading to induced resistance, which encompasses include systemic acquired resistance and induced systemic resistance. The identification of the proper inputs by a number of recognition receptors is necessary for the execution of plant defense mechanisms. According to recent investigations, several recognition proteins in the plant immune system have been shown to identify PAMPs such bacterium flagellin, fungal chitin, and polysaccharides [39, 72, 76]. PAMPs increase the disease resistance of crop plants by enhancing their defensive mechanisms. Damage-associated molecular patterns (DAMPs), which are endogenous plant-derived triggers, activate the plant defense mechanism [10].

Immunity produced by a PAMP serves as a person's first line of protection against germs that could infect them and cause illness (PTI). After PTI signaling has suppressed the host detection system, the secondary line of defense, aptamer immunity, is activated when prion virulent effector molecules manage to bypass the PTI and begin plant infection [10, 109, 110]. After the second line of defense has been formed, the effector-triggered immunity (ETI) pathway keeps signaling downstream defense-related genes by identifying attack-specific effector molecules using internal nucleotide-binding domains and leucine-rich repeat domains [165].

This is due to the fact that SAR spreads a single distant signal among unharmed plant portions, stimulating the first and secondary lines of defense and encouraging elevated defensive mechanisms far from the sites of infection [160]. Both induced systemic resistance and systemic acquired resistance are crucial plant immune responses that protect plants from phytopathogenic fungi, and PR genes and NPR1 have long been thought to play a key role in regulating these responses [32, 88, 89]. Even in the dearth of the cell cycle genes, NPR1 can continue to operate correctly. It was discovered that plants with PR genes also had SAR intercellular communication.

When it comes to plant pathways, SAR is among the strongest. All plant tissues are capable of expressing SAR during the initial stages of a pathogen infection [85, 92, 93]. This signal transduction system activates in response to phytopathogen infections, protecting plants by triggering hypersensitive reactions such as necrotic lesion growth at the infection site [33]. When pathogens infect plants, a chemical known as an "elicitor" activates immunological processes, increasing the plant's defenses against infection [37, 96–98]. It was found that retinine is a more potent elicitor than previously thought; it upregulates NPR1 and SA-related genes, leading to increased protein translation involved in defense and protein production in disease. Reticine is obtained from citrus fruit peel extract (A). In contrast to the commercial elicitor benzothiadiazole, systemic treatment of salicylic acid and hydrogen peroxide caused adverse reactions even against tobacco mosaic virus (TMV) [101, 103, 156].

Effectiveness of SAR stimulants, such as salicylic acid, methyl jasmonate, benzothiadiazole, and aminobutyric acid in avoiding *Pseudoperonospora cubensis* and *Phytophthora infestans*, respectively, in muskmelon and potatoes. A plan was made to apply a fungal mixture once every five weeks. It was shown that salicylic acid was more effective in both settings [28, 117, 123]. A wide spectrum of rhizobacterial elicitors can trigger SAR responses against various pathogens, as discovered by Rudrappa [132]. Acetoin, a rhizobacterial elicitor produced by *Bacillus subtilis*, contains salicylic acid and nuclear protein ribosomal protein 1 (NPR1), a combination that has been shown to provide systemic acquired resistance to *Pseudomonas syringae* pv. tomato DC3000 in the model plant Arabidopsis thaliana [132, 135, 136]. Significant upregulation of SAR-related genes was detected in barley in response to cis-jasmone, -aminobutyric acid, and acibenzolar-S methyl, all of which have been shown to be effective in activating the plant defense system against *Rhynchosporium commune* [139].

Berberine is a chemical that is produced by the tobacco mosaic virus infected plant Nicotiana tabacum. Other effects of berberine include raising hydrogen peroxide levels and promoting hypersensitive reactions. Additionally, the antibacterial properties of berberine offer extra protection from a number of diseases [74]. In order to demonstrate systemic acquired resistance to a variety of phytopathogens, a number of biotic inducers may activate a signaling pathways pathway in response to infection. SARs therefore offer the highest protection against phytopathogens.

9.3 Pathways for Multiple Stress Response Modulation

When biotic and abiotic stressors are present together, responses are more complex. A plant's ability to adapt to stress and defend itself from a variety of threats is enhanced by the accumulation of particular metabolites [124, 161]. Metabolic reprogramming in favor of resistance is triggered by signaling from stress responses such as callose synthesis, ion influx alterations, reactive oxygen species (ROS), and plant hormones [35].

9.4 Species of Reactive Oxygen

Immediately after detecting stress, ROS production spikes [22]. Signaling molecules are one of ROS's primary functions in cells [65, 148, 154]. The plant limits ROS generation to protect its tissues [101, 104, 144, 146]. In stressed organisms, ROS are known to be extremely toxic and destructive substances. Xia [157] showed that while larger levels of reactive oxygen species (ROS) can still promote cell death, it is the lower levels of ROS that are more important for controlling the plant's response to stress. Signaling is a major role for reactive oxygen species (ROS) under biotic stress. Potentially less oxidative damage from abiotic stress could result [84]. Additionally, ROS may prevent cross-tolerance [40]. Initial treatment with the vascular pathogen Verticillium spp., ROS help Arabidopsis thaliana adapt to stress by encouraging the creation of de novo xylem and a boost in water flow [40]. Reactive oxygen species (ROS) production can also function as a secondary messenger by altering the architecture of proteins, turning on defense genes (such as respiratory burst oxidase homologous D), and enhancing cellular communication [143].

Different strategies exist for ROS to react to biotic and abiotic stressors [118]. That the transcription factor Zat12 is involved in both biotic and abiotic stress was first shown by Davletova [53]. Most likely, ROS scavenging is under the command of Zat12. Reactive oxygen species (ROS), according to Atkinson [29], are the main mechanism enabling bridge spanning biotic and abiotic stress response networks. ROS-sensitive transcription factors in Arabidopsis can detect ROS production and activate genes related to the plant's stress response [109]. Gechev postulated several variables as potential sources of the stress response, including gene transcription, antioxidants, dehydrins, and proteins linked to low-temperature heat shock, and pathogenicity [69].

Modifications in ROS signaling that take place following the administration of some medications are associated with conditioning for stress tolerance [84]. Brassinosteroid-treated cucumber plants yield more hydrogen peroxide and are more resilient to biotic and abiotic stress [69]. Citrus plants were dosed with H2O2 to boost their salt tolerance, which resulted in a slight rise in the fraction of oxidizing and S-nitrosylated proteins. Meanwhile, untreated plants have a worse ability to withstand stress [84].

9.4.1 Mitogen-Activated Protein Kinases Cascades

Mitogen-activated protein kinase (MAPK) circuits start as soon as stress stimuli are identified. They regulate the body's response to a demanding situation [161]. Mitogen-activated protein kinases (MAPKs), a class of signal transducers, play multiple roles in cells that respond both to abiotic and biotic stress [42, 146]. There is proof connecting specific kinases to both kinds of stress. Numerous stress responses include different adenosine monophosphate kinases (MAPKs), which suggests that they might influence interaction between biotic and abiotic stimuli [163]. For instance, GhMPK6a negatively controls biotic and abiotic stress in cotton [91]. When a pathogen infects a host, SA increases the activity of the MAPK pathway, which in turn triggers the production of the PR gene and consequent protective responses [159]. After being phosphorylated by MAP kinase 3 in Arabidopsis, VIP1 enters the nucleus and indirectly triggers PR1 [125]. A pathogen-associated molecular pattern (PAMP) which has been demonstrated to activate MAPK cascades is flagellin [47]. The formation of the regulatory network that combats infections is crucial.

A variety of abiotic stressors also control MAPKs like MPK3, MPK4, and MPK6, among others. Stress response crosstalk is closely regulated by MAPK pathways [25]. It is likely that MPK3 and MPK6 play a key role in regulating tolerance to upcoming stresses because they are required for the complete activation of defense responses. The results of ABA-induced OsMPK5 kinase activity include increased tolerance to biotic and abiotic stimuli and OsMPK5 overexpression. As a modulator of abiotic stress tolerance and a positive promoter of resilience to the necrotrophic brown spot infection *Cochliobolus miyabeanus*, OsMPK5 appears to play a pivotal role in the rice stress response [36]. To increase their innate immunity to UVB, injury, and infections, tomatoes activate MPK1 and MPK2 [79]. According to Zhou [166], MAPK signaling collaborates with reactive oxygen species (ROS) and auxin biogenesis inhibitor (ABA) signaling pathways to promote cross-acclimation to abiotic and biotic stress.

9.4.2 Importance of Hormone Signaling During Stress Interaction

A variety of hormones control different types of stress, permitting diverse forms of protection against a variety of external stresses. According to Cramer *et al.* [50], the primary hormone involved in recognizing various abiotic stressors is ABA. A surge in ABA concentration, according to Xiong [158], reshapes the network that controls how plants react to abiotic stress. The main way that biotic stress responses are controlled is through this relationship, despite the fact that SA and acid JA/ET are hostile to one another and the other stress hormones [92]. According to studies, ABA levels may rise after a health worry [64, 147]. For instance, researchers found that after a plant contracted Pst DC 3000, its ABA levels increased, inhibiting the plant's other defense mechanisms [54].

However, recent data suggests that ABA boosts tolerance to biotic stress [94]. This molecule is known as adenosine monophosphate (ABA), and it is problematic because, depending on the situation (such as the type and time of the stressor [94]. Additionally, SA/JA/ethylene and ABA frequently behave antagonistically when abiotic and biotic factors interact, rendering plants more susceptible to disease and herbivore attack [130]. Abiotic stress causes the stomata to close, which is an "unplanned secondary impact" that defends the crop from biotic invaders. The plant is shielded from biotic and abiotic stress in this situation [100]. ABA can be observed to have an impact on the pathogen during infection [29]. As infection takes longer to establish and delays plant response, the first effect of ABA on the interplay involving various biotic and abiotic stresses is only related to abiotic stress [146]. Kim [83] found that shutting stomata in relation to ABA aids plants in conserving water and maintaining a high water content. Because ABA can prevent the induction of SA, JA, and ethylene, it is probable that they are not implicated in this stage. It will be necessary to adapt future disease prevention strategies to this new reality. Here, we concentrate on the virus's internal manifestations. An essential inducible defense that stops the invasion of microbes is calereous acid [67].

Ton [149] makes the case that an effective ABA signaling pathway is necessary for the buildup of callose in infected plants. Depending on the situation, elevated ABA levels might either promote or prevent callus formation. Therefore, the way an organism reacts to biotic stress may be affected by ABA changes brought on by past stress. It might improve the resistance phenotype by encouraging callose development or other additional defense mechanisms [100]. The defense response is controlled by third-phase activation of SA, JA, and ethylene buildup [113]. Therefore, the precise role of ABA as a regulator of the interaction between abiotic and biotic stress is highly dependent on the timing of the stress perception, whether an infection affects a plant that has already been exposed to abiotic stress or whether an infected plant becomes further exposed to abiotic stress. This claim is supported by the evidence [81].

Numerous studies have looked into the role of SA in the cooperative connection between pathogens and plants. It is widely known that the antagonistic actions of the plant hormones ABA and SA contribute in plant resistance against stress [34]. Miura and Tada's research demonstrated that SA is equally crucial to plant adaptations to drought stress as ABA [145]. Additionally, SA enhanced the drought resistance of barley [34].

9.4.3 Molecular Response and Cross-Tolerance Transcription Modulators

Transcription factors are in charge of this reorganization of the regulatory network after a stress is identified. In order to help plants take a strong defensive position, it is essential to alter the expression of specific genes.

There is strong evidence to suggest that so many genes have numerous meanings and can help plants develop stress resistance [145]. The presence or absence of specific phytohormones, including as ABA, SA, JA, and ethylene, has an impact on the expression of genes involved to defense. for instance, the botrytis susceptible1 (bos1) gene is regulated by both ABA and JA, and hence BOS1 mutant plants are less vulnerable to osmotic stress. By facilitating the molecular link between ABA generated in reaction to drought stress and SA released in response to pathogen infection, the Arabidopsis gene called MYB96 plays a crucial role in plant resistance during pathogen infection [137]. OsMAPK5 is a kinase-active enzyme that positively regulates rice's response to drought, salt, and heat, as well as the plant's resilience to disease. The stimulation of tomato SlAIM1 boosts the plant's resistance to abiotic stress and Botrytis cinerea infection [138]. Strangely, the plant's ability to resist illness is ensured by the synthesis of several PR genes as a result to abiotic stress [137]. Elevated expression is a sign of plant harm [138]. The cold-responsive PR genes are activated by the overexpression of a number of transcription factors, protecting plants from both cold stress and disease. A chain of transcription factors is triggered by abiotic stress, and they in turn trigger the production of an abundance of PR proteins.

C-repeat binding factors (CBF), no apical meristem ATAF, and cup-shaped cotyledon (NAC) proteins regulate the main abiotic stress signaling pathways that enhance stress tolerance [119].

According to research by Snider [142], Stress from cold has been proven to increase CBF and other PR proteins. By inducing an upregulation of the PR1 gene, mutant Arabidopsis, which overproduces the NAC signaling pathway NTL6 in response to cold stress is well able to ward against pathogen attack [82]. Tsutsui [152] demonstrated that Arabidopsis' resistance to cold and illnesses is facilitated by the DREB transcription factor, which regulates the plant's responsiveness to cross-tolerance between biotic and abiotic stress.

Although opinions on how ABA interacts with plant diseases vary among experts, it has been shown that, in certain conditions, it increases tolerance to biotic stress after a plant has been subjected to abiotic stress. It is because TF overexpression upregulates the PR genes, making them more active.

Recently, it has been proposed that the redox-responsive transcription factor 1 (RRTF1) and WHIRLY1 protein play a part in the transmission of plastid-nucleus communication [66]. WHIRLY1 can recognize plastid redox changes and transmit that information to the nucleus independently of NPR1. As a result of its role in assisting the body's adaptation to new stimuli, scientists claim that this enzyme is the most crucial component of retrograde transmission. In a similar line, it may be possible to prepare distant leaves for defense by using RRTF1, which is generated in response to abiotic and biotic stimuli.

9.4.4 Pesticides

Around the turn of the 20th century, increased population demands, especially in the United States, required an increase in agricultural output. In 1900, there were only 1.5 billion people on Earth; by 2000, there were 6.1 billion. The United Nations predicts that by 2050, there will be anywhere from 9.4 and 10 billion individuals on Earth, a rise of one billion over the following 10 years if current population growth rates are maintained [55]. To satisfy the demands of a fast-growing global population, there was a compelling need to significantly increase food production during the 20th century. This was achieved by the use of fertilizers and other agricultural inputs in the 20th century. Around the turn of the twentieth century, urea and other inorganic phosphate compounds were first used, and since then, their use has only grown [70]. People first began utilizing guano and other biofertilizers in the latter half of the nineteenth century. Phosphates had a significant role in the "green revolution" in agricultural output by increasing crop diversity and yields. Grain production surged by a ratio of 10 for every acre of farmland, resulting in a global surplus of food [43]. With an R2 of 0.97 from 1900 to 1988, phosphate production expanded along with world population all through the 20th century [78].

Food production increased as artificial crop storage agents became available in the 1940s. It increased dramatically from the previous year, jumping from 0.2 million tons to approximately 5 million tons in 2000. There was a growth of almost 11 percent between 1950, when output was only 0.2 million tons, to 2000, when it was over 5 million tons. Chemicals used to protect crops from pests have many different ingredients, including plant hormones, neonicotinoids, organochlorine pesticides, pyrethroids, organophosphorus, carbamates, and, more lately, biopesticides. To protect crops from a range of hazards, countless distinctive pesticides and insecticides have indeed been developed over time. Insecticides, fungicides, and herbicides all saw an increase in sales, but the industry for herbicides saw the biggest jump. This issue has been made worse by the high cost of chemicals, labor, and pesticide application. It has also been made worse by the wide variety of pests that can be seen in any climate or region. The overall average pesticide application rates, which ranged from 6.5 to 60 kg/ha, were found in South American and Asian countries. When compared to the West and the Americas, pesticide use in Asia's cities and rural areas has not increased as much. According to the International Food Policy Research Institute and the International Monetary Fund, herbicide use has remained comparatively low in recent years despite the extensive use of insecticides across Asia [77].

DDT and other synthetic pesticides created in the twentieth century for agricultural purposes were widely used to treat ticks, parasites, and other pests throughout the century. For instance, the original rationale behind the development of DDT was its application in agriculture. Despite being prohibited, it is frequently utilized in South Asia as a pesticide to eradicate domestic pests, malaria vectors, and fish [151]. Pesticides can be applied in agricultural settings using a range of techniques, including as hand-held sprayers, truck-mounted applicator, and aerial spraying. Some people have claimed to be satisfied after using pesticides, which runs counter to evidence on the impacts and risks of synthetic chemicals on human health and wellness. Farmworkers, locals, and their families participated in this study at the time after pesticide use.

High chemical sensitivity and poor chemical control, as well as the almost 355,000 yearly fatalities related to unintentional poisoning, have been linked by numerous previous studies. Additionally, they discovered a connection between these toxicities, reckless handling of potentially dangerous substances, and hypersensitivity [105]. The aim of the study was to estimate the amount of pesticides in the environment that are causing a mass extinction of animal species. There were fish, birds, bees, frogs, even a few smaller animals. Statistics were kept on both the number of fatalities and the murderous techniques [52].

Plant protection pesticides, that were commonly believed to have been released into the environment soon after synthetic chemicals were first developed, were initially held responsible for domestic and international sickness [77]. Many people think that part of the DDT that is sprayed on crops eventually makes its way into waterways, where it is swiftly transformed to DDE, bioaccumulates in aquatic food systems, and eventually finds its way back to the environment and human bodies. Today, endosulfan manufacturing decisions are made via a thorough, evidence-based process. Scientific research to enhance food production, food security, and environmental security is prioritized in this plan as well as the aforementioned goals [77].

By stopping the emergence of plant infections with resistance, synthetic pesticides prevent illnesses from spreading again. Thankfully, farmers are increasingly using insecticides. According to Parween and Jan, 2019 [122]. Plants absorb and store chemicals used in synthetic insecticides after being treated with them. People's health continues to decline even after they quit eating these crops because of the high quantities of hazardous chemical deposits in them [75]. Plants absorb manmade pesticides after treatment and retain their active ingredients. The high concentration of harmful chemical residues in these crops meant for human consumption is causing people to have ongoing health problems [121].

9.5 Biopesticides

Recent years have seen a rise in the need for biopesticides, especially in third world nations where numerous synthetic pesticides are banned because to concerns about human and environmental health: organophthaloids, halogenated hydrocarbons, carbamates, and organophosphates. While synthetic pesticides can have a short-term impact on pests and diseases, but they also put people, animals, and the environment at danger by contaminating food sources. In order to combat disease and pests in agriculture, various synthetic pesticides have indeed been produced. Synthetic pesticides are used in agriculture to combat soil-borne pests and diseases. The Environmental Protection Agency warns that synthetic pesticides pose risks to humans and nontarget species alike. Synthetic pesticides can be harmful to both humans and other animals, according to the EPA. They also mess with the ecosystem's natural balance. The widespread use of chemical pesticides has far-reaching biological effects, including disruption of cell signaling, inhibition of enzyme activity, changes in pH, disruption of electrolysis, stimulation of osmotic and membrane breakdown, and the formation of transmembrane pH gradients. In addition, they release potentially dangerous elements, such as free-radical, which can weaken tissue and even demolish proteins and DNA [63]. Parkinson's disease and other forms of neurodegeneration, diabetes type 2, disturbance of hormones, cancer, and obesity are only some of the diseases that have been related to the use of synthetic pesticides. In this category, Parkinson's disease is one of the most well discussed cases.

The use of synthetic pesticides has been linked to an increased risk of illness. The mechanisms by which these chemicals work and the increased rate with which people have been exposed to them throughout time are some potential explanations for this link [27, 135, 144].

Unfortunately, the usage of most pesticides is rapidly declining, according to the pesticide industry. However, because the technique is still common, people's bodies continue to absorb pesticides from what they eat. Additionally, it has been observed that pesticide-exposed workers frequently get intoxicated as a result of the toxins throughout their shifts [61, 62]. The lower risk to environmental and human health is the main advantage of natural pesticides over synthetic ones. This does not, however, give you permission to use them recklessly or to disregard the repercussions of your choices. Products that have been licensed to be employed as biopesticides may cause harm to the general public. Although copper is necessary for plant and animal life, it can be dangerous for aquatic species in high amounts and over time.

Microcystis aeruginosa, Gracilaria coronopifolia, Chrysanthemum spp., and other hazardous plant, microalgae, and algae species can look quite similar to cyanide [63]. Biopesticides are pesticides with active components derived from natural chemicals or microorganisms rather than synthetic ones, as their name suggests. They are known as "biopesticides," and they are applied to kill insects in a variety of contexts. Plant-incorporated preservatives, microbial biopesticides, and plant biopesticides are the three primary types of plant-based pesticides. Microbial biopesticides are a comparatively new family of insecticides that are efficient against insects and diseases.

"Microbials" are microorganisms that are employed as biopesticides and are effective against both disease and pests. Many illnesses and pests have been effectively combated with the help of microbial biopesticides. A pathogenic microbial biopesticide that also works against the other diseases and insect species can be used to treat some fungal infections [45].

Targeting Microbes with Biopesticides

Insect destruction is followed by fungus growth. Species like *Metarhizium anisopliae* and *Beauveria bassiana*, both harmful to insects, are examples of environmental entomopathogenic fungus. The World Health Organization claims that B. bassiana spores quickly multiply when in touch with their insect hosts, penetrate the cuticle, and form new colonies deep into the host's tissues. The bassiana spores spread to nearby insects after the insect host dies. A fungus known as bassiana could help to either suppress or eradicate the bug population. The white mold growing on the decaying carcass releases new spores into the air, adding to the general filth. Insects are killed when M. anisopliae mold spores land on them, germinate, and then release hyphae that pierce the cuticle. A few days after the first touch, the fungus starts to infect and disseminate throughout the insect's body. For a long time, soil-borne diseases have been prevented from spreading by using microbes like *Pseudomonas* and *Trichoderma* spp [134].

UC Berkeley research indicates that trichoderma is a form of filamentous fungus that thrives in decomposing organic matter like wood and soil. In addition to *T. virens, T. viride,* and *T. harzianum,* many more *Trichoderma species* have been identified to have biocontrol abilities. Mycoparasitism involves the release of cell wall-degrading enzymes such proteases, chitinases, and glucanases and is one biocontrol-worthy method for competing for finite resources. Antibiosis can be brought on by a wide variety of antibiotics, including but not limited to 6-pentylpyrone, massoilactone, heptelidic acid, peptaibols, viridin, and harzianic acid.

Antibiotics like heptelidic acid and harzianic acid are among the most popular choices. Bacteria produce the antibiotic heptelidic acid, which is used to treat illnesses brought on by different pathogens [57].

Phytophthora, Candida albicans, Fusarium, and *Sclerotia* are just a few of the many fungal infections that *Trichoderma* is successful against. Cotton crop failure due to *Fusarium* wilt is caused by *Fusarium* sp. Similar to other crops, maize is affected by rhizoctonia and pyrthium species' diseases. Cucumbers were able to develop a resistance against the fungus *Colletotrichum* sp., which causes anthracnose, in this way. *Pseudomonas aeruginosa* infections are distinguished by the synthesis of antibiotic derivatives and medicines, such as pyoluteorin (Plt), 2,4-diacetylaminoglucinol (DAPG), phenazine-1-pyrrolnitrin (Prn), carboxylic acid (PCA), and systemic resistance (ISR) [58].

Whenever microbial biopesticides are to be used widely, microorganisms including bacteria, fungi, and algae must be incorporated into formulations. According to the Global Biopesticide Trade Association, there has been a bacterial epidemic in the biopesticide sector, notably as among species of *Bacillus thuringiensis* that are frequently moved over international borders and used to manage pest populations in plantations [71]. Plaguing the insect's gut and causing holes to form is how this bacterium kills parasites. As a result, the insect develops an anion disequilibrium, which impairs its ability to digest and finally results in death. Because they are less toxic to nontarget insects, birds, and mammals than traditional insecticides, these pesticides are thought to be less hazardous to the environment than those pesticides [167–177].

Despite this, research indicates that the use of microalgae as biopesticides can successfully halt the spread of a variety of diseases that afflict plants. The ability of this bacteria to produce a wide variety of biocompounds, such as phenolic compounds, among others, has been demonstrated through extensive research. The results of these studies show that these bacteria can produce a variety of chemicals apart from those already identified. Research has been conducted on a variety of components, including terpenes, growth regulators, phenolic compounds, and others [49, 133].

9.5.1 Natural Chemical Pesticides

Pesticides made from living organisms like bacteria, fungi, or insects can be used with confidence. They have no negative effects on the species that create them. The purpose of these devices is to put an end to the proliferation of unwanted pests. These substances could control plant growth and act as pheromones either to attract or repel pests, which are both advantageous for growth (PGR).

It is impossible to determine if a particular biopesticide poses any risk because so few countries have regulatory authorities responsible for analyzing metabolites. As a result, determining what type of biopesticide is suitable to use may be difficult [87].

Since their discovery, auxin-type plant growth regulators (PGRs) have proven to be efficient biological control agents and herbicides. They do so for good reason. It's one of the top biological control and herbicides on the market right now. Consider the difference between the effects of PGR and marijuana on action selectivity. It's possible that marijuana's strength is increased by the rapid detoxifying process. When present in modest amounts, these substances have a biofertilizer impact and promote cell development, cell division, and elongation. High marijuana exposure causes symptoms of intoxication and developmental flaws in plants, including decreased respiratory, perspiration, and carbon absorption. Weeds eventually succumb to the effects of the damage these anomalies bring to their blood vessels and membranes [63].

9.5.2 Pesticides Made From Plant Substance

Crops receive pesticide spraying to prevent the spread of insects and other pests (chemical compounds and plant extracts). Insect pest control is a common use for pesticides. Terpenes, phenolics, and essential oils are examples of secondary metabolites that can be used to defend plants against a variety of dangers [86]. Given that they are quasi in the environment and nontoxic to humans, essential oils are among the most effective insecticides for agricultural usage. Along with destroying insects and mites, these compounds could be used to destroy fungi, bacteria, and other environmental contaminants.

The antioxidant properties of essential oils aid in the prevention of disease-causing microorganisms in plant cultures. These microbes' proteins and DNA contain prooxidants, which contribute to cytotoxicity, ROS production, and membrane and organelle destruction [42].

The procedure used to create this type of biopesticide, the amount of material purchased, how old the plant was that produced the oil was extracted can all affect how effective the finished product is. One factor among several that can influence the toxicity of a biological insecticide is the morphological and physiological age of the plant used to make the oil. Despite their value, a number of problems have prevented agricultural pesticides from being used widely. Pesticides have issues such as deterioration over time, complex extraction combinations, extraction techniques, or active component composition, and purifying issues [133].

9.5.3 Biopesticides' Constrains

Biopesticides provide a number of benefits over conventional methods of disease and pest control, including less negative effects on the environment and a higher possibility of food safety. According to a growing body of research, plant-isolated biopesticides face significant activity issues and need extremely high dosages of their component parts to be efficacious in field settings [59]. That's because plants include a wide variety of different bioactive compounds that could alter their chemical properties, from which they are extracted. Environmental harm is also caused by the discharge of organic compounds for use as solvents to extricate pesticides. It has been observed that biopesticides have a shorter shelf life and are extremely biodegradable. For a certain subset of pests in the field, microbial pesticides might be more effective than botanical pesticides, but this comes at the cost of their poor performance against a number of insect species. Because microbial pesticides' effectiveness is decreased by factors including moisture loss, heat, light, and UV radiation, crop harm endures [120].

Plants' bioactive chemical content and their preferred habitat are both affected by their growing conditions. The number and composition of plants' active compounds are also impacted by plant diversity and variation, which changes how resistant plants are to disease [38]. However, many extraction techniques can be used to produce plant extracts of differing quality. Maintaining the proper ratios of active and inert ingredients all through formulation process may be difficult. The absence of established methods for planning and measuring effectiveness in the field, where time is limited, is another issue [108]. Despite positive *in vitro* results, real-world outcomes can be unpredictable; this is partly because the raw materials have a limited shelf life and, in some cases, are of poor quality and/or were improperly prepared. Only after carefully analyzing the host crops and their propensity for spread may predatory biopesticides be employed. On small acreages, hand application might not be feasible due to the time needed to ensure proper crop coverage and exposure length. Chemistry, hazard, packaging, and composition data are necessary for product registration, although this isn't always the situation in the pharmaceutical sector [111].

9.5.4 Biopesticide Action Mechanism

The different modes of action of biopesticides permit for their usage in a variety of contexts, including farming. Microorganisms produce harmful pesticides through a variety of methods, such as parasitism, abiotic stress, and predators. Numerous studies have shown how effective botanical insecticides are at getting rid of pests and preventing the spread of diseases. During the attack, the victim is either drugged or parasitized to death. Pests are drawn to the treatment area by the delivery of the therapy, where they are then eliminated or sterilized. Asteraceae plant extracts prevent plant pathogenic fungal hyphal growth and structural alterations in the mycelia [114].

Asteraceae plants include terpenoids, coumarin alkaloids, and flavonoids, which render them completely toxic to fungi. Numerous substances can change the way the inner membrane and the compartments inside it are constructed [51]. Therefore, when bioactive substances are in contact with the membrane surface of fungal cells, they may cause partitioning and penetration. The partition of the cell will be compromised as a result, allowing the inmates to escape. Additionally, it has been shown that plant bioactive chemicals can rupture the cytoplasmic membrane, release the contents of the cell, stop cell proliferation, and ultimately kill the cell [60].

9.6 Microbial Biopesticides

9.6.1 Bactericides and Fungicides

These biopesticides frequently block or prevent translation, which prevents the creation of proteins [20]. They attach to bacterial 50S ribosomes and prevent chain lengthening and peptide transfer. By preventing aminoacyl tRNA from binding to the 30S and 70S ribosomal subunits complexes, they sporadically have the ability to stop translation [11]. The interaction of streptomycin with the 30S ribosomal subunit results in two adverse effects: inhibition of peptidyltransferase activity and disruption of protein synthesis [31].

Additionally, by altering the plasma membrane's permeability and allowing substances like amino acids and electrolytes to leach out, they can limit chitin synthase activity and amino acid and electrolyte metabolism [20]. When avermectins and emamectins come into contact with neuron terminals, they trigger the release of GABA, which blocks the electrical transmission of nerve impulses. In addition, GABA can activate GABA-gated channels to open Cl ion channels [56]. The mitochondrial membrane may become permeable due to polynactins, enabling potassium ions to escape. Herbicides work by blocking glutamine synthase, leading to an excess of ammonia (bilanafos), which prevents plants from phosphorylating [116].

Because of their high effectiveness, species specificity, and environmental acceptability, the use of pesticides made with the assistance of microorganisms and their byproducts is generally recognized in the field of pest management [61]. Given their established importance, the industrial sector's marketing and monetization of these products has much opportunity for improvement [61].

Bacillus thuringiensis, B. popilliae, obligatory pathogens, potential pathogens, and facultative pathogens are the four species of bacteria as being used as biopesticides [13]. These microbes include *Pseudomonas aeruginosa* and *E. coli*, for instance. The bacteria most frequently utilized in industrial settings are those that generate spores. *Bacillus thuringiensis* and *Bacillus sphaericus,* the two most common bacteria, are highly specialized, safe, and effective at halting and controlling pest populations [109]. The cry genes code for a family of crystalline proteins called Cry, which are generated by *Bacillus thuringiensis* in sporulation crystals. Depending on the strain, Cry proteins can be anywhere from 65 to 145 kDa in size and have three structural domains linked together by a single linker. There is a single family of cry proteins, which is further subdivided into roughly 50 subgroups [109]. Koul has provided a comprehensive analysis of the Cry protein's supplementary features and mechanism [112]. The only remaining techniques of exterminating insects are lethal septicemia and hunger. The bi mosquito larvicidal toxin, BinB (51.4kDa) and BinA (41.9kDa), is present in a product made from Bacillus sphaericus that is frequently used to control mosquitoes [22].

9.6.2 Disease-Fighting Biopesticides Derived From Arbuscular Mycorrhizal Fungi

When present on plant roots, arbuscular mycorrhizal fungi (AMF) significantly boost growth and harvest success [115]. By enhancing the plants' resilience to dangerous organisms, they help farmers maintain disease-free crops. AMF's composition and prevalence are influenced by factors like crop and soil type, application of fertilizers, and tillage [11]. Plants have evolved a variety of both direct and indirect defense strategies to stave off herbivory. To deter herbivores, they may, for example, create compounds of the same type as nicotine, gossypol, and others. Changing gene expression patterns and, in turn, affecting the nutritional state of the crop, colonization by AMF on agricultural plants has been demonstrated to be incredibly beneficial in enhancing hosts' defense capabilities [16].

9.6.3 Biological Pesticides Have Advantages Over the Chemical Ones

The chemical pesticides, which have hitherto been the industry norm, are no match for biopesticides in many ways. They can replace chemical pesticides in insect control since they are more effective, less hazardous to the environment, and only kill the targeted pests [19]. Due to its value in environmentally friendly farming practices, the usage of biopesticides has recently increased [105, 151].

IPM programs heavily rely on biopesticides since they are very effective at low doses, biodegrade rapidly without leaving bothersome residues, and are therefore environmentally preferable to conventional pesticides [151]. Even though biopesticides have many potential applications, they have not been used nearly as widely as they should. Support is given for the following: The price of making insecticides is affected by the following factors:

First, it is expensive to find, produce, and receive approval for new biological agents. Additionally, the shelf life of biopesticides is finite. Fieldwork can only be carried out to a certain extent due to local climate and environmental conditions like warmth, moisture, soil, and so forth. The biopesticides are extremely specialized, which means that they are only effective against the particular illnesses that they were created to eradicate. For various diseases and pests in the field, different biological agents are required. These techniques are time-consuming, costly, and uncomfortable to employ, and some diseases and pests cannot be treated using them.

Conventional chemical pesticides have a number of drawbacks that biopesticides do not. Table 9.1 shows the attributes of conventional chemicals and biopesticides.

Table 9.1 The attributes of conventional chemicals and biopesticides.

Conventional chemical	Biopesticides
Pesticides Biopesticides.	Put into manufacturing only substances found in nature and found in living things.
They are not ecofriendly and contribute to pollution levels.	They are safe for the environment and pose no threats to it.
Often kills things it's not supposed to	Avoid harming unintended organisms.
Inefficient in terms of cost.	As opposed to chemical fertilizers, it is more cost effective and affordable.
As the dosage is increased, resistant microorganisms will eventually emerge.	Resistance in pests is not an issue.
Profitability at a very high level.	Not well received by consumers.
Pollute the ground and the water supply.	Not allowed to pollute water supplies.
Promote bioaccumulation.	Have no potential for bioaccumulation.

9.7 Alleviation of Biotic Stress

9.7.1 *Bacillus subtilis* Reduces Biotic Stress in Plants

Bacillus species are strong prospects for long-term environmental survival because endospores can endure harsh circumstances for an extended period. A vast variety of secondary metabolites that support crop growth and disease resistance can also be produced by them. The physiological mechanisms by which *Bacillus* species increase plant stress tolerance have received little attention from researchers [86]. The use of bacterial agents to increase crop yield and resistance to disease is safe and has no adverse impacts on the surrounding ecosystem, according to organic farming guidelines [95]. *B. subtilis* can lessen reliance on synthetic pesticides and insecticides in contemporary farming. Chemical fungicides and insecticides kill beneficial soil microorganisms for plant growth. Therefore, as chemical pesticides are finally phased out, *B. subtilis* and other advantageous bacteria could be used to enhance the use of other microbial pesticides [99]. Bt toxin and *Bacillus thuringiensis* (Bt) provide adaptable pest management [112]. Bt prevents insect larval development and promotes crop growth [27]. Pest colonies can also be lowered with the aid of bacteria like *Bacillus subtilis, Bacillus amyloliquefaciens*, and *Bacillus cereus* [54]. Beneficial rhizospheric bacteria (PGPR) like B use nanoparticles encapsulated with aluminum, gold, or silver. *Paenibacillus, B. subtilis, P. fluorescens, P. putida*, and others to encourage plant growth and prevent fungal development As a result, nanobiofertilizers must be considered. Researchers have found that using encapsulated nanobiofertilizers to deliver fertilizer to individual cells reduces the overall quantity of fertilizer needed and eliminates runoff pollution [104]. Herbicides like pendimethalin and metalochlor have had compounds like polypropylene sulphonate and polyallylamine hydrochloride added to them to make a delayed release of the active ingredient to a particular region, significantly improving the effectiveness and efficiency of weed control [84].

9.7.2 Phytonutrient-Producing Rhizobia; PGPR

This is a fascinating new field of study because plants' innate defensive mechanisms are frequently activated when they are exposed to PGPR strains or when they become infected with the plant pathogen. Medical professionals refer to this syndrome as "induced systemic resistance" (ISR). Plants can develop greater resistance to a variety of diseases and pests when they are stimulated in the proper way; this process is referred to as "induced resistance." Improved resistance that develops in response to an inducing agent during pathogen infection is known as inducible systemic resistance (ISR) or systemic acquired resistance (SAR) [128]. Originally proposed by Chester [46], the concept of "acquired physiological immunity" in plants has since been variously referred to as "systemic acquired resistance" [131], "translocated resistance," and "plant immunization" [152].

9.7.3 Aspects That Improve Plant Growth

Root-invading PGPR bacteria encourage plant growth, reduce damage from insects and diseases, and ultimately aid in the survival of plants. Both directly and indirectly, bacteria that invade the root surface promote plant growth. Joe Kloepper came up with the term "PGPR" in the 1980s to designate this large group of microorganisms. Because of the durability of their endospores, Bacillus species have good long-term chances of surviving in the environment. A vast variety of secondary metabolites that support plant growth and resistance to disease can also be produced by them. The physiological mechanisms that increase plant tolerance in relation to Bacillus species presence are poorly understood [127]. The use of infectious pathogens to increase crop yield and resistance to disease is fully safe and has no detrimental impacts on the ecology, according to organic farming regulations [54]. Utilizing *B. subtilis* could reduce the need for artificial pesticides and insecticides in modern farming. Chemical fungicides and pesticides eliminate beneficial soil microbes, which are necessary for plant growth. Therefore, as chemical pesticides are finally phased out, *B. subtilis* and related beneficial bacteria could be used to increase the use of other microbial pesticides [73]. The use of Bacillus thuringiensis (Bt) and Bt toxin offers a broad array of pesticide control alternatives. Bt prevents insect larval development and encourages crop growth [27]. Additionally used in pest control are the strains of bacteria *Bacillus subtilis, Bacillus amyloliquefaciens,* and *Bacillus cereus* [67]. Beneficial rhizospheric bacteria (PGPR), such as *B. subtilis, P. fluorescens, P. putida,* and *Paenibacillus,* use coated aluminum, gold, or silver nanoparticles to encourage plant growth and prevent the growth of fungi. Nanobiofertilizers are thus a practical choice. According to studies, encapsulated nanobiofertilizers deliver fertilizer to individual cells, lowering the amount of fertilizer required and preventing runoff pollution [106].

9.7.4 Fungi and Bacteria That Promote Plant Development; Rhizobacteria

Exciting new study opportunities arise when plants are exposed to strains of plant growth-promoting rhizobacteria (PGPR) or when the plant pathogen infects a plant. This is referred to as "induced systemic resistance" in the medical field (ISR). By supplying the right stimuli, plants can develop their resistance to a wide range of viruses and pests; this process is known as "induced resistance." An increase in pathogen resistance that occurs after infection and exposure to an inducing factor is known as inducible systemic resistance (ISR) or systemic acquired resistance (SAR) [126]. First proposed by Chester [46], the idea of "acquired physiological immunity" in plants has gone by a number of other names since then, including "systemic acquired resistance," "translocated resistance," and "plant immunization" [152].

9.7.5 Nematodes

PGPRs lead to a systemic resistance against nematode pests [160]. *P. fluorescens* inhibited *Heterodera schachtii*, a sugar beet cyst nematode, from penetrating the roots as quickly [160]. Cotton that has been exposed to B. subtilis is also resistant to *Meloidogyne incognita* and *Meloidogyne arenaria* [160]. For parasitic nematodes that plague plants, PGPR has shown effective as a biological control agent, especially against sugarcane.

9.7.6 PGPR as Endophytes

There have been reports of success employing the 227 beet and potato cyst nematode to reduce these pests. *Hirschmanniella oryzae*, a root-eating worm, can damage rice crops. This can be avoided by applying PGPR to rice seed, either alone or in combination with chitin and neem cake [130]. The prevalence of the root-knot nematode *M. incognita* on tomatoes was diminished after treatment with *P. fluorescens* strain Pf1. At the same time, there was a decrease in the number of galls and egg masses in the ground [130]. Similar results were obtained when *P. chitinolytica* was applied to tomato fields in an effort to eradicate root-knot nematodes [143]. Many different plant species, both monocots (like rice and maize) and dicots (including Arabidopsis, beans, carnations, tobacco, radishes, cucumbers, and tomatoes), have shown that PGPR-mediated ISR is successful.

9.8 Consideration and Forecasting

A plant's developmental stage and environmental variables have significant effects on the plant's reaction to abiotic and biotic stress [164]. It has been demonstrated that various infections and types of abiotic stress have diverse impacts on the expression of the immune system, indicating that a wide variety of stressor combinations may cause phenotypic injury. A plant's complex responses to biotic and abiotic stress are made possible by the collaboration of multiple signaling systems. When many stressors are exposed, the concentration of numerous signaling molecules increases, which frequently results in cross-tolerance.

Stress leads plants to emit defensive chemicals as a "signal." Even while only some of these elements may be implicated in a plant's defense reaction to a specific stress combination, those that are shared by both stressors are crucial in defending the plant and enhancing its resistance.

Plants have evolved a variety of defense mechanisms to handle combinatorial stress. For instance, a technology based on "omics" is one such method. Transcriptomics, proteomics, and metabolomics have revealed the methods by which plants overcome adversity. These investigations also reveal potential target genes, proteins, or metabolites to enhance plant responses and generate tolerance.

Even though the issue has been addressed in recent publications [29, 95], relatively little is documented about the "omics" characterization of abiotic and biotic stress combinations. Although many crops and model plants have whole genome sequences available, protein and metabolite libraries are still very uncommon, making it more challenging to combine these results. Whole-plant statistics could be misleading because various species of plants and even cultivars may behave differently. Based on which organ is addressed while working with a plant, the outcome can be very different. Another option is to use molecular engineering to insert particular genes into crop plants.

To improve the potato plant's resilience to biotic and abiotic stress, a gene encoding a short antimicrobial peptide was modified and injected into the plant [73]. One choice is to make an effort to persuade powerful regulators. By increasing flavonoid synthesis, the detrimental effects of biotic and abiotic stress can be lessened [110, 150]. An additional example is supplied by polyamines. These substances have long been known to have a role in mediating resistance to infectious pathogens, and now we now know that they do so when it comes to resilience to abiotic stress [155]. Genetic modification that increases polyamine accumulation can boost resistance to a range of stressors [80]. Priming is another strategy that can promote tolerance among different groups. A number of substances have the potential to improve management of a variety of stresses in the field since they have been shown to properly equip plants for biotic and abiotic difficulties in the lab [126]. The ultimate goal is always to maintain or improve plant productivity, output, and performance.

The initial step in the plant cell's response is the identification of biotic and abiotic stress. It is generally known that Ca++ ions serve as crucial transducers (messengers). The use of MAPK (mitogen-activated protein kinase) complexes has been shown for both. MAPKs play a critical role in controlling Ca2+-ROS crosstalk and signaling in response to a specific stress. Given that both forms of stresses have been highlighted as being necessary for ROS, it is expected that ROS will be essential for the incorporation of both stresses throughout cross-tolerance. Plants' capacity to resist stress is greatly influenced by hormonal cues. Salicylic acid (SA) and jasmonate/ethylene (JA/ET) play a key role in the plant's response to biotic stress, whereas abscisic acid has a similar role in the plant's reaction to abiotic stress (ABA). Endocrine pathways engage with one another in intricate ways, but it is unclear what kind of interaction occurs when both biotic and abiotic stress are present. Callose deposition is encouraged by ABA signaling activation, which benefits in both preinvasion protection and wound healing. Positive effects result from the ABA and JA/ET signaling symbiosis. If SA signaling is turned on, the existence of a pathogen can result in the suppression of ABA responses. ABA signaling promotes infection transmission outward from the site of original infection by inhibiting signals that lead to systemic acquired resistance. Antimicrobial resistance is facilitated by the interaction of SA, JA, and ET signals. Hormones, secondary metabolites, priming agents, and other cytoplasmic chemicals promote the expression of a phenotype known as cross-tolerance by increasing the expression of transcription factors (TF), pathogenesis-related (PR), defense, heat shock protein (HSP), and other genes associated with stress resistance.

References

1. Moustafa Farag, M., Almoneafy, A., Mahmoud, A., Elkelish, A., Arnao, M.B., Li, L., Ai, S., Melatonin and its protective role against biotic stress impacts on plants. *Biomolecules*, 10, 1, 54, 2019.
2. Keswani, C., Singh, S.P., García-Estrada, C., Mezaache Aichour, S., Glare, T.R., Borriss, R., Rajput, V.D., Minkina, T.M., Ortiz, A., Sansinenea, E.,

Biosynthesis and beneficial effects of microbial gibberellins on crops for sustainable agriculture. *J. Appl. Microbiol.*, 132, 3, 1597–1615, 2021.

3. Mertens, D., Boege, K., Kessler, A., Koricheva, J., Thaler, J.S., Whiteman, N.K., Poelman, E.H., Predictability of biotic stress structures plant defence evolution. *Trends Ecol. Evol.*, 36, 5, 444–456, 2021.
4. Meena, M., Zehra, A., Dubey, M.K., Aamir, M., Gupta, V.K., Upadhyay, R.S., Comparative evaluation of biochemical changes in tomato (Lycopersicon esculentum Mill.) infected by Alternaria alternata and its toxic metabolites (TeA, AOH, and AME). *Front. Plant Sci.*, 7, 1408, 2016.
5. Adetuyi, B.O., Oluwole, E.O., Dairo, J.O., Biochemical effects of shea butter and groundnut oils on white albino rats. *Int. J. Chem. Chem. Processes (IJCC)*, 1, 8, 1–17, 2015.
6. Batiha, G.E.B., Awad, D.A., Abdelazeem, M.A., Nyamota, R., Wahed, M.I., Shah, M.A., Amin, M.N., Adetuyi, B.O., Hetta, H.F., Cruz-Marins, N., Koirala, N., Ghosh, A., Sabatier, J., Diary-derived and egg white proteins in enhancing immune system against COVID-19. *Front. Nutr. (Nutritional Immunology)*, 8, 629440, 2021.
7. Ogunlana, O.O., Adetuyi, B.O., Rotimi, M., Esalomi, I., Adeyemi, A., Akinyemi, J., Ogunlana, O., Adetuyi, O., Adebisi, O., Okpata, E., Baty, R., Batiha, G., Hypoglycemic activities of ethanol seed extract of Hunteria umbellate (Hallier F.) on Streptozotocin-induced diabetic rats. *Clin. Phytosci.*, 7, 1, 1–9s, 2021.
8. James-Okoro, P.O., Iheagwam, F.N., Sholeye, M.I., Umoren, I.A., Adetuyi, B.O., Ogundipe, A.E., Braimah, A.A., Adekunbi, T.S., Ogunlana, O.E., Ogunlana, O.O., Phytochemical and *in vitro* antioxidant assessment of Yoyo bitters. *World News Nat. Sci.*, 37, 1–17, 2021.
9. Didunyemi, M.O., Adetuyi, B.O., Oyebanjo, O.O., Morinda lucida attenuates acetaminophen-induced oxidative damage and hepatotoxicity in rats. *J. Biomed. Sci.*, 8, 1–9, 2019, https://www.jbiomeds.com/biomedical-sciences/morinda-lucida-attenuates-acetaminopheninduced-oxidative-damage-and-hepatotoxicity-in-rats.php?aid=24482.
10. Abdul Malik, N.A., Kumar, S., Nadarajah, K., Elicitor and receptor molecules: Orchestrators of plant defense and immunity. *Int. J. Mol. Sci.*, 21, 3, 963, 2020.
11. Adeleke, D.A., Olajide, P.A., Omowumi, O.S., Okunlola, D.D., Taiwo, A.M., Adetuyi, B.O., Effect of monosodium glutamate on the body system. *World News Nat. Sci.*, 44, 1–23, 2022.
12. Adetuyi, B.O., Adebayo, P.F., Olajide, P.A., Atanda, O.O., Oloke, J.K., Involvement of free radicals in the ageing of cutaneous membrane. *World News Nat. Sci.*, 43, 11–37, 2022.
13. Adetuyi, B.O., Adebisi, O.A., Adetuyi, O.A., Ogunlana, O.O., Toloyai, T.Y., Egbuna, C., Uche, C.J., Khan, J., Adumanya, O.C.U., Patrick-Iwuanyanwu,

K.C., Ficus exasperata attenuates acetaminophen-induced hepatic damage via NF-κB signaling mechanism in experimental rat model. *BioMed. Res. Int.*, 2022, Article number 6032511, 2022.

14. Adetuyi, B.O., Adebisi, O.A., Awoyelu, E.H., Adetuyi, O.A., Ogunlana, O.O., Phytochemical and toxicological effect of ethanol extract of Heliotropium indicum on liver of male albino rats. *Lett. Appl. NanoBiosci. (LIANB)*, 10, 2, 2085–2095, 2020, https://doi.org/10.33263/LIANBS102.20852095.
15. Adetuyi, B.O. and Farombi, E.O., 6 Gingerol, an active constituent of ginger, attenuates Lipopolysaccharide-induced Oxido-inflammation, cognitive deficits, neuroplasticity and amyloidogenesis in rat. *J. Food Biochem.*, 45, 4, 1–11, 2021, https://doi.org/10.1111/jfbc.13660.
16. Adetuyi, B.O., Dairo, J.O., Didunyemi, M.O., Anti-hyperglycemic potency of Jatropha Gossypiifolia in alloxan induced diabetes. *Biochem. Pharmacol. (Los Angel)*, 4, 5, 193, 2015.
17. Adetuyi, B.O., Odine, G.O., Olajide, P.A., Adetuyi, O.A., Atanda, O.O., Oloke, J.K., Nutraceuticals: Role in metabolic disease, prevention and treatment. *World News Nat. Sci.*, 42, 1–27, 2022.
18. Adetuyi, B.O., Ogundipe, A.E., Ogunlana, O.O., Egbuna, C., Estella, O.U., Mishra, A.P., Akram, M., Achar, R.R., Banana peel as a source of nutraceuticals, in: *Food and Agricultural Byproducts as Important Source of Valuable Nutraceuticals*, 1st ed, p. 400, Springer, Berlin, 2022.
19. Adetuyi, B.O., Okeowo, T.O., Adetuyi, O.A., Adebisi, O.A., Ogunlana, O.O., Oretade, O.J., Marraiki, N., Beshbishy, A.M., Welson, N.N., Batiha, G.E.S., Ganoderma lucidum from red mushroom attenuates formaldehyde-induced liver damage in experimental male rat model. *Biology*, 9, 313, 2020, https://doi.org/10.3390/biology9100313.
20. Adetuyi, B.O., Olajide, P.A., Adetuyi, O.A., Oloke, J.K., Preventive phytochemicals of cancer as speed breakers in inflammatory signaling. *Res. J. Life Sci., Bioinform. Pharm. Chem. Sci.*, 8, 1, 30–61, 2022.
21. Adetuyi, B.O., Olajide, P.A., Awoyelu, E.H., Adetuyi, O.A., Adebisi, O.A., Oloke, J.K., Epidemiology and therapeutic measure for COVID-19; a review. *Afr. J. Reprod. Health* (Special Edition COVID-19), 24, 2, 142, June 2020. https://www.ajrh.info/index.php/ajrh/article/view/2300.
22. Adetuyi, B.O., Oluwole, E.O., Dairo, J.O., Chemoprotective potential of ethanol extract of ganoderma lucidum on liver and kidney parameters in plasmodium Beghei-induced mice. *Int. J. Chem. Chem. Processes (IJCC)*, 1, 8, 29–36, 2015.
23. Adetuyi, B.O., Omolabi, F.K., Olajide, P.A., Oloke, J.K., Pharmacological, biochemical and therapeutic potential of Milk Thistle (Silymarin): A review. *World News Nat. Sci.*, 37, 75–91, 2021.
24. Adetuyi, B.O., Toloyai, P.Y., Ojugbeli, E.T., Oyebanjo, O.T., Adetuyi, O.A., Uche, C.Z., Olisah, M.C., Adumanya, O.C.U., Jude, C., Chikwendu, J.K., Akram, M., Awuchi, C.G., Egbuna, C., Neurorestorative roles of microgliosis

and astrogliosis in neuroinflammation and neurodegeneration. *Scicom J. Med. Appl. Med. Sci.*, 1, 1, 1–5, 2021.
25. Andreasson, E. and Ellis, B., Convergence and specificity in the Arabidopsis MAPK nexus. *Trends Plant Sci.*, 15, 106–113, 2010.
26. Apel, K. and Hirt, H., Reactive oxygen species: Metabolism; oxidative stress; and signal transduction. *Ann. Rev. Plant Biol.*, 55, 373–399, 2004.
27. Arena, J.P., Liu, K.K., Paress, P.S., Frazier, E.G., Cully, D.F., Mrozik, H., Schaeffer, J.M., The mechanism of action of avermectins in Caenorhabditis elegans: Correlation between activation of glutamate-sensitive chloride current, membrane binding, and biological activity. *J. Parasitol.*, 81, 2, 286–294, 1995.
28. Astha, and Sekhon, P.S., Efficacy of SAR elicitors and fungicides against late blight of potato and downy mildew of muskmelon. *Agric. Res. J.*, 544, 529–533, 2017.
29. Atkinson, N. and Urwin, P.E., The interaction of plant biotic and abiotic stresses: From genes to the field. *J. Exp. Bot.*, 63, 3523–3544, 2012.
30. Atkinson, N.J., Lilley, C.J., Urwin, P.E., Identification of genes involved in the response of Arabidopsis to simultaneous biotic and abiotic stresses. *Plant Physiol.*, 162, 2028–2041, 2013.
31. Awoyelu, E.H., Oladipo, E.K., Adetuyi, B.O., Senbadejo, T.Y., Oyawoye, O.M., Oloke, J.K., Phyloevolutionary analysis of SARS-CoV-2 in Nigeria. *New Microbes New Infect.*, 36, 100717, 2020, https://doi.org/10.1016/j.nmni.2020.100717.
32. Backer, R., Naidoo, S., van den Berg, N., The nonexpressor of pathogenesis-related genes 1 (NPR1) and related family: Mechanistic insights in plant disease resistance. *Front. Plant Sci.*, 10, 102, 2019.
33. Balint-Kurti, P., The plant hypersensitive response: Concepts, control and consequences. *Mol. Plant Pathol.*, 20, 8, 1163–1178, 2019.
34. Bandurska, H. and Stroiński, A., The effect of salicylic acid on barley response to water deficit. *Acta Physiol. Plant*, 27, 379–386, 2005.
35. Bartoli, C.G., Casalongué, C.A., Simontacchi, M., Marquez-Garcia, B., Foyer, C.H., Interactions between hormone and redox signaling pathways in the control of growth and cross-tolerance to stress. *Environ. Exp. Bot.*, 94, 73–88, 2013.
36. Beckers, G.J.M., Jaskiewicz, M., Conrath, U., Mitogen-activates protein kinases 3 and 6 are required for full priming of stress responses in Arabidopsis thaliana. *Plant Cell*, 21, 944–953, 2009.
37. Bektas, Y. and Eulgem, T., Synthetic plant defense elicitors. *Front. Plant Sci.*, 5, 804, 2015.
38. Bharadwaj, A. and Sharma, S., Role of Arbuscular Mycorrhizae in pest management. *Int. J. Ecol. Environ. Sci.*, 32, 193–205, 2006.
39. Boller, T. and Felix, G.A., renaissance of elicitors: Perception of microbe-associated molecular patterns and danger signals by pattern-recognition receptors. *Annu. Rev. Plant Biol.*, 60, 379–406, 2009.

40. Bowler, C. and Fluhr, R., The role of calcium and activated oxygens as signals for controlling cross-tolerance. *Trends Plant Sci.*, 5, 241–246, 2000.
41. Brader, G., Djamei, A., Teige, M., Palva, E.T., Hirt, H., The MAP kinase kinase MKK2 affects disease resistance in Arabidopsis. *Mol. Plant Microbe Interact.*, 20, 589–596, 2007, 2017.
42. Bravo, A., Gill, S.S., Soberón, M., Mode of action of Bacillus thuringiensis Cry and Cyt toxins and their potential for insect control. *Toxicon*, 49, 423–435, 2007.
43. Carvalho, F.P., Agriculture, pesticides, food security and food safety. *Environ. Sci. Policy*, 9, 7-8, 685–692, 2006.
44. Chandran, H., Meena, M., Swapnil., P., Plant growth-promoting rhizobacteria as a green alternative for sustainable agriculture. *Sustanibility*, 13, 19, 10986, 2021.
45. Chen, J., *Biopesticides: Global markets to 2022*, Report Code CHM029G, BCC Research, Wellesley, MA, USA, 2018.
46. Chester, K.S., The problem of acquired physiological immunity in plants. *Rev. Biol.*, 8, 3, 275–324, 1933.
47. Chinchilla, D., Zipfel, C., Robatzek, S., Kemmerling, B., Nurnberger, T., Jones, J.D.G., Felix, G., Boller, T.A., Flagellin-induced complex of the receptor FLS2 and BAK1 initiates plant defence. *Nature*, 448, 497–500, 2007.
48. *Consumption of chemical pesticides in various States/UTs during 2014–2015 to 2019–2020*, Department of Agriculture, Cooperation and Farmers Welfare, Ministry of Agriculture and Farmers Welfare, Government of India, 2022.
49. Costa, J.A.V., Freitas, B.C.B., Cruz, C.G., Silveira, J., Morais, M.G., Potential of microalgae as biopesticides to contribute to sustainable agriculture and environmental development. *J. Environ. Sci. Health Part B*, 54, 366–375, 2019.
50. Cramer, G.R., Urano, K., Delrot, S., Pezzotti, M., Shinozaki, K., Effects of abiotic stress on plants: A systems biology perspective. *BMC Plant Biol. 2011*, 11, 163–177, 2011.
51. da Silva, I.F., de Oliveira, R.G., Soares, I.M., da Costa, A.T., Ascêncio, S.D., de Oliveira Martins, D.T., Evaluation of acute toxicity, antibacterial activity, and mode of action of the hydroethanolic extract of Piper umbellatum L. *J. Ethnopharmacol.*, 151, 1, 137–143, 2014.
52. Dannon, H.F., Dannon, A.E., Douro-kpindou, O.K., Zinsou, A.V., Houndete, A.T., Toffa-Mehinto, J., Elegbede, I.A.T.M., Olou, B.D., Tamo, M., Toward the efficient use of Beauveria bassiana in integrated cotton insect pest management. *J. Cotton Res.*, 3, 24, 2020.
53. Davletova, S., Schlauch, K., Coutu, J., Mittler, R., The zinc-finger protein Zat12 plays a central role in reactive oxygen and abiotic stress signaling in Arabidopsis. *Plant Physiol.*, 139, 847–856, 2005.
54. De Torres-Zabala, M., Truman, W., Bennett, M.H., Lafforgue, G., Mansfield, J.W., Egea, P.R., Bogre, L., Grant, M., Pseudomonas syringae pv. tomato

hijicks the Arabidopsis abscisic acid signaling pathway to cause disease. *EMBO J.*, 26, 1434–1443, 2007.

55. Desa, U., United nations department of economic and social affairs, population division, in: *World Population Prospects: The 2015 Revision, Key Findings and Advance Tables*, Technical Report: Working Paper No ESA/P/WP 241, 2015.
56. Didunyemi, M.O., Adetuyi, B.O., Oyewale, I.A., Inhibition of lipid peroxidation and *in-vitro* antioxidant capacity of aqueous, acetone and methanol leaf extracts of green and red Acalypha wilkesiana Muell Arg. *Int. J. Biol. Med. Res.*, 11, 3, 7089–7094, 2020.
57. Duan, C.B., Du, Y., Hou, X., Yan, N., Dong, W., Mao, X., Zhang, Z., chemical basis of the fungicidal activity of tobacco extracts against Valsa mali. *Molecules*, 21, 1743, 2016.
58. Elgar, M.A., Zhang, D., Wang, Q., Wittwer, B., Thi Pham, H., Johnson, T.L., Freelance, C.B., Coquilleau, M., Insect antennal morphology: The evolution of diverse solutions to odorant perception. *Yale J. Biol. Med.*, 91, 457–469, 2018.
59. El-Sayed, A., Bengtsson, M., Rauscher, S., Lofqvist, J., Witzgall, P., Multicomponent sex pheromone in Codling Moth (Lepidoptera: Tortricidae). *Environ. Entomol.*, 28, 775–779, 1999.
60. El-Wakeil, N.E., Botanical pesticides and their mode of action. *Gesunde Pflanzen*, 65, 4, 125–149, 2013.
61. Farombi, E.O., Abolaji, A.O., Adetuyi, B.O., Awosanya, O.O., Fabusoro, M., Neuroprotective role of 6-Gingerol-rich fraction of Zingiber officinale (Ginger) against acrylonitrile-induced neurotoxicity in male Wistar rats. *J. Basic Clin. Physiol. Pharmacol. (JBCPP)*, 30, 3, 2018.
62. Feduchi, E., Cosín, M., Carrasco, L., Mildiomycin: A nucleoside antibiotic that inhibits protein synthesis. *J. Antibiot.*, 38, 415–419, 1985.
63. FICCI, A report on indian agrochemical industry, in: *A Federation of Indian Chambers of Commerce and Industry Report*, p. 45, FICCI, New Delhi, India, 2016, Available online: https://ficci.in/spdocument/23103/agrochemical-ficci.pdf, (Accessed on 3 May 2021).
64. Flors, V., Ton, J., van Doorn, R., Jakab, G., García-Agustín, P., Mauch-Mani, B., Interplay between JA, SA and ABA signalling during basal and induced resistance against Pseudomonas syringae and Alternaria brassicicola. *Plant J.*, 54, 81–92, 2008.
65. Foyer, C. and Noctor, G., Redox homeostasis and antioxidant signaling: A metabolic interface between stress perception and physiological responses. *Plant Cell*, 17, 1866–1875, 2005.
66. Foyer, C.H., Karpinska, B., Krupinska, K., The functions of whirly1 and redox-responsivetranscription factor 1 in crosstolerance responses in plants: A hypothesis. *Philos. Trans. R. Soc. Lond. B Biol. Sci.*, 369, 20130226, 2014.
67. Garcia-Andrade, J., Ramirez, V., Lopez, A., Vera, P., Mediated plastid RNA editing in plant immunity. *PLoS Pathog.*, 9, e1003713, 2013.

68. García-Andrade, J., Ramirez, V., Flors, V., Vera, P., Arabidopsis ocp3 mutant reveals amechanism linking ABA and JA to pathogen-induced callose deposition. *Plant J.*, 67, 783–794, 2011.
69. Gechev, T.S., van Breusegem, F., Stone, J.M., Denev, I., Laloi, C., Reactive oxygen species as signals that modulate plant stress responses and programmed cell death. *Bioessays*, 28, 1091–1101, 2006.
70. Gilland, B., Nitrogen, phosphorus, carbon and population. *Sci. Prog.*, 98, 4, 379–390, 2015.
71. Gomiero, T., Food quality assessment in organic vs. conventional agricultural produce: Findings and issues. *Appl. Soil Ecol.*, 123, 714–728, 2018.
72. Gowthami, L., Role of elicitors in plant defense mechanism. *J. Pharmacogn. Phytochem.*, 7, 6, 2806–2812, 2018.
73. Goyal, R.K., Hancock, R.E.W., Mattoo, A.K., Misra, S., Expression of an engineered heterologous antimicrobial peptide in potato alters plant development and mitigates normal abiotic and biotic responses. *PLoS One*, 8, e77505, 2013.
74. Guo, W., Yan, H., Ren, X., Tang, R., Sun, Y., Wang, Y., Feng., J., Berberine induces resistance against tobacco mosaic virus in tobacco. *Pest Manag. Sci.*, 76, 5, 1804–1813, 2020.
75. Gwinn, K.D., Bioactive natural products in plant disease control, in: *Studies in Natural Products Chemistry*, vol. 56, R. Attaur (Ed.), pp. 229–246, Elsevier Inc, Amsterdam, The Netherlands, 2018.
76. Hakim, S., Naqqash, T., Nawaz, S., Laraib, I., Siddique, M.J., Zia, R., Mirza, M.S., Imran, A., Rhizosphere engineering with plant growth-promoting microorganisms for agriculture and ecological sustainability Front. *Sustain. Food Syst.*, 5, 16, 2021.
77. Halder, J., Rai, A.B., Kodandaram, M.H., Compatibility of neem oil and different entomopathogens for the management of major vegetable sucking pests. *Natl. Acad. Sci. Lett.*, 36, 19–25, 2013.
78. Hendrix, C.S., *Applying hubbert curves and linearization to rock phosphate*, vol. 4, pp. 1–12, Peterson Institute for International Economics, 1750 Massachusetts Avenue NW, Washington, D.C., U.S., 2011.
79. Holley, S.R., Yalamanchili, R.D., Moura, D.S., Ryan, C.A., Stratmann, J.W., Convergence of signaling pathways induced by systemin; oligosaccharide elicitors; and ultraviolet-B radiation at the level of mitogen-activated protein kinases in Lycopersicon peruvianum suspension-cultured cells. *Plant Physiol.*, 132, 1728–1738, 2003.
80. Hussain, S.S., Ali, M., Ahmad, M., Siddique, K.H., Polyamines: Natural and engineered abiotic and biotic stress tolerance in plants. *Biotechnol. Adv.*, 29, 300–311, 2011.
81. Jensen, M.K., Hagedorn, P.H., de Torres-Zabala, M., Grant, M.R., Rung, J.H., Collinge, D.B., Lyngkjaer, M.F., Transcriptional regulation by an NAC (NAM-ATAF1;2-CUC2) transcription factor attenuates ABA signalling for efficient

basal defence towards Blumeria graminis f. sp.hordei in Arabidopsis. *Plant J.*, 56, 867–880, 2008.

82. Kim, S.Y., Kim, S.G., Kim, Y.S., Seo, P.J., Bae, M., Yoon, H.K., Park, C.M., Exploring membrane-associated NAC transcription factors in Arabidopsis: Implications for membrane biology in genome regulation. *Nucleic Acids Res.*, 35, 203–213, 2007.
83. Kim, T.H., Hauser, F., Ha, T., Xue, S., Bohmer, M., Nishimura, N., Munemasa, S., Hubbard, K., Peine, N., Lee, B., Chemical genetics reveals negative regulation of abscisic acid signalingby a plant immune response pathway. *Curr. Biol.*, 21, 990–997, 2011.
84. Kissoudis, C., van de Wiel, C., Visser, R.G.F., van der Linden, G., Enhancing crop resilience to combined abiotic and biotic stress through the dissection of physiological and molecular crosstalk. *Front. Plant Sci.*, 5, e207, 2014.
85. Klessig, D.F., Choi, H.W., Dempsey, D.M.A., Systemic acquired resistance and salicylic acid: Past, present, and future. *Mol. Plant Microbe Interact.*, 31, 9, 871–888, 2018.
86. Koul, O., Microbial biopesticides: Opportunities and challenges. *CAB Rev. Perspect. Agric. Vet. Sci. Nutr. Nat. Resour.*, 6, 1–26, 2011.
87. Kumar, M., Prasanna, R., Bidyarani, N., Babu, S., Mishra, B.K., Kumar, A., Adak, A., Jauhari, S., Yadav, K., Singh, R., Evaluating the plant growth promoting ability of thermotolerant bacteria and cyanobacteria and their interactions with seed spice crops. *Sci. Hortic.*, 164, 94–101, 2013.
88. Kumar, S., Chandra, R., Behera, L., Keswani, C., Sansinenea, E., Dual Trichoderma consortium mediated elevation of systemic defense response against early blight in potato. *Eur. J. Plant Pathol.*, 162, 681–696, 2021.
89. Kumari, P., Meena, M., Gupta, P., Dubey, M.K., Nath, G., Upadhyay, R.S., Plant growth promoting rhizobacteria and their biopriming for growth promotion in mung bean (Vigna radiata (L.) R Wilczek). *Biocatal. Agric. Biotechnol.*, 16, 163–171, 2018.
90. Kumari, P., Meena, M., Upadhyay., R.S., Characterization of plant growth promoting rhizobacteria (PGPR) isolated from the rhizosphere of Vigna radiata (mung bean). *Biocatal. Agric. Biotechnol.*, 16, 155–162, 2018.
91. Li, Y., Zhang, L., Wang, X., Hao, L., Chu, X., Guo, X., Cotton GhMPK6a negatively regulates osmotis tolerance and bacterial infection in transgenic Nicotiana benthaminana and plays apivotal role in development. *FEBS J.*, 280, 5128–5144, 2013.
92. Liu, C., Ruan, Y., Lin, Z., Wei, R., Peng, Q., Guan, C., Ishii, H., Antagonism between acibenzolar-S-methyl-induced systemic acquired resistance and jasmonic acid-induced systemic acquired susceptibility to Colletotrichum orbiculare infection in cucumber. *Physiol. Mol. Plant Pathol.*, 72, 141–145, 2008.
93. Lucas, J.A., García-Cristobal, J., Bonilla, A., Ramos, B., Gutierrez-Manero, J., Beneficial rhizobacteria from rice rhizosphere confers high protection

against biotic and abiotic stress inducing systemic resistance in rice seedlings. *Plant Physiol. Biochem.*, 82, 44–53, 2014.
94. Luna, E., Pastor, V., Robert, J., Flors, V., Mauch-Mani, B., Ton, J., Callose deposition: A multifaceted plant defense response. *Mol. Plant Microbe Interact.*, 24, 183–193, 2011.
95. Marco, F., Altabella, T., Alcázar, R., Cuevas, J.C., Bortolotti, C., González, M.E., Carrasco, P., Transcriptome analysis of polyamine overproducers reveals activation of plant stress responses and related signalling pathways tolerance in plants, in: *Omics and Plant Abiotic Stress Tolerance*, vol. 1, pp. 82–90, 2011.
96. Mauch-Mani, B. and Mauch, F., The role of abscisic acid in plant-pathogen interactions. *Curr. Opin. Plant Biol.*, 8, 409–414, 2005.
97. Meena, M., Swapnil, P., Zehra, A., Aamir, M., Dubey, M.K., Patel, C.B., Upadhyay, R.S., Virulence factors and their associated genes in microbes, in: *New and Future Developments in Microbial Biotechnology and Bioengineering*, H.B. Singh, V.K. Gupta, S. Jogaiah (Eds.), pp. 181–208, Elsevier, Amsterdam, United State, 2019.
98. Meena, M., Swapnil, P., Divyanshu, K., Kumar, S., Harish, Y.N., Tripathi, A., Zehra, A., Marwal, R., Upadhyay, S., PGPR-mediated induction of systemic resistance and physiochemical alterations in plants against the pathogens: Current perspectives. *J. Basic Microbiol.*, 60, 10, 828–861, 2020.
99. Meena, M., Gupta, S.K., Swapnil, P., Zehra, A., Dubey, M.K., Upadhyay., R.S., Alternaria toxins: Potential virulence factors and genes related to pathogenesis. *Front. Microbiol.*, 8, 1451, 2017.
100. Melotto, M., Underwood, W., Koczan, J., Nomura, K., He, S.Y., Plant stomata function in innate immunity against bacterial invasion. *Cell.*, 126, 969–980, 2006.
101. Meng, L., Wong, J.H., Feldman, L.J., Lemaux, P.G., Buchanan, B.B.A., Membrane-associated thioredoxin required for plant growth moves from cell to cell; suggestive of a role in intercellular communication. *Proc. Natl. Acad. Sci. U. S. A*, 107, 3900–3905, 2010.
102. Mhlongo, M.I., Piater, L.A., Madala, N.E., Labuschagne, N., Dubery, A., The chemistry of plant–microbe interactions in the rhizosphere and the potential for metabolomics to reveal signaling related to defense priming and induced systemic resistance. *Front. Plant Sci.*, 9, 112, 2018.
103. Miller, G., Suzuki, N., Ciftci-Yilmaz, S., Mittler, R., Reactive oxygen species homeostasis and signalling during drought and salinity stresses. *Plant Cell Environ.*, 33, 453–467, 2010.
104. Mittler, R., Vanderauwera, S., Suzuki, N.N., Miller, G., Tognetti, V.B., Vandepoele, K., Gollery, M., Shulaev, V., van Breusegem, F., ROS signaling: The new wave? *Trends Plant Sci.*, 16, 1360–1385, 2011.
105. Mordue, A.J., Morgan, E.D., Nisbet, A.J., Azadirachtin, a natural product in insect control, in: *Comprehensive Molecular Insect Science*, L.I. Gilbert (Ed.), pp. 117–135, Elsevier, Amsterdam, The Netherlands, 2005.

106. Mossa, A.-T.H., Green pesticides: Essential oils as biopesticides in insect-pest management. *J. Environ. Sci. Technol.*, 9, 5, 354, 2016.
107. Moustafa-Farag, M., Almoneafy, A., Mahmoud, A., Elkelish, A., Arnao, M.B., Li, L., Ai, S., Melatonin and its protective role against biotic stress impacts on plants. *Biomolecules*, 10, 1, 54, 2019.
108. Mukerji, K.G. and Ciancio, A., Mycorrhizae in the integrated pest and disease management, in: *General Concepts in Integrated Pest and Disease Management. Integrated Management of Plants Pests and Diseases*, vol. 1, Springer, Dordrecht, Switzerland, 2007.
109. Munir, N., Hasnain, M., Waqif, H., Adetuyi, B.O., Egbuna, C., Olisah, M.C., Gelling agents, micro and nanogels in food system applications, in: *Application of Nanotechnology in Food Science, Processing and Packaging*, pp. 153–167, 2022.
110. Nakabayashi, R., Yonekura-Sakakibara, K., Urano, K., Suzuki, M., Yamada, Y., Nishizawa, T., Matsuda, F., Kojima, M., Sakakibara, H., Shinozaki, K., Enhancement of oxidative and drought tolerance in Arabidopsis by over accumulation of antioxidant flavonoids. *Plant J.*, 77, 367–379, 2014.
111. Narware, J., Yadav, R.N., Keswani, C., Singh, S.P., Singh, H.B., Silver nanoparticle-based biopesticides for phytopathogens: Scope and potential in agriculture, in: *Nano-Biopesticides Today and Future Perspectives*, pp. 303–314, Academic Press, Cambridge, MA, USA, 2019.
112. Nazir, A., Itrat, N., Shahid, A., Mushtaq, Z., Abdulrahman, S.A., Egbuna, C., Adetuyi, B.O., Khan, J., Uche, C.Z., Toloyai, P.Y., Orange peel as a source of nutraceuticals, in: *Food and Agricultural Byproducts as Important Source of Valuable Nutraceuticals*, 1st ed, p. 400, Springer, Berlin, 2022.
113. Nishimura, M.T., Stein, M., Hou, B.H., Vogel, J.P., Edwards, H., Somerville, S.C., Loss of a callose synthase results in salicylic acid-dependent disease resistance. *Science*, 301, 969–972, 2003.
114. Nuzhat, T. and Vidyasagar, G., Antifungal investigations on plant essential oils. A review. *Int. J. Pharm. Pharm. Sci.*, 2, 19–28, 2014.
115. Ogunlana, O.O., Adetuyi, B.O., Adekunbi, T.S., Adegboye, B.E., Iheagwam, F.N., Ogunlana, O.E., Ruzu bitters ameliorates high–fat diet induced non-alcoholic fatty liver disease in male wistar rats. *J. Pharm. Pharmacogn. Res.*, 9, 3, 251–26, 2021.
116. Ogunlana, O.O., Adetuyi, B.O., Esalomi, E.F., Rotimi, M.I., Popoola, J.O., Ogunlana, O.E., Adetuyi, O.A., Antidiabetic and antioxidant activities of the twigs of Andrograhis paniculata on Streptozotocin-induced diabetic male rats. *BioChem*, 1, 3, 238–249, 2021.
117. Ogunlana, O.O., Ogunlana, O.E., Adekunbi, T.S., Adetuyi, B.O., Adegboye, B.E., Iheagwam, F.N., Anti-inflammatory mechanism of Ruzu bitters on diet-induced nonalcoholic fatty liver disease in male wistar rats. *Evid. Based Complement. Altern. Med.*, 1, 1–8, 2020 2020, https://doi.org/10.1155/2020/5246725.

118. Ogunlana, O.O., Ogunlana, O.E., Popoola, J.O., Adetuyi, B.O., Adekunbi, T.S., David, O.L., Adeleye, O.J., Udeogu, S.A., Adeyemi, A.O., Twigs of Andrographis paniculata (Burn. F) Nees attenuates carbon tetrachloride (CCl4) induced liver damage in wistar albino rats. *RASAYAN J. Chem.*, 14, 4, 2598–2603, 2021.
119. Oh, S.J., Song, S.I., Kim, Y.S., Jang, H.J., Kim, S.Y., Kim, M., Kim, Y.K., Nahm, B.H., Kim, J.K., Arabidopsis CBF3/DREB1A and ABF3 in transgenic rice increased tolerance to abiotic stress without stunting growth. *Plant Physiol.*, 138, 341–351, 2005.
120. Olowe, O.M., Olawuyi, O.J., Sobowale, A.A., Odebode, A.C., Role of arbuscular mycorrhizal fungi as biocontrol agents against Fusarium verticillioides causing ear rot of Zea mays L. (Maize). *Curr. Plant Biol.*, 15, 30–37, 2018.
121. Parker, K.M., Barragán, B.V., van Leeuwen, D.M., Lever, M.A., Mateescu, B., Sander, M., Environmental fate of RNA interference pesticides: Adsorption and degradation of double-stranded RNA molecules in agricultural soils. *Environ. Sci. Technol.*, 53, 3027–3036, 2019.
122. Parween, T. and Jan, S., Pesticides and environmental ecology, in: *Ecophysiology of Pesticides*, pp. 1–38, Academic Press, Cambridge, MA, USA, 2019.
123. Pastori, G.M. and Foyer, C.H., Common components, networks, and oathways of cress-tolerance to stress. The central role of "redox" and abscisic acid-mediated controls. *Plant Physiol.*, 129, 460–468, 2002.
124. Pastori, G.M. and Foyer, C.H., Common components, networks, and oathways of cress-tolerance to stress. The central role of "redox" and abscisic acid-mediated controls. *Plant Physiol.*, 129, 460–468, 2002.
125. Pitzschke, A., Djamel, A., Teige, M., Hirt, H., VIP1 response elements mediate mitogen-activatedprotein kinase 3-induced stress gene expression. *Proc. Natl. Acad. Sci. U. S. A.*, 106, 18414–18419, 2009.
126. Prime-A-Plant Group, Conrath, U., Beckers, G.J., Flors, V., García-Agustín, P., Jakab, G., Mauch, F., Newman, M.A., Pieterse, C.M., Poinssot, B., Priming: Getting ready for battle. *Mol. Plant Microbe Interact.*, 19, 1062–1071, 2006.
127. Rajput, R.S., Ram, R.M., Vaishnav, A., Singhn, H.B., Microbe-based novel biostimulants for sustainable crop production, in: *Microbial Diversity in Ecosystem Sustainability and Biotechnological Applications*, vol. 2, T. Satyanarayana, S.K. Das, B.N. Johri (Eds.), pp. 109–144, Soil & Agroecosystems, Springer, Singapore, 2019.
128. Ranglová, K., Lakatos, G.E., Câmara Manoel, J.A., Grivalský, T., Suárez Estrella, F., Acién Fernández, F.G., Molnár, Z., Ördög, V., Masojídek, J., Growth, biostimulant and biopesticide activity of the MACC-1 Chlorella strain cultivated outdoors in inorganic medium and wastewater. *Algal Res.*, 53, 102136, 2021.
129. Reglinski, T. and Walters., D., Induced resistance for plant disease control, in: *Disease Control in Crops: Biological and Environmentally Friendly Approaches*,

D. Walters (Ed.), p. 62–92, John Wiley & Sons, Blackwell Publishing Ltd, Hoboken, New Jersey, United States, 2009.

130. Robert-Seilaniantz, A., Navarro, L., Bari, R., Lones, J.D.G., Pathological hormone imbalances. *Curr. Opin. Plant Biol.*, 10, 372–379, 2007.

131. Romera, F.J., García, M.J., Lucena, C., Martínez Medina, A., Aparicio, M.A., Ramos, J., Alcántara, E., Angulo, M., Pérez-Vicente, R., Induced systemic resistance (ISR) and Fe deficiency responses in dicot plants. *Front. Plant Sci.*, 10, 287, 2019.

132. Rudrappa, T., Biedrzycki, M.L., Kunjeti, S.G., Donofrio, N.M., Czymmek, K.J., Paul, W.P., Bais, H.P., The rhizobacterial elicitor acetoin induces systemic resistance in Arabidopsis thaliana. *Commun. Integr. Biol.*, 3, 2, 130–138, 2010.

133. Saberi, F., Marzban, R., Ardjmand, M., Pajoum, S.F., Tavakoli, O., Optimization of culture media to enhance the ability of local Bacillus thuringiensis var. tenebrionis. *J. Saudi Soc. Agric. Sci.*, 19, 468–475, 2020.

134. Schorderet, W.S., Kaminski, K.P., Perret, J.-L., Leroy, P., Mazurov, A., Peitsch, M.C., Ivanov, N.V., Hoeng, J., Antiparasitic properties of leaf extracts derived from selected Nicotiana species and Nicotiana tabacum varieties. *Food Chem. Toxicol.*, 132, 110660, 2019.

135. Schuwirth, B.S., Day, J.M., Hau, C.W., Janssen, G.R., Dahlberg, A.E., Cate, J.H., Vila-Sanjurjo, A., Structural analysis of kasugamycin inhibition of translation. *Nat. Struct. Mol. Biol.*, 13, 879–886, 2006.

136. Segarra, G., Van der Ent, S., Trillas, I., Pieterse, C.M.J., MYB72, a node of convergence in induced systemic resistance triggered by a fungal and a bacterial beneficial microbe. *Plant Biol.*, 11, 1, 90–96, 2019.

137. Seo, P.J. and Park, C.M., MYB96-mediated abscisic acid signals induce pathogen resistance response by promoting salicylic acid biosynthesis in Arabidopsis. *New Phythol.*, 186, 471–483, 2010.

138. Seo, P.J., Kim, M.J., Park, J.Y., Kim, S.Y., Jeon, J., Lee, Y.H., Kim, J., Park, C.M., Cold activation of a plasma membrane-tethered NAC transcription factor induces a pathogen resistance response in Arabidopsis. *Plant J.*, 61, 661–671, 2010.

139. Shen, Y., Li, J., Xiang, J., Wang, J., Yin, K., Liu., Q., Isolation and identification of a novel protein elicitor from a Bacillus subtilis strain BU412. *AMB Express*, 9, 1, 117, 2019.

140. Singh, H.B. and Vaishnav, A., New and future developments in microbial biotechnology and bioengineering, in: *Sustainable Agriculture: Revitalization through Organic Products*, pp. 1–366, Elsevier, Amsterdam, United State, 2022.

141. Singh, H.B. and Vaishnav, A., New and future developments in microbial biotechnology and bioengineering, in: *Sustainable Agriculture: Advances in Microbe-Based Biostimulants*, pp. 1–374, Elsevier, Amsterdam, United State, 2022.

142. Snider, C.S., Hsiang, T., Zhao, G., Griffith, M., Role of ice nucleation and antifreeze activities in pathogenesis and growth of snow molds. *Phytopathology*, 90, 354–361, 2000.
143. Spoel, S.H. and Loake, G.J., Redox-based protein modifications: The missing link in plant immune signalling. *Curr. Opin. Plant Biol.*, 14, 358–364, 2011.
144. Svidritskiy, E., Ling, C., Ermolenko, D.N., Korostelev, A.A., Blasticidin S inhibits translation by trapping deformed tRNA on the ribosome. *Proc. Natl. Acad. Sci. U. S. A.*, 110, 12283–12288, 2013.
145. Swindell, W.R., The association among gene expression responses to nine abiotic stress treatments in Arabidopsis thaliana. *Genet. Soc. Am.* 174, 1811–1824, 2003, 2006.
146. Teige, M., Scheikl, E., Eulgem, T., Roczi, F., Ichimura, K., Shinozaki, K., Dangl, J.L., Hirt, H., The MKK2 pathway mediates cold and salt stress signaling in Arabidopsis. *Mol. Cell*, 15, 141–152, 2004.
147. Ton, J., Ent, V.D.S., Hulten, V.M., Pozo, M., Oosten, V.V., Loon, L.C.V., Mauch-Mani, B., Turlings, T.C.J., Pieterse, C.M.J., Priming as a mechanism behind induced resistance against pathogens; insects and abiotic stress. *IOBC/wprs Bull.*, 44, 3–13, 2009.
148. Ton, J., Flors, V., Mauch-Mani, B., The multifaceted role of ABA in disease resistance. *Trends Plant Sci.*, 14, 310–317, 2009.
149. Ton, J., Jakab, G., Toquin, V., Flors, V., Iavicoli, A., Maeder, M.N., Metraux, J.P., Mauch-Mani, B., Dissecting the β-aminobutyric acid induced priming phenomenon in Arabidopsis. *Plant Cell.*, 17, 987–999, 2005.
150. Treutter, D., Significance of flavonoids in plant resistance: A review. *Environ. Chem. Lett.*, 4, 147–157, 2006.
151. Tripathi, A.K., Upadhyay, S., Bhuiyan, M., Bhattacharya, P.R., A review on prospects of essential oils as biopesticide in insect-pest management. *Pharmacogn. Phytother.*, 1, 52–63, 2009.
152. Tsutsui, T., Kato, W., Asada, Y., Sako, K., Sato, T., Sonoda, Y., Kidokoro, S., Yamaguchi-Shinozaki, K., Tamaoki, M., Arakawa, K., DEAR1; a transcriptional repressor of DREB protein thatmediates plant defense and freezing stress responses in Arabidopsis. *J. Plant Res.*, 122, 633–643, 2009.
153. Vallad, G.E. and Goodman, R.M., Systemic acquired resistance and induced systemic resistance in conventional agriculture. *Crop Sci.*, 44, 6, 1920–1934, 2004.
154. Venegas-Molina, J., Proietti, S., Pollier, J., Orozco-Freire, W., Ramirez-Villacis, D., Leon-Reyes, A., Induced tolerance to abiotic and biotic stresses of broccoli and Arabidopsis after treatment with elicitor molecules. *Sci. Rep.*, 10, 1, 1–17, 2020.
155. Walters, D., Resistance to plant pathogens: Possible roles for free polyamines and polyamine catabolism. *New Phytol.*, 159, 109–115, 2003.
156. Wang, D., Liu, B., Ma, Z., Feng, J., Yan, H., Reticine, A., A new potent natural elicitor: Isolation from the fruit peel of Citrus reticulate and induction

of systemic resistance against tobacco mosaic virus and other plant fungal diseases. *Pest Manag. Sci.*, 77, 354–364, 2021.

157. Xia, X.J., Wang, Y.J., Zhou, Y.H., Tao, Y., Mao, W.H., Shi, K., Asami, T., Chen, Z., Yu, J.Q., Reactive oxygen species are involved in brassinosteroid-induced stress tolerance in cucumber. *Plant Physiol.*, 158, 1034–1045, 2012.
158. Xiong, L., Schumaker, K.S., Zhu, J.K., Cell signaling during cold; drought; and salt stress. *Plant Cell*, 14, 165–183, 2002.
159. Xiong, L. and Yang, Y., Disease resistance and abiotic stress tolerance in rice are inversely modulated by an abscisic acid-inducible mitogen-activated protein kinase. *Plant Cell*, 15, 745–759, 2003.
160. Zaim, S., Belabid, L., Bayaa, B., Bekkar, A.A., Biological control of chickpea fusarium wilts using rhizobacteria "PGPR", in: *Microbial-Mediated Induced Systemic Resistance in Plants*, D.K. Choudhary and A. Varma (Eds.), pp. 147–162, Springer, New York, NY, USA, 2016.
161. Zehra, A., Meena, M., Dubey, M.K., Aamir, M., Upadhyay, R.S., Synergistic effects of plant defense elicitors and Trichoderma harzianum on enhanced induction of antioxidant defense system in tomato against Fusarium wilt disease. *Bot. Stud.*, 58, 44, 2017.
162. Zhang, D., Martyniuk, C.J., Trudeau, V.L., SANTA domain: A novel conserved protein module in Eukaryote with potential involvement in chromatin regulation. *Bioinformatics*, 22, 20, 2459-62, 2006.
163. Zhang, S. and Liu, Y., Activation of salicylic acid-induced protein kinase; a mitogen-activated protein kinase; induces multiple defense responses in tobacco. *Plant Cell.*, 13, 1877–1889, 2001.
164. Zhang, Y., Butelli, E., de Stefano, R., Schoonbeek, H.J., Magusin, A., Pagliarani, C., Wellner, N., Hill, L., Orzaez, D., Granell, A., Anthocyanins double the shelf life of tomatoes by delaying over ripening and reducing susceptibility to graymold. *Curr. Biol.*, 23, 1094–1100, 2013.
165. Zhang, L. and Gleason, C., Enhancing potato resistance against root-knot nematodes using a plant-defence elicitor delivered by bacteria. *Nat. Plants*, 6, 6, 625–629, 2020.
166. Zhou, J., Xia, X.J., Zhou, Y.H., Shi, K., Chen, Z., Yu, J.Q., RBOH1-dependent H2O2 production and subsequent activation of MPK1/2 play an important role in acclimation-induced cross-tolerance in tomato. *J. Exp. Bot.*, 65, 595–607, 2014.
167. Adetunji, C.O., Titilayo, O., Mathew, J.T., Inobeme, A., Singh, K.R.B., Olaniyan, O.T., Singh, J., Singh, R.P., *Carbon dots: Next-generation materials for biomedical applications*, pp. 8–12, IOP Publishing, Bristol, England, 2022.
168. Farid, A., Muzammal, M., Ghazanfar, S., Ismail, S., Ullah, A., Dauda, W.P., Adetunji, C.O., Shah, M., Dadrasnia, A., Idrees, M. *et al.*, Green cleaning activity of *Bacillus salmalaya* 139Sl: A novel strain for removing common household stains. *Biomass Convers. Biorefin.*, 12, 1–10, 2022, https://doi.org/10.1007/s13399-022-03147-z.

169. Wali, S., Zahra, M., Okla, M.K., Wahidah, H.A., Tauseef, I., Haleem, K.S., Farid, A., Maryam, A., AbdElgawad, H., Adetunji, C.O., Akhtar, N., Akbar, S., Rehman, W., Yasir, H., Shakira, G., Brassica oleracea L. (Acephala Group) based zinc oxide nanoparticles and their efficacy as antibacterial agent. *Braz. J. Biol.*, 84, 1–8, 2022, https://doi.org/10.1590/1519-6984.259351.
170. Adetunji, C.O., Mathew, J.T., Singh, K.R.B., Inobeme, A., Nayak, V., Olaniyan, O.T., Singh, J., Singh, R.P., Zero-dimensional carbon nanomaterials for targeted drug delivery and photothermal therapy applications, in: *Zero-Dimensional Carbon Nanomaterials: Fundamentals and Applications*, pp. 1–10, IOP Publishing, Bristol, England, 2022.
171. Adetunji, C.O., Mathew, J.T., Singh, K.R.B., Inobeme, A., Nayak, V., Olaniyan, O.T., Singh, J., Singh, R.P., Zero-dimensional carbon nanomaterials for targeted drug delivery and photothermal therapy applications, in: *Zero-Dimensional Carbon Nanomaterials: Fundamentals and Applications*, pp. 1–12, IOP Publishing, Bristol, England, 2022.
172. Adetunji, C.O., Michael, O.S., Nwankwo, W., Anani, O.A., Adetunji, J.B., Olayinka, A.S., Akram, M., Biogenic nanoparticles based drugs derived from medicinal plants: A sustainable panacea for the treatment of malaria, in: *Green Synthesis in Nanomedicine and Human Health*, Boca Raton, Florida, United State, pp. 103–122, CRC Press, 2021.
173. Adetunji, O., Michael, O.S., Akram, M., Oseni, K., Anani, O.A., Nwankwo, W., Anwar, H., Adetunji, J.B., Olayinka, A.S., Bottom of formrecent advances in the utilization of bioengineered plant-based nanoparticles: A sustainable nanobiotechnology for the management of extensively drug-resistant tuberculosis, in: *Green Synthesis in Nanomedicine and Human Health*, pp. 149–166, CRC Press, Boca Raton, Florida, United State, 2021.
174. Adetunji, C.O., Michael, O.S., Akram, M., Oseni, K., Temidayo, A.K., Anani, O.A., Olayinka, A.S., Nwankwo, W., Ghaffar, I., Adetunji, J.B., *Green synthesis in nanomedicine and human health*, pp. 81–100, CRC Press, Boca Raton, Florida, United State, 2021.
175. Islam, S., Thangadurai, D., Adetunji, C.O., Michael, O.S., Nwankwo, W., Kadiri, O., Anani, O.A., Makinde, S., Adetunji, J.B., Nanomaterials: Applications in biomedicine and biotechnology, in: *Handbook of Nanomaterials and Nanocomposites for Energy and Environmental Applications*, O.V. Kharissova, L.M. Torres-Martínez, B.I. Kharisov (Eds.), Springer, Cham, 2021, https://doi.org/10.1007/978-3-030-36268-3_4.
176. Thangadurai, D., Prakash, L., Sangeetha, J., Mohammad Said Al-Tawaha, A.R., David, M., Islam, S., Adetunji, C.O., Adetunji, J.B., Sustainable synthesis of greener nanomaterials: Principles, processes, and products, in: *Handbook of Nanomaterials and Nanocomposites for Energy and Environmental Applications*, O.V. Kharissova, L.M.T. Martínez, B.I. Kharisov (Eds.), Springer, Cham, 2021, https://doi.org/10.1007/978-3-030-11155-7_30-1.

177. Adetunji, C.O. and Varma, A., Biotechnological application of Trichoderma: A powerful fungal isolate with diverse potentials for the attainment of food safety, management of pest and diseases, healthy planet, and sustainable agriculture, in: *Trichoderma: Agricultural Applications and Beyond*, Soil Biology, vol. 61, C. Manoharachary, H.B. Singh, A. Varma (Eds.), Springer, Cham, 2020, https://doi.org/10.1007/978-3-030-54758-5_12.

10

The Influence of Nanopesticides on the Social Economy, Its Bioeconomy Perspectives in Attaining Sustainable Development Goals

Abere Benjamin Olusola[1]* and Charles Oluwaseun Adetunji[2]

[1]Department of Economics Edo State University Uzairue, Iyamho – Uzairue Edo State, Nigeria
[2]Applied Microbiology, Biotechnology and Nanotechnology Laboratory, Department of Microbiology, Edo University Iyamho, Auchi, Edo State, Nigeria

Abstract
The goal of this research was to examine how nanopesticides impacted the social economy and bioeconomy in terms of achieving socioeconomic development objectives. Plant protection using nanopesticides is a new study topic that brings up new possibilities for producing nanoscale active chemicals. In the advancement of other practical and safer insecticide/biopesticides, nanopesticide-based formulations have a promising future. Nanotechnology research has resulted in the development of many nanoformulations for crop protection, including nanoinsecticides, nanoherbicides, nanofungicides, and nanonematicides. Nanopesticides are made to increase solubility, restrict active ingredient release, and prevent degradation, among other things. Other chemical modifications to the carrier molecule been categorized as organic polymer-based formulations, lipid-based formulations, nanosized metals, and metal oxides, clay-based nanomaterials, and so on. On this page, we will go through some of the most significant nanoformulations. The effective use of nanopesticides in agriculture has been discovered to solve problems, such as rising global food demand, climate change, hunger and malnourishment, food loss and waste, biodiversity loss, and overexploitation of common assets, all of which are posing a fresh threat to world's agriculture food production scheme, as stated in the sustainable development goals.

**Corresponding author*: abere.benjamin@edouniversity.edu.ng

Charles Oluwaseun Adetunji and Julius Kola Oloke (eds.) Handbook of Agricultural Biotechnology: Volume I Nanopesticides, (279–292) © 2024 Scrivener Publishing LLC

Keywords: Nanopesticides, social economy, bioeconomy perspectives, sustainable development goals

10.1 Introduction

Nanopesticides similar to nanoemulsions, nanocapsules (e.g., containing polymers), and inorganic finagled nanoparticles (ENPs, for illustration, essence, essence oxides, and nanoclays) have been more prognosticated for the agrochemical business due to the rapid-fire growth of nanotechnologies [19]. Nanopesticides containing nanoscale active constituents (AIs) have broad-diapason insecticidal, fungicidal, and herbicidal characteristics [35]. Also, nanopesticides can ameliorate AI solubility, release operation, splint adherence, and stability, performing in increased efficacy and lifetime as well as a drop in the volume of AIs used. As a result, nanopesticides can be used to maintain or boost agrarian productivity while demanding reduced operation rates, potentially reducing environmental enterprises [12] likewise, the worldwide fungicide assiduity is anticipated to rise from$ 75 billion in 2013 to$ 90 billion in 2023, with nanopesticides having the eventuality to give multibillion-bone advantages [7].

Despite this, only many marketable synthetic nanopesticides have been produced (e.g., Dupont's Kocide 3000 and Vive Crop Protection's AZteroid FC) due to three crucial technological challenges (i.e., effective conveyance at the field scale, regulation and safety enterprises, and consumer acceptance) [1]. The study of nanopesticides is presently generally concentrated on their styles of action, environmental fortune, and ecological consequences [20]. Still, there are significant knowledge gaps in nanogovernance and socioprofitable aspects, especially in public stations and amenability to pay (WTP) for nanopesticides, which are both critical in adding nanopesticides use and request share [10]. Large tablets per unit crop, drift hazards, functional troubles, remainders in the terrain, shops, and marketable food, as well as on-target foliage and nontarget species, are all downsides of traditional fungicides [13]. As a result, they must be replaced with a pest operation plan able of prostrating the forenamed restrictions. Traditional germicides have faults, which nanofungicides can help to overcome [21].

Nanofungicides are factory protection products that use nanotechnology to induce the active element or carrier patch [15]. The major thing of nanofungicide development is to reduce the environmental troubles associated with a fungicide's active element while adding the chemical's efficacity [17]. In this script, the advantage is the extremely bitsy flyspeck size. One nanometer equals one billionth of a cadence, while a nanoparticle can be 1 to 100 nm in size [9]. Even though these patches are bitsy, they have a big face area, which means further germicides are exposed to the bugs. Nanoparticles can access due to their extremely bitsy size and structure [2]. In nanopesticides phrasings, the active element will be present, exactly as its in traditional fungicide phrasings. Surfactants, as well as the carrier patch, these nanoparticles have lesser solubility of active constituents, increased expression stability, progressive active component release, and bettered mobility due to their lower flyspeck size and larger face area [22]. Nanoparticles, rather than bulk chemicals, are anticipated to enhance the medium of action against pests. Systemic parcels, invariant splint content, and enhanced soil rates are all features of nanophrasings that contribute to their agrarian use [34].

10.2 Literature Review

Pesticides are often regarded as the most effective way to manage insect pests, diseases, and weeds. Because of their simple availability, potency, and convenience of administration, pesticides are often utilized in agricultural circumstances to boost crop output and quality [3]. Agriculture now faces considerable issues such as environmental pollution, insect resistance, bioaccumulation, and health hazards as a result of extensive pesticide usage, all of which must be addressed and for which swift answers are required [25]. One approach to do this is to reduce the number of pesticides sprayed on crop or used in the preservation of a stockpiled product, for example. In this sense, nanotechnology is proving to be an enticing tool for achieving the goal of reducing pesticide use by providing new means to synthesize and transport active chemicals in pesticides, as well as novel active compounds known as nanopesticides [14].

Nanotechnologies concern information on a Nanoscale and have shown to be very promising in the development of novel pest control approaches [30]. The constraints of the present pest will be transcended by nanotechnologies. Increasing the active component's stability, lowering the needed pesticide dose, and conserving agri-inputs [16]. Modern nanobased formulations for specific targets should be stable, active in a variety of conditions, easy to make, manufacture, and rather have a unique mechanism of activity [16]. Due to the low toxicity of nanopesticides formulations, good biocompatibility, and increased bioavailability, their usage is limited since the usage of nanotechnologies or nanoparticles (NPs) necessitates research on the potential risks of soil and water pollution [28]. We need to build nanoformulations with certain features, such as early breakdown without harmful residues and precise active ingredient release, to maximize the potential advantages by making insecticides more potent [31].

Nanoformulations, unlike traditional pesticide formulations, are meant to enhance the solubility of intractable or deficiently soluble agile chemicals while also allowing for manage and targeted biocide release [11]. As a result, a lesser quantity of active ingredients per area may be appropriate for the application, and active ingredient delivery may be endured for long periods. Because of the decreased dose, production costs, nontarget impacts, and phytotoxicity are all minimized [29]. In addition, until the active component is released, controlled-release formulations must remain inert. The most often manufactured controlled-release formulations include nanocapsules, nanospheres, nanogels, and micelles [27]. A mix of physical and chemical procedures is used to create them. The use of a polymer matrix for nanoencapsulation can improve the diffusion of hydrophobic active components in aqueous results, permitting for guarded discharge with great choosiness while preserving biocidal action [26].

The active component's release profile is influenced by the chemical characteristics of the polymeric matrix, the firmness of chemical connections, and the amount of the biocide molecule [43–51]. When the polymer containing the active component comes into touch with water and receives the necessary stimulus, it undergoes diffusion or dissolution [34]. Encapsulation high technology is extensively employed in agricultural operations because hydrophobic and hydrophilic bioactive substances may be contained in controlled release formulations made with encapsulation technology, such as liposomes and micelles, formed from lecithin [4]. Pesticide reduction, strengthening the firmness of unstable basic materials, lowering the strong odors of unrestricted chemicals, and assuring biocompatibility with carrier systems are all possible with these [38]. Because of its biodegradability, nontoxicity, and adsorption capabilities, chitosan is anticipated to be a useful carrier for regulated distribution. Encapsulated active chemicals can be discharged in controlled amounts while staying safe from the environment by using a chitosan matrix as a protected tank [36]. Controlled release minimized evaporation, degradation, and leaching losses [19]. Polymer-encapsulated nanoformulations have several advantages over normal formulations, including the longer activity of active compounds with a short half-life. Although because nanoformulations have significantly different qualities than standard bulk pesticides, rising health risks ranging from inhalation to skin penetration are currently being debated [4].

10.3 Nanopesticides Categories

Depending on their intended purpose, pesticide nanoformulations are classified as (1) formulations that expand the solubility of water-insoluble active constituents, (2) formulations that moderate the discharge amount of active constituents, and (3) formulations that can attain targeted conveyance and increased chemical firmness [20]. Nanoformulations founded on the chemical creation of nanocarriers, for example, are polymer-based preparations. Lipid-based formulations, nanosized metals and metal oxides, clay-based nanomaterials, silica nanoparticles, and other subjects are discussed on this page [33]. Nanoemulsions have several advantages, including bioavailability, chemical stability, and controlled release methods. Nanoemulsions can be classed as (1) thermodynamically steady nanoemulsions or (2) kinetically stable nanoemulsions, depending on the amount and kind of surfactant utilized [39]. When a nonpolar insecticide is partly dissolved in the aqueous stage and a highly dispersing surfactant is existent in proportions greater than critical micelle concentrations, thermodynamically solid nanoemulsions are generated (CMC). Because of their chemical properties and ease of manufacture, they are frequently recommended for commercial formulations [23]. These nanoemulsions, however, are not suited for application since they require a substantial amount of surfactant and have restricted micelle integration. They also have a variety of drawbacks, including high cost and phytotoxicity [18]. Kinetically steady nanoemulsions are created once the pesticide is nearly irresolvable in the aqueous stage and the surfactant molecules' aggregation into micelles is reduced owed to the feebly resisting surface-active agent [24]. The key benefit is that a larger spectrum of surface-active agent may be used but the main disadvantage is that the pesticide is still present.

10.4 Formulations of Nanopesticides

Nanotechnology research has resulted in the development of many nanoformulations for crop protection, including nanoinsecticides, nanoherbicides, nanofungicides, and nanonematicides. Nanopesticides are made to increase solubility, restrict active ingredient release, and prevent degradation, among other things [32]. Organic polymer-based formulations, lipid-based formulations, nanosized metals, and metal oxides, clay-based nanomaterials, and other chemical nature carrier molecule alterations have all been made to achieve these goals [51]. This article discusses some of the most important nanoformulations. Nanoemulsions are oil-in-water (O/W) mixtures in which the chemical's active component is spread as nanosized vapor in water, with surface-active agent molecules limited to the pesticide-water confluence [39]. Nanoemulsions are characterized as thermodynamically or kinetically stable depending on the amount and kind of surfactants used. Because the insecticide is partly soluble in the aqueous stage, a thermodynamically stable nanoemulsion arises when the surfactant, pesticide, and water constituents are mixed, and a stable emulsion forms spontaneously [37]. Due to the in-solubility of the agile constituent, the pesticide and surfactant primarily form a two-phase method; however, constant shearing drives them to mix, and pesticide sprayer in the nanoemulsion will endure spread for a long time, making them kinetically stable [46]. Neem oil in water nanoemulsion using Tween 20 as the surfactant has been created for pest management, for example. Insecticides are dispersed as hard nanosized particles in an aqueous means to create nanosuspensions, also known as nanodispersions [43]. The polar portions of the pesticide extend into the flowing result, while the nonpolar parts interact with the hard pesticide in nanodispersions, which are made up of surfactant molecules on the particle surface [45].

Researchers have generated aqueous dispersions of nanopermethrin, novaluron, and cypermethrin, for example [40]. Nanoparticles made of polymers: Polymer-based pesticide nanocarriers are commonly employed to deliver active chemicals to a specific location in a regulated and progressive manner [48]. They can also be used as a protective reservoir to aid in the dispersion of aqueous media [41]. This includes nanoencapsulation, nanospheres, nanogels, nanofibers, and other nanomaterials. An inner core chamber limits the hydrophobic or hydrophilic active component and is bounded by a polymer coating or membrane in a heterogeneous reservoir-type structure. This formulation can protect the active component in the neem-azadiractin combination [50].

Neonicotinoid pesticides, such as acetamipirid and imidacloprid, for example, have been developed as controlled-release nanoformulations. Nanospheres are vesicular formations in which bioactive components are distributed uniformly during the polymer matrix. For example, nanospheres consisting of bifenthrin nanoparticles are stabilized by polymers. Nanogels, also known as hydrogel nanoparticles, is a form of nanogel. These are made by crosslinking hydrophilic groups onto polymeric particles to increase their water absorption. A good example is chitosan nanogel. Electrospinning and thermally induced phase separation are used to create nanofibers. For the control of a variation of lepidopteran insect pests, researchers developed electrospun nanofibers burdened with the chemical (Z)-9-dodecenyl acetate, a component of pheromone that stays lodged in the polymer matrix [42].

10.5 Biopesticides

Biopesticides are pesticides that come from biological forms such as animals, plants, bacteria, and other microorganisms. They may be a potential option for long-term insect pest management due to their nontoxic method of action. Microorganisms and the compounds they generate are biopesticides. Microorganisms' active substance that acts as a biopesticide and, if registered, can be used as a crop protection agent. Bacterial growth is facilitated by the sapping process [44]. Furthermore, understanding the mechanisms of action of biomultiple pesticides is critical throughout the fungi–host relationship. Numerous microbes compete with one another in this process. A bioimpact pesticide is made up of a variety of distinct action mechanisms. Antibiosis and *Trichoderma harzianum*, for example, are competitors [47]. As a result, plants develop resistance. In January 2015, most tomato and sweet pepper producers implemented integrated pest management for the majority of insect pests and plant diseases for virtually the whole year. It's possible that eight novel biopesticides were employed as test cases for a 'Green Deal' registration procedure [5] Nanopesticides' Influence on the Establishment of Sustainable Development Goals.

Agriculture and food (agrifood) is a $7.8 trillion worldwide business that employs over 40% of the world's workforce [6]. Unfortunately, this enormous organization is a big contributor to greenhouse gas emissions, accounting for around a third of global carbon emissions from the stock chain, which includes food production, transportation, and marketing. Factors like increased global food consumption, climate change, poverty and hunger, food loss and waste, biodiversity loss, and overexploitation of common assets are all bringing new encounters to the world's agriculture food production schemes [49]. That is why we must protect the food and agriculture sectors' long-term survival. (1) Increasing food system productivity, employment, and value addition, (2) guarding and improving natural assets, (3) enhance living and creating general economic upturn, (4) improving the flexibility of publics, societies, and ecosystems, and (5) adjusting governance to new challenges are the five main principles of having a viable food and agriculture system. Personalities, families, communities, and society as a whole are all affected by poverty, which is a complex, multifaceted socioeconomic issue. Poverty, according to the International Covenant on Economic, Social, and Cultural Rights, the United Nations, is a social circumstance noticeable by a persistent or prolonged lack of essential resources, capabilities, choices, security, and strengths that would allow a realistic standard of living and additional civil, cultural, economic, political, and social rights. Scarcity might make it challenging to thoroughly use one's human rights [53].

As a result of economic constraints, many wealthy countries are currently dealing with the problem of poor diet. In contrast to poor nations, where food shortages can be acute and head to life-threatening malnourishment, food shortages in the United States are unusual. In wealthy nations such as Israel, the difficult is characterized as "a lack of regular access to a sufficient quantity of nutritious food that allows for a healthy, active life and the attainment of one's developmental potential." Several aspects of poverty, as well as the trajectory of the poverty rate, impact nutrition insecurity. In the Western world, the most significant barrier to obtaining food security is a household's financial ability to buying food [52].

Pesticides are chemicals that are used to prevent, repel, or even kill pests of any kind, whether they are plants or animals. Herbicides, insecticides, fungicides, disinfectants, and chemicals are pesticides that kill and prevent a wide range of pests. Plant and animal pests, as well as bacteria and mold development. However, they are well known for endangering human health and the environment, also creating major environmental health threats. Only 0.1 percent of pesticides are used to effectively single out pests, whereas the remaining 99.9% pollute the environment, endangering the food chain and human health [5]. Moreover, insecticide overuse drives bugs to develop and become resistant to current insecticides, resulting in the development of new pesticides that are more potent, lethal, and damaging to humans and the environment.

However, it is critical to note that we will never be able to eliminate pesticide use in agriculture due to the 1,000-fold increase in the number of people killed by bugs [8]. Biopesticides are insecticides that are less harmful to the environment. Biopesticides are claimed to be less effective and hence less popular, even though they may have less detrimental impacts on the environment [12]. As a result, nanopesticides are being developed. They not only keep pests at bay, but they also help to save the environment by dropping the use of harmful pesticides. Nanopesticides can save water and energy, according to George *et al.* [12] since they are used in lesser amounts and do not need to be applied as often as normal pesticides. It has also been suggested that enhancing the efficiency of nanopesticides leads to greater yields. We were able to minimize manufacturing costs by decreasing waste and labor.

10.6 Conclusion

Nanopesticides are critical to achieving the sustainable development goals because they help to achieve the five prime principles of a sustainable food and agriculture system: (1) growing production, employment, and value inclusion in food systems, (2) guarding and improving natural means, (3) enhancing livings and creating complete economic development, (4) improving the flexibility of persons, societies, and ecosystems, and (5) adaptation.

References

1. Amenta, V., Aschberger, K., Arena, M., Bouwmeester, H., Botelho Moniz, F., Brandhoff, P., Gottardo, S., Marvin, H.J.P., Mech, A., QuirosPesudo, L. *et al.*, Regulatory aspects of nanotechnology in the agri/feed/food sector in EU and non-EU countries. *Regul. Toxicol. Pharmacol.*, 73, 463–476, 2015.
2. Brown, J. and Kuzma, J., Hungry for information: Public attitudes toward food nanotechnology and labeling. *Rev. Policy Res.*, 30, 512–548, 2013.
3. Chatterjee, A.K., Chakraborty, R., Basu, T., Mechanism of antibacterial activity of copper nanoparticles. *Nanotechnology*, 25, 135101, 2014.
4. Certo, S.T., Busenbark, J.R., Woo, H., Semadeni, M., Sample selection bias and Heckman models in strategic management research. *Strateg. Manag. J.*, 37, 2639–2657, 2016.
5. Carson, R.T. and Hanemann, W.M., Chapter 17 contingent valuation, in: *Handbook of Environmental Economics*, vol. 2, pp. 821–936, Elsevier, Amsterdam, The Netherlands, 2005.
6. Cawley, J., Contingent valuation analysis of willingness to pay to reduce childhood obesity. *Econ. Hum. Biol.*, 6, 281–292, 2008.
7. Kidd, J., Westerhoff, P., Maynard, A., Survey of industrial perceptions for the use of nanomaterials for in-home drinking water purification devices. *NanoImpact*, 22, 100320, 2021.
8. D'Silva, J. and Bowman, D.M., To label or not to label?—It's more than a nano-sized question. *Eur. J. Risk Regul.*, 1, 420–427, 2010.
9. Eng, J., Sample size estimation: How many individuals should be studied? *Radiology*, 227, 309–313, 2003.
10. Frewer, L.J., Bergmann, K., Brennan, M., Lion, R., Meertens, R., Rowe, G., Siegrist, M., Vereijken, C., Consumer response to novel agri-food technologies: Implications for predicting consumer acceptance of emerging food technologies. *Trends Food Sci. Technol.*, 22, 442–456, 2011.

11. Ganesh Pillai, R. and Bezbaruah, A.N., Perceptions and attitude effects on nanotechnology acceptance: An exploratory framework. *J. Nanopart. Res.*, 19, 41, 2017.
12. George, S., Kaptan, G., Lee, J., Frewer, L., Awareness on adverse effects of nanotechnology increases negative perception among public: Survey study from Singapore. *J. Nanopart. Res.*, 16, 2751, 2014.
13. Gene Denovo Omic Share Tools, https://www.omicshare.com/tools (accessed on 29 May 2021).
14. Heckman, J.J., Sample selection bias as a specification error. *Econometrica*, 47, 153–161, 1979.
15. Hofmann, T., Lowry, G.V., Ghoshal, S., Tufenkji, N., Brambilla, D., Dutcher, J.R., Gilbertson, L.M., Giraldo, J.P., Kinsella, J.M., Landry, M.P. *et al.*, Technology readiness and overcoming barriers to sustainably implement nanotechnology-enabled plant agriculture. *Nat. Food*, 1, 416–425, 2020.
16. Juster, F.T. and Smith, J.P., Improving the quality of economic data: Lessons from the HRS and AHEAD. *J. Am. Stat. Assoc.*, 92, 1268–1278, 1997.
17. Karina Gallardo, R. and Wang, Q., Willingness to pay for pesticides' environmental features and social desirability bias: The case of apple and pear growers. *J. Agric. Resour. Econ.*, 38, 124–139, 2013.
18. Khan, J., Jan, A., Lim, K.H., Shah, S.A., Khanal, A.R., Ali, G., Household's perception and their willingness to pay for pesticidesfree fruits in Khyber Pakhtunkhwa (Kp) Province of Pakistan: A double-bounded dichotomous choice contingent valuation study. *Sarhad J. Agric.*, 35, 1266–1271, 2019.
19. Kookana, R.S., Boxall, A.B.A., Reeves, P.T., Ashauer, R., Beulke, S., Chaudhry, Q., Cornelis, G., Fernandes, T.F., Gan, J., Kah, M. *et al.*, Nanopesticides: Guiding principles for regulatory evaluation of environmental risks. *J. Agric. Food Chem.*, 62, 4227–4240, 2014.
20. Kah, M., Beulke, S., Tiede, K., Hofmann, T., Nanopesticides: State of knowledge, environmental fate, and exposure modeling. *Crit. Rev. Environ. Sci. Technol.*, 43, 1823–1867, 2013.
21. Kah, M., Kookana, R.S., Gogos, A., Bucheli, T.D., A critical evaluation of nanopesticides and nanofertilizers against their conventional analogues. *Nat. Nanotechnol.*, 13, 677–684, 2018.
22. Klaine, S.J., Koelmans, A.A., Horne, N., Carley, S., Handy, R.D., Kapustka, L., Nowack, B., von der Kammer, F., Paradigms to assess the environmental impact of manufactured nanomaterials. *Environ. Toxicol. Chem.*, 31, 3–14, 2012.
23. Lee, C.J., Scheufele, D.A., Lewenstein, B.V., Public attitudes toward emerging technologies: Examining the interactive effects of cognitions and affect on public attitudes toward nanotechnology. *Sci. Commun.*, 27, 240–267, 2005.
24. Macoubrie, J., Nanotechnology: Public concerns, reasoning and trust in government. *Public Underst. Sci.*, 15, 221–241, 2006.

25. Pandey, S., Giri, K., Kumar, R., Mishra, G., Raja Rishi, R., Nanopesticides: Opportunities in crop protection and associated environmental risks. *Proc. Natl. Acad. Sci. India Sect. B Biol. Sci.*, 16, 108–129, 2018.
26. Petrescu-Mag, R.M., Banatean-Dunea, I., Vesa, S.C., Copacinschi, S., Petrescu, D.C., What do romanian farmers think about the effects of pesticides? perceptions and willingness to pay for bio-pesticides. *Sustainability*, 11, 3628, 2019.
27. StataCorpStata, *Version: Stata base reference manual: Release*, StataCorp LLC, College Station, TA, USA, 2019.
28. Scheerder, J., Vos, S., Taks, M., Expenditures on sport apparel: Creating consumer profiles through interval regression modelling. *Eur. Sport Manag. Q.*, 11, 251–274, 2011.
29. Skordoulis, M., Ntanos, S., Arabatzis, G., Socioeconomic evaluation of green energy investments. *Int. J. Energy Sect. Manag.*, 14, 871–890, 2020.
30. Sodano, V., Gorgitano, M.T., Verneau, F., Vitale, C.D., Consumer acceptance of food nanotechnology in Italy. *Br. Food J.*, 118, 714–733, 2016.
31. Stokes, E., Regulating nanotechnologies: Sizing up the options. *Leg. Stud.*, 29, 281–304, 2009.
32. Aaron O., Statista, *Distribution of the workforce across economic sectors in China from 2010 to 2020* (accessed on 1 November 2021), China, 2023.
33. Vidogbéna, F., Adégbidi, A., Tossou, R., Assogba-Komlan, F., Martin, T., Ngouajio, M., Simon, S., Parrot, L., Zander, K., Consumers' willingness to pay for cabbage with minimized pesticide residues in Southern Benin. *Environments*, 2, 449–470, 2015.
34. Vo, N.X., Huyen Nguyen, T.T., Van Nguyen, P., Tran, Q.V., Vo, T.Q., Using contingent valuation method to estimate adults' willingness to pay for a future coronavirus 2019 vaccination. *Value Health Reg. Issues*, 24, 240–246, 2021.
35. Walker, G.W., Kookana, R.S., Smith, N.E., Kah, M., Doolette, C.L., Reeves, P.T., Lovell, W., Anderson, D.J., Turney, T.W., Navarro, D.A., Ecological risk assessment of nano-enabled pesticides: A perspective on problem formulation. *J. Agric. Food Chem.*, 66, 6480–6486, 2018.
36. Wang, W., Jin, J., He, R., Gong, H., Tian, Y., Farmers' willingness to pay for health risk reductions of pesticide use in China: A contingent valuation study. *Int. J. Environ. Res. Public Health*, 15, 625, 2018.
37. Wang, Z., Gong, Y., Mao, X., Exploring the value of overseas biodiversity to Chinese netizens based on willingness to pay for the African elephants' protection. *Sci. Total Environ.*, 637–638, 600–608, 2018.
38. Yue, C., Zhao, S., Cummings, C., Kuzma, J., Investigating factors influencing consumer willingness to buy GM food and nano-food. *J. Nanopart. Res.*, 17, 283, 2015.
39. Yang, M., *The study on pesticide use behavior of greenhouse vegetable growers: Taking Shandong as an example*, China Agricultural University, Beijing, China, 2019.

40. Yu, J. and Wu, J., The sustainability of agricultural development in China: The agriculture–environment nexus. *Sustainability*, 10, 1776, 2018.
41. Zhou, G. and Hu, W., Public acceptance of and willingness-to-pay for nanofoods in the U.S. *Food Control*, 89, 219–226, 2018.
42. Zobar, Z.A.M., Ali, A., Admi, F., Suleiman, M.R., Ahmad, K., A review of nanopestcides for plants protection synthesized using the supramolecular chemistry of layered hydroxide hosts. *Int. J. Environ. Res. Public Health*, 10, 1077, 2021.
43. Islam, S., Thangadurai, D., Adetunji, C.O., Micheal, O.S., Nwankwo, W., Kadiri, O., Anani, O.A., Makinde, S., Adetunji, J.B., Nanostructured biomaterials for targeted drug delivery, in: *Biogenic Nanomaterials*, pp. 83–108, Springer, Singapore, 2022.
44. Adetunji, C.O., Olaniyan, O.T., Okeke, N.E., Production of next-generation biodiesel from high yielding strains of microorganisms: Recent advances, in: *Sustainable Green Chemical Processes and Their Allied Applications. Nanotechnology in the Life Sciences*, A.A. Inamuddin (Ed.), Springer, Cham, 2020, https://doi.org/10.1007/978-3-030-42284-4_2.
45. Jhala, Y.K., Panpatte, D.G., Adetunji, C.O., Vyas, R.V., Shelat, H.N., Management of biotic and abiotic stress affecting agricultural productivity using beneficial microorganisms isolated from higher altitude agro-ecosystems: A remedy for sustainable agriculture, in: *Microbiological Advancements for Higher Altitude Agro-Ecosystems & Sustainability. Rhizosphere Biology*, R. Goel, R. Soni, D. Suyal (Eds.), Springer, Singapore, 2020, https://doi.org/10.1007/978-981-15-1902-4_7.
46. Adetunji, C.O., Samuel, M.O., Adetunji, J.B., Oluranti, O.I., Corn silk and health benefits, in: *Medical Biotechnology, Biopharmaceutics, Forensic Science and Bioinformatics*, pp. 167–178, Springer, Singapore, 2022.
47. Olaniyan, O.T., Adetunji, C.O., Adetunji, J.B., Oloke, J.K., Olaleye, O.O., Hefft, D.I., Ubi, B.E., Soy-based food products consumed in Africa: A panacea to mitigate food insecurity and health challenges, in: *Fermentation and Algal Biotechnologies for the Food, Beverage and Other Bioproduct Industries*, pp. 87–103, Springer, Singapore, 2022.
48. Adetunji, C.O. and Oyeyemi, O.T., Antiprotozoal activity of some medicinal plants against Entamoeba histolytica, the causative agent of amoebiasis, in: *Medical Biotechnology, Biopharmaceutics, Forensic Science and Bioinformatics*, pp. 341–358, Springer, Singapore, 2022.
49. Adetunji, C.O., Panpatte, D.G., Jhala, Y.K., *Microbial rejuvenation of polluted environment*, vol. 3, Springer, Singapore, 2021.
50. Adetunji, C.O., Michael, O.S., Nwankwo, W., Anani, O.S., Adetunji, J.B., Olayinka, A.S., Akram, M., Based drugs derived from medicinal plants, in: *Green Synthesis in Nanomedicine and Human Health*, p. 103, 2021.

51. Adetunji, J.B., Adetunji, C.O., Olaniyan, O.T., Masajuwa, F.U., Islam, I., Thangadurai, D., Olaleye, O.O., Hefft, D.I., Dauda, W.P., Ubi, B.E., Nutritional and health benefits of nutraceutical beverages derived from cocoa and other caffeine products: A comprehensive review, in: *Fermentation and Algal Biotechnologies for the Food, Beverage and Other Bioproduct Industries*, pp. 105–118, Springer, Singapore, 2022.
52. Kashyap, P.L., Kumar, S., Jasrotia, P., Singh, D.P., Singh, G.P., Nanotechnology in wheat production and protection, in: *Environmental Nanotechnology*, Vol. 4, N. Dasgupta, S. Ranjan, E. Lichtfouse (Eds.), pp. 165–194, Springer, Cham, 2020.
53. Adetunji, C.O., Olaniyan, O.T., Adetunji, J.B., Osemwegie, O.O., Ubi, B.E., African mushrooms as functional foods and nutraceuticals, in: *Fermentation and Algal Biotechnologies for the Food, Beverage and Other Bioproduct Industries*, pp. 233–251, Springer, Singapore, 2022.

11

Application of Nanotechnology for the Production of Biopesticides, Bioinsecticides, Bioherbicides, Mosquitoe Repellants and Biofungicides

Olorunsola Adeyomoye[1*], Charles Oluwaseun Adetunji[2], Olugbemi T. Olaniyan[3], Juliana Bunmi Adetunji[4], Olalekan Akinbo[5], Babatunde Oluwafemi Adetuyi[6], Abel Inobeme[7], John Tsado Mathew[8] and Shakirat Oloruntoyin Ajenifujah-Solebo[9]

[1]Department of Physiology, University of Medical Sciences, Ondo City, Nigeria
[2]Applied Microbiology, Biotechnology and Nanotechnology Laboratory, Department of Microbiology, Edo University Iyamho, Auchi, Edo State, Nigeria
[3]Laboratory for Reproductive Biology and Developmental Programming, Department of Physiology, Rhema University, Aba, Nigeria
[4]Department of Biochemistry, Osun State University, Osogbo, Nigeria
[5]Centre of Excellence in Science, Technology, and Innovation, Directorate of Knowledge Management and Programme Evaluation, Cape Town, South Africa
[6]Biochemistry Unit, Department of Natural Sciences, Faculty of Pure and Applied Sciences, Precious Cornerstone University, Ibadan, Nigeria
[7]Department of Chemistry, Edo State University Uzairue, Edo, Nigeria
[8]Department of Chemistry, Ibrahim Badamasi Babangida, Niger, Nigeria
[9]National Biotechnology Development Agency, Umar Musa Yardua Way, Lugbe P.M.B., Wuse, Abuja, Nigeria

Abstract

Nanotechnology is one intriguing technology, which has the potential to increase agricultural output through the use of pesticides, herbicides, and nanofertilizers. Mosquitoes, which are known to spread certain dreadful diseases, are a serious menace to people all over the world. Conventional insecticides, insect growth regulators, and other microbial control agents are used to restrict the growth and

**Corresponding author*: oadeyomoye@unimed.edu.ng; adeyomoyeshola@yahoo.com

Charles Oluwaseun Adetunji and Julius Kola Oloke (eds.) Handbook of Agricultural Biotechnology: Volume I Nanopesticides, (293–310) © 2024 Scrivener Publishing LLC

reproduction of mosquitoes. However, the usage of these typical chemicals and conventional pesticides ultimately has a detrimental effect on both the environment and human health, which raises serious concerns. By reducing the use of dangerous chemicals and encouraging new ideas like the use of biological controls, new tactics are being introduced using new environmentally friendly technology in plant protection for better human health and a safer environment. This particular method addresses the need to reduce pesticide use while switching to new biobased agrochemicals. As a result, stakeholders, like farmers, seed firms, agronomists, breeders, and consumers, are paying more attention to biological control measures. These biobased remedies offer enormous potential for eradicating dangerous plants and pathogenic germs without degrading the environment and human health.

***Keywords*:** Nanotechnology, biocontrol, biobased solution, human health, environment

11.1 Introduction

The emergence of smart and active packaging, nanosensors, nanopesticides, and nanofertilizers is particularly helpful in transforming the traditional food and agricultural sectors [21]. For the purpose of enhancing agricultural growth, food quality and safety, and environmental monitoring, numerous novel nanomaterials have been created [12, 16]. The efficacy and durability of agrochemicals are facilitated by the distinctive architectures of nanomaterials, such as high specific surface area, concentrated size distribution, and great biocompatibility. Additionally, the application of suitable nanomaterials during plant growth phases or stressful situations efficiently accelerates plant growth and improves stress resistance [30]. New nanotools and nanobiotechnologies offer a new platform for molecular crop monitoring and modification. Agrochemical additives are carried by nanoparticles, which are nanostructures with two to three dimensions. Compared to using free insecticides, encapsulating nanoinsecticide has advantages because of their special characteristics [55]. The demand for food has increased due to the world's fast growing population, which has increased the need of agrochemicals to bridge the gap between food production and consumption. Conventional agronanochemicals have the problem that their active ingredient is less effective at reaching the goal and that additional factors, including leaching and evaporation, also have a harmful influence on the environment and life, including humans [39]. Fortunately, nanotechnology platforms improve the bioavailability of conventional pesticides through a variety of kinetics, mechanisms, and pathways on their intended target species. This enables them to conveniently avoid biological and other unwelcome resistances and thereby increase their effectiveness. Phytotoxic allelochemical-containing plants or specific disease-carrying microorganisms that can control weed populations are the sources of bioherbicides. Although bioherbicides are highly effective in stopping the germination and growth of weed seeds, relatively few *in vitro* research on the physiological changes they cause in weeds have been described [20]. Nanotechnology-based products like nanoinsecticides, nanopesticides, and nanofertilizers, have exceptional functionalities of nanoparticles and are utilized in sustainable agriculture [56–65]. The transmission of various deadly diseases by mosquitoes, including dengue, West Nile fever, chikungunya, yellow fever, and filaria, is responsible for millions of fatalities each year throughout the world [26]. Although synthetic insecticides are frequently used to control a range of vector-borne diseases, they have a number of disadvantages, such as a negative impact on unintended species, detrimental environmental effects, and the development of resistance in vector species by altering the target location. Without harming the environment, plant extracts, phytochemicals, and their nanoformulations can be employed as insect growth regulators, larvicides, oviposition attractants, and repellents [17]. In addition to being far less expensive, environmentally benign, biodegradable, readily available, and nontoxic to organisms other than the intended target, these plant-based treatments offer broad-spectrum resistance against a range of mosquito species.

11.2 Nanotechnology

Even though researchers have been studying nanoscience phenomena for many years, technological advancements in the second half of the 20th century gave researchers useful tools that made it possible for them to study and create materials in the nanoscale size range and contributed to the formalization of nanotechnology as a scientific field [43]. The environment and nanotechnology are already interacting. At the nanoscale, matter is manipulated by scientists and engineers; the industrial sector uses these nanoscale processes and outcomes in commercially available goods. Either these products are directly applied to the environment or they indirectly enter the ecosystem. Advances in nanoscience and nanotechnology has made life easier in practically every field of study [6]. Because of the arrangement of their atoms at the 1 to 100 nm scale, structures, gadgets, and systems in the field of nanoscience and nanotechnology have new features and functionalities. Beginning in the early 21st century, this topic has drawn more attention from the public and generated controversy. As a result, commercial uses of nanotechnology have started to emerge [5]. Nanotechnology is advantageous to almost all fields of study, including physics [31], materials science, chemistry, biology, computer science, and engineering. The current application of nanotechnology to better human health, particularly in the fight against cancer, is interesting [22]. In order to enhance therapeutic efficacy while reducing side effects, innovative materials and devices that interact with the body at the molecular level can be used in a wide variety of highly selective and targeted applications [53]. Quantum dots, carbon nanotubes, liposomes, and superparamagnetic nanoparticles are some of the nanotechnologies that have received the most attention. While this is happening, other upcoming platforms are creating the groundwork for a completely new scientific paradigm that might fundamentally change the direction of research. The manipulation of nanomaterials via dielectrophoresis is one of the main study disciplines that could potentially help micro or nanoscience for a number of applications, such as microfluidics, nanomachinery, and biosensors [30]. Innovation and development of fundamental theories, approaches, or applications will have a significant impact on the associated field as a whole. Enhancing precision, flexibility, and scalability specifically could enable further applications in microresearch or nanoresearch for manipulating nanomaterials using dielectrophoresis [29]. Numerous issues, such as climate change, stagnant crop yields, and rising pesticide resistance, beset the agricultural industry. Agricultural systems have already entered a condition of production saturation due to the explicit use of agrochemicals, the introduction of high-yielding cultivars, and genetically engineered seeds. Therefore, to boost crop yield in line with the growing demand for sustainability, advanced nanotechnology technologies must be used [48]. Some nutritional issues may be resolved because to the improved solubility, absorption, and target specificity of nanofertilizers made from materials including silver, copper, gold, zinc, and iron oxides. Pest control uses nanopesticides such spinosad-loaded chitosan, silica encapsulating fipronil, and sodium alginate encapsulating imidacloprid, while herbicide and heavy metal detection uses fluorescent nanosensors, carbon and graphene nanodots. Hard and waste water's quality might be greatly improved with nanofiltration using graphem, cellulose, and cyclodextrin to remove salts, heavy metals, and organic contaminants, making it suitable for irrigation. Numerous difficulties in agriculture, forestry, and the environment, including urbanization, energy shortages, and the sustainable use of resources, could be positively impacted by nanotechnology [44].

However, new risks to the environment and to human health could result from applications that leverage nanotechnology. This poses issues for agricultural employees who may be exposed to these xenobiotics largely while doing their job duties. Nanopesticides (nanoformulations of conventional active ingredients or inorganic nanomaterials) and nanofertilizers (nano-sized nutrients, nanolayered fertilizers, or engineered nanomaterials based on metal oxide or carbon per se) can deliver targeted/controlled release of agrochemicals to achieve their maximum biological efficacy without overdosing. Environmental toxins can be found and eliminated using nanosensors and nanoremediation techniques. To analyze the potential dangers for nanoagriculture, more research is necessary due to the limited understanding regarding the biosafety of nanomaterials, adverse effects, destiny, and acquired biological reactivity after dispersion into the environment [40]. Crop yield and ensuring food security for a growing population depend on plant nutrition. The foundation of any agricultural economy is the application of chemical and biological fertilizers and pesticides. Chemical pesticides and fertilizers have unforeseen side effects, though. Their use has a negative impact on the ecosystem and ecological equilibrium. Despite significant drawbacks related to their use, biofertilizers and biopesticides mitigate some of the negative environmental consequences of chemical pesticides and fertilizers. Nanotechnology's most recent advancements provide promise for sustainable agriculture. Concerns about both agriculture and the environment must be addressed in order to practice sustainable agriculture. Both nanopesticides and nanofertilizers have gradual and sustained release capabilities in addition to being environmentally friendly [23]. They can be modified to meet the unique requirements of the crop. Additionally, nanofertilizers have higher stress tolerance, making them extremely valuable in the face of climate change. Additionally, nanofertilizers and nanopesticides are used in extremely small doses, which lowers the associated transport costs and has a beneficial economic impact. They are used in ways that go beyond, say, the usage of nanoparticles (NPs) in high quantities, which harm plant pathogens. Nanopesticides and nanofertilizers based on polymers provide a number of benefits. There is a bad side to the usage of nanomaterials in agriculture. Metal/metal oxide nanoparticles, which can enter human bodies and accumulate through biomagnification, are frequently used in nanotechnology [2]. NPs can reach dangerous quantities in soil and run into rivers and other water bodies, and their removal represents a significant economic burden even though it is uncertain how they may affect human health. However, nanoformulations must first undergo a risk-benefit analysis before being used in sustainable agriculture.

11.3 Formulation and Delivery of Biopesticides Using Nanotechnology

Pesticides are intended to eliminate and control weeds and pests. They are divided into many groups according to where they came from and the kind of pests they are meant to control. In agricultural areas, chemical pesticides like fungicides, insecticides, and herbicides are frequently utilized [25]. However, overuse of these agrochemicals threatens endangered species, bird habitats, and insect pollinator populations, among other aspects of the environment. Chemical pesticides can cause cancer and other health issues when taken, including issues with the skin, eyes, and nervous system [34]. Numerous techniques based on surface adsorption, membrane filtering, and biological degradation have been created in the past to lower the pesticide level. However, some drawbacks of these approaches are their poor response times, lack of specificity, and sensitivity. Nanotechnology has recently become a supporting tool for the identification and removal of pesticides. Due to its low use and indiscriminate application, pesticide-related pollution has become a major issue on a global scale. Nanotechnology has showed considerable promise in promoting sustainable agriculture. Due to its particular features of low toxicity, great thermal and chemical stability, and especially size tunability and varied functionality, mesoporous silica-based nanomaterials have attracted a lot of attention nowadays for enhancing pesticide efficacy and safety. The numerous functions of mesoporous silica in the secure use of pesticides utilizing nanotechnology cannot be overemphasized based on the description of the structure and synthesis of several forms of mesoporous silica nanoparticles [11]. Nanocarriers are used to deliver pesticides over time and in controlled environments. They serve as a pesticide adsorbent for enriching or removing pesticides in aqueous media, support for catalysts for the breakdown of pesticide pollutants, and support for sensors for pesticide detection. A form of nanomaterial called mesoporous silica nanoparticles has lately become popular in smart delivery systems. Multifunctional MSNs with bright luminescent centers that make it simpler to monitor MSNs in biological systems and adaptable structural features that have a high drug loading capacity and controlled release are still highly desired, even though research into the use of these nanoparticles as pesticide carriers has advanced. With a major reduction in the environmental load of pesticides, the process of bioencapsulation has gained attention, particularly as an environmentally beneficial form of pest control. Conventional pesticide application may be replaced by the creation of effective nanocarrier systems and nanoencapsulated insecticides for sustainable agriculture. To increase the selectivity, specificity, and longevity of encapsulated pesticides while minimizing adverse environmental effects on ecosystems and people, much more research is required in this area to produce safe and promising pesticide delivery systems [18]. The pesticide formulation with nanoencapsulation has improved solubility, permeability, and stability with gradual release characteristics. These qualities are mostly obtained by either preventing the encapsulated active ingredients from premature deterioration or by prolonging the duration of their activity against pests [24]. Reduced pesticide dosage and human exposure from nanoencapsulated pesticide formulation is environmentally benign for crop protection [54]. However, the adoption of nanoencapsulation materials for pesticide administration may be hampered by the lack of understanding of the production mechanism and the absence of a cost–benefit analysis. With the aid of further investigation into the behavior of these materials and their final fate in the environment, a regulatory framework for their commercialization may be devised [32]. More than 90% of pesticides escape into the environment and remain in agricultural products during application due to the flaws in conventional pesticide formulation, such as the use of toxic solvent, inadequate dispersion, dust entrainment, etc. The effectiveness and safety of these insecticides could be improved because to recent advances in nanotechnology [27, 42]. The major objective of the development of nanoparticle-based pesticides is the exact release of essential and sufficient amounts of their active ingredients in response to environmental cues and biological demands through controlled release mechanisms.

11.4 Application of Nanotechnology for Bioinsecticide Production

To create systems for pest management in agriculture, nanotechnology is used in conjunction with insecticides. The primary categories of insecticides have been described, along with a variety of carriers and their possible applications [13]. Additionally to essential oils derived from specific plants, insecticides can also be based on active principles separated from plant extracts. Before agricultural pest management technologies may be used in the future, the systems benefits and major technological difficulties may have been described. Nanotechnology-based insecticides have a lot of potential to boost agricultural output while minimizing negative effects on the environment and human health [14]. Due of their high residue concentration, synthetic insecticides can pollute the environment when used frequently. New pest management strategies are also necessary due to the fact that many insect pests are becoming more resistant to insecticides [3]. In the creation of novel nanoagrochemicals like nanoinsecticides and nanofertilizers, nanotechnology has the potential to be a fruitful field of contemporary agriculture [41, 45]. While nanoparticles themselves can be used as insecticides, they can also transport other pesticides. Studies indicates that the toxicity and penetration into the insect body increase with decreasing nanoparticle size [10]. However, there is a paucity of literature on the behavior of insects. Nanoparticles also affect cuticle integrity and color on the surface, affecting insect development and reproduction. Internal immunological responses and changed gene expression result in altered protein, lipid, and carbohydrate metabolism as well as cellular toxicity [15, 52]. In contrast, some nanoparticles can promote growth and development while reducing the effects of off-target toxicity in insects. This is because they can disrupt nutrient intake, produce reactive oxygen species, and alter metabolic activity.

11.5 Application of Nanotechnology for Bioherbicide Production

In the course of agriculture, weeds have significantly reduced agricultural productivity. Herbicides are crucial to maintaining agricultural output [46]. But a heavy dependence on herbicides has resulted in both environmental issues and the emergence of herbicide resistance. New herbicides with safer toxicity profiles and novel mechanisms of action are thus urgently needed. Due to their unique physical and chemical characteristics, nanoparticles can pass through cell barriers and have an impact on living things [8]. Their usage in agriculture as well as in health, electronics, chemical, and physical sciences has been observed as a result of their capacity to break the cell walls. Using different nanoproducts like nanofertilizers, nanoherbicides, nanofungicides, nanopesticides, etc., in agriculture can assist increase crop growth and yield. It is feasible to distribute nanoparticles at the ideal concentration by incubating seeds, roots, pollen, isolated cells, and protoplasts as well as through foliar spraying, irrigation with nanoparticles, direct injection, hydroponic treatment, and biolistic delivery [49]. Once in contact with plant cells, nanoparticles are taken up via plasmodesmatal or endocytotic pathways and are subsequently transported by apoplastic or symplastic pathways. Atrazine, one of the most widely used herbicides, has been linked to lasting contamination of surface and groundwater; new formulations created using nanotechnology may hold the key to a potential cure. Atrazine nanoencapsulated in poly(caprolactone) (NC+ATZ) was used to make a highly effective herbicide formulation [35]. A complete structural investigation of the interaction between the formulation and *Brassica juncea* plants was conducted in order to gauge the uptake of nanoatrazine by the leaf and the structural changes generated in the leaves. After postemergence therapy, NC+ATZ remained attached to the leaf and primarily entered the mesophyll tissue through the hydathode regions. The stability of the nanoparticles for 120 days was determined using an analysis of the particle size, potential, polydispersity, pH, and encapsulation efficiency [14]. SLN showed great physicochemical stability and high encapsulation effectiveness. Release kinetics tests showed that the use of SLN changed the herbicide release patterns in water. In tests of herbicidal activity employing preemergent and postemergent treatments of the target species, the effectiveness of formulations incorporating nanoparticles of a herbicide was shown [14]. Tests on *Zea mays* showed that the formulations had no impact on the growth of nontarget species. Cytotoxicity tests show that the addition of SLN reduces the toxicity of herbicides. Utilizing less pesticides is now possible because to new nanoparticle compositions, offering a more environmentally friendly means of controlling weeds in agriculture.

11.6 Nanobiotechnology as an Emerging Approach to Combat Malaria

As carriers of several illnesses, including parasites and viruses, mosquitoes pose a serious hazard to millions of humans and animals worldwide. Failure to follow application directions and resistance to insecticides like organophosphates and microbial control agents are two main causes of insecticide treatment failure [4]. Therefore, research into safer, less expensive, and more powerful insecticides with new mechanisms of action is urgently needed to enhance mosquito control. Recently, the possibility of using biosynthesized nanoparticles as a malaria cure and a strategy to fight malaria vectors has been explored [7]. Understanding the characteristics and effectiveness of biogenic nanoparticles against the main malaria vectors, such as avian malaria, may help to better understand the human malaria vectors. One of the main infectious diseases that plagues people on a regular basis is malaria, which is made more difficult by the development of drug-resistant parasite strains and the accessibility of medications with a nonspecific pharmacodynamic action that are known to cause greater side effects [36]. The disease has spread to epidemic levels in tropical nations where sanitation standards are subpar and environmental factors encourage the growth of the parasite and its transmission, particularly through female anopheles. It is obvious that in order to address the issues, it is necessary to create and develop more potent medications that can combat parasite strains that are resistant to existing medications [1]. The development of next-generation antimalarial medications and drug delivery technologies could deliberately take into account a variety of therapeutic targets provided by the parasite's molecular biology and its targeting of host cellular motifs [38]. Recent studies show that a vaccine against the malaria parasite could be created, yet it can evade immune monitoring due to a sophisticated parasite mutation. Consequently, the created vaccine was unable to offer complete protection against malaria infection [51]. The dose and dosage regimen used in the conventional method of antimalarial drug treatment frequently results in contraindications and thereby jeopardizes the efficacy of the treatment. Numerous regimens and modules that could use nanotechnology specifically to treat this difficult infection are now available because to recent developments in contemporary biological sciences, material sciences, and drug delivery. Nanotechnology has numerous biomedical uses, including the ability to treat malaria. A large amount of activity has been demonstrated by silver nanoparticles (AgNPs) against the malaria pathogen *P. falciparum* and vector (female Anopheles mosquito) [33].

11.7 Application of Nanotechnology for Biofungicide Production

About 70 to 80 percent of agricultural yield losses are due to fungi. Agrochemicals, which are currently used to treat plant diseases are necessary, but these have detrimental impacts on both humans and the environment [28]. By enhancing the solubility of fungicides with poor water solubility, prolonging shelf life, and reducing toxicity in a sustainable and environmentally friendly manner, nanotechnology can help reduce the negative effects of fungicides [47]. Despite the fact that employing nanoparticles has several benefits, very few nanoparticle-based products have been made in large enough quantities for use in agriculture. Numerous issues, including insufficient field trials and underuse of pest host systems, may contribute to the lack of commercial usage. Nanotechnology has advanced in several fields, and the best approach to keep up with them is to comprehend the key components of the study and fill in the knowledge gaps to speed up technological advancements that can justify them. Ginger (*Zingiber officinale* Rosc.) is cultivated as a tropical plant for both culinary and medicinal applications and is one of the nanoproducts in commercial quantities all over the world. Its market value is increased by the fact that it is one of the most significant spices that is regularly used in food. However, fungi like *Pythium* and *Fusarium* spp. are responsible for the prevalent disease of ginger known as soft rot (rhizome rot) [37]. It is the most harmful illness to ginger, and it can drastically limit production by up to 90%. Although the use of chemical fungicides is thought to be an efficient means of preventing soft rot of ginger, the environment and public health are seriously endangered by the widespread use of fungicides [9]. Therefore, it is crucial to create alternative, economically and environmentally sound management strategies for diseases like this one that cause ginger soft rot. Nanotechnology is a promising strategy that is currently being investigated and may be utilized to control Pythia and Fusarium. In addition, fungus could be used for the biological production of silver nanoparticles due to its capacity to create a cap from fungal biomolecules that is stable and with increased biological activity [19, 50].

11.8 Conclusion and Future Perspectives

Over the past decade, interest in studying how to employ nanotechnology in agriculture has grown significantly, with a focus on the creation of novel nanoagrochemicals. Through the use of nanoparticles, the application of nanotechnology in agriculture can considerably improve the efficacy of agricultural inputs, providing a key pathway for sustaining the sustainable growth of agroecosystems. For possible use in agriculture, agrochemical encapsulated nanocarriers, nanopesticides, nanofertilizers, and nanoherbicides have all been created as nanoplant growth boosters. The limited bioavailability, simple photolysis, and organic solvent pollution of conventional agrochemicals are significantly improved by the use of these nanomaterials and nanotechnology in agrochemicals. Future studies and regulations are required to encourage intelligent design and the creation of more environmentally friendly agrochemicals.

References

1. Anamika, J., Nikhar, V., Laxmikant, G., Priya, S., Sonal, V., Vyas, S.P., Nanobiotechnological modules as molecular target tracker for the treatment and prevention of malaria: Options and opportunity. *Drug Deliv. Transl. Res.*, *10*, 4, 1095–1110, 2020.
2. Arora, S., Murmu, G., Mukherjee, K., Saha, S., Maity, D., A comprehensive overview of nanotechnology in sustainable agriculture. *J. Biotechnol.*, *355*, 21–41, 2022.
3. Awad, M., Ibrahim, E.S., Osman, E., II, Elmenofy, W.H., Mahmoud, A., Atia, M., Moustafa, M., Nano-insecticides against the black cutworm Agrotis ipsilon (Lepidoptera: Noctuidae): Toxicity, development, enzyme activity, and DNA mutagenicity. *PLoS One*, *17*, 2, e0254285, 2022.
4. Barabadi, H., Alizadeh, Z., Rahimi, M.T., Barac, A., Maraolo, A.E., Robertson, L.J., Masjedi, A., Shahrivar, F., Ahmadpour, E., Nanobiotechnology as an emerging approach to combat malaria: A systematic review. *Nanomed.: Nanotechnol. Biol. Med.*, *18*, 221–233, 2019.
5. Bassi, M., Santinello, I., Bevilacqua, A., Bassi, P., Nanotecnologie: Una grande rivoluzione che parte dal piccolo [Nanotechnology: A big revolution from the small world]. *Urologia*, *80*, 1, 46–55, 2013.
6. Bayda, S., Adeel, M., Tuccinardi, T., Cordani, M., Rizzolio, F., The history of nanoscience and nanotechnology: From chemical-physical applications to nanomedicine. *Molecules*, Basel, Switzerland, *25*, 1, 112, 2019.
7. Benelli, G., Plant-mediated biosynthesis of nanoparticles as an emerging tool against mosquitoes of medical and veterinary importance: A review. *Parasitol. Res.*, *115*, 1, 23–34, 2016.
8. Bombo, A.B., Pereira, A., Lusa, M.G., de Medeiros Oliveira, E., de Oliveira, J.L., Campos, E., de Jesus, M.B., Oliveira, H.C., Fraceto, L.F., Mayer, J., A Mechanistic View of Interactions of a Nanoherbicide with Target Organism. *J. Agric. Food Chem.*, *67*, 16, 4453–4462, 2019.
9. Boxi, S.S., Mukherjee, K., Paria, S., Ag doped hollow TiO2 nanoparticles as an effective green fungicide against Fusarium solani and Venturia inaequalis phytopathogens. *Nanotechnology*, *27*, 8, 085103, 2016.
10. Bricarello, P.A., Barros, G.P., Seugling, J., Podestá, R., Veleirinho, M.B., Mazzarino, L., Data of insecticide action of a nanoemulsion of Citronella essential oil on *Cochliomyia hominivorax* blowfly. *Data Br.*, *38*, 107391, 2021.
11. Cao, L., Zhang, H., Zhou, Z., Xu, C., Shan, Y., Lin, Y., Huang, Q., Fluorophore-free luminescent double-shelled hollow mesoporous silica nanoparticles as pesticide delivery vehicles. *Nanoscale*, *10*, 43, 20354–20365, 2018.
12. Chaud, M., Souto, E.B., Zielinska, A., Severino, P., Batain, F., Oliveira-Junior, J., Alves, T., Nanopesticides in agriculture: Benefits and challenge in agricultural productivity, toxicological risks to human health and environment. *Toxics*, *9*, 6, 131, 2021.

13. de Oliveira, J.L., Campos, E.V., Bakshi, M., Abhilash, P.C., Fraceto, L.F., Application of nanotechnology for the encapsulation of botanical insecticides for sustainable agriculture: Prospects and promises. *Biotechnol. Adv.*, *32*, 8, 1550–1561, 2014.
14. de Oliveira, J.L., Campos, E.V., Gonçalves da Silva, C.M., Pasquoto, T., Lima, R., Fraceto, L.F., Solid lipid nanoparticles co-loaded with simazine and atrazine: Preparation, characterization, and evaluation of herbicidal activity. *J. Agric. Food Chem.*, *63*, 2, 422–432, 2015.
15. Duarte, J.L., Maciel de Faria Motta Oliveira, A.E., Pinto, M.C., Chorilli, M., Botanical insecticide-based nanosystems for the control of Aedes (Stegomyia) aegypti larvae. *Environ. Sci. Pollut. Res. Int.*, *27*, 23, 28737–28748, 2020.
16. Fincheira, P., Tortella, G., Seabra, A.B., Quiroz, A., Diez, M.C., Rubilar, O., Nanotechnology advances for sustainable agriculture: Current knowledge and prospects in plant growth modulation and nutrition. *Planta*, *254*, 4, 66, 2021.
17. Ghayempour, S. and Montazer, M., Micro/nanoencapsulation of essential oils and fragrances: Focus on perfumed, antimicrobial, mosquito-repellent and medical textiles. *J. Microencapsul.*, *33*, 6, 497–510, 2016.
18. Grillo, R., Abhilash, P.C., Fraceto, L.F., Nanotechnology applied to bio-encapsulation of pesticides. *J. Nanosci. Nanotechnol.*, *16*, 1, 1231–1234, 2016.
19. Guilger-Casagrande, M., Germano-Costa, T., Pasquoto-Stigliani, T., Fraceto, L.F., Lima, R., Biosynthesis of silver nanoparticles employing Trichoderma harzianum with enzymatic stimulation for the control of Sclerotinia sclerotiorum. *Sci. Rep.*, *9*, 1, 14351, 2019.
20. Hasan, M., Ahmad-Hamdani, M.S., Rosli, A.M., Hamdan, H., Bioherbicides: An eco-friendly tool for sustainable weed management. *Plants (Basel, Switzerland)*, *10*, 6, 1212, 2021.
21. He, X., Deng, H., Hwang, H.M., The current application of nanotechnology in food and agriculture. *J. Food Drug Anal.*, *27*, 1, 1–21, 2019.
22. Huynh, K.H., Lee, K.Y., Chang, H., Lee, S.H., Kim, J., Pham, X.H., Lee, Y.S., Rho, W.Y., Jun, B.H., Bioapplications of nanomaterials. *Adv. Exp. Med. Biol.*, *1309*, 235–255, 2021.
23. Iavicoli, I., Leso, V., Beezhold, D.H., Shvedova, A.A., Nanotechnology in agriculture: Opportunities, toxicological implications, and occupational risks. *Toxicol. Appl. Pharmacol.*, *329*, 96–111, 2017.
24. Khandelwal, N., Barbole, R.S., Banerjee, S.S., Chate, G.P., Biradar, A.V., Khandare, J.J., Giri, A.P., Budding trends in integrated pest management using advanced micro- and nano-materials: Challenges and perspectives. *J. Environ. Manag*, *184*, Pt 2, 157–169, 2016.
25. Kong, X.P., Zhang, B.H., Wang, J., Multiple roles of mesoporous silica in safe pesticide application by nanotechnology: A review. *J. Agric. Food Chem.*, *69*, 24, 6735–6754, 2021.

26. Kumar, D., Kumar, P., Singh, H., Agrawal, V., Biocontrol of mosquito vectors through herbal-derived silver nanoparticles: Prospects and challenges. *Environ. Sci. Pollut. Res.*, *27*, 21, 25987–26024, 2020.
27. Kumar, S., Nehra, M., Dilbaghi, N., Marrazza, G., Hassan, A.A., Kim, K.H., Nano-based smart pesticide formulations: Emerging opportunities for agriculture. *J. Control. Release: Off. J. Controlled Release Soc.*, *294*, 131–153, 2019.
28. Kutawa, A.B., Ahmad, K., Ali, A., Hussein, M.Z., Abdul Wahab, M.A., Adamu, A., Ismaila, A.A., Gunasena, M.T., Rahman, M.Z., Hossain, M., II, Trends in nanotechnology and its potentialities to control plant pathogenic fungi: A review. *Biology*, *10*, 9, 881, 2021.
29. Lemley, M.A., Patenting nanotechnology. *Stanford Law Rev.*, *58*, 2, 601–630, 2005.
30. Liu, C., Zhou, H., Zhou, J., The applications of nanotechnology in crop production. *Molecules (Basel, Switzerland)*, *26*, 23, 7070, 2021.
31. Liu, L., Chen, K., Xiang, N., Ni, Z., Dielectrophoretic manipulation of nanomaterials: A review. *Electrophoresis*, *40*, 6, 873–889, 2019.
32. Luiz de Oliveira, J., Ramos Campos, E.V., Fraceto, L.F., Recent developments and challenges for nanoscale formulation of botanical pesticides for use in sustainable agriculture. *J. Agric. Food Chem.*, *66*, 34, 8898–8913, 2018.
33. Najafzadeh, M., Bakhshandeh, H., Zamani, Z., Movahedi, M., Atyabi, S.M., Preparation, characterization, and evaluation of eosin B-loaded nano-liposomes for growth inhibition of Plasmodium falciparum. *Parasitol. Res.*, *121*, 1, 383–393, 2022.
34. Nuruzzaman, M., Rahman, M.M., Liu, Y., Naidu, R., Nanoencapsulation, nano-guard for pesticides: A new window for safe application. *J. Agric. Food Chem.*, *64*, 7, 1447–1483, 2016.
35. Pereira, A.E., Grillo, R., Mello, N.F., Rosa, A.H., Fraceto, L.F., Application of poly(epsilon-caprolactone) nanoparticles containing atrazine herbicide as an alternative technique to control weeds and reduce damage to the environment. *J. Hazard. Mater.*, *268*, 207–215, 2014.
36. Rahman, K., Khan, S.U., Fahad, S., Chang, M.X., Abbas, A., Khan, W.U., Rahman, L., Haq, Z.U., Nabi, G., Khan, D., Nano-biotechnology: A new approach to treat and prevent malaria. *Int. J. Nanomed.*, *14*, 1401–1410, 2019.
37. Rai, M., Ingle, A.P., Paralikar, P., Anasane, N., Gade, R., Ingle, P., Effective management of soft rot of ginger caused by Pythium spp. and Fusarium spp.: Emerging role of nanotechnology. *Appl. Microbiol. Biotechnol.*, *102*, 16, 6827–6839, 2018.
38. Rai, M., Ingle, A.P., Paralikar, P., Gupta, I., Medici, S., Santos, C.A., Recent advances in use of silver nanoparticles as antimalarial agents. *Int. J. Pharm.*, *526*, 1–2, 254–270, 2017.
39. Rajput, V.D., Singh, A., Minkina, T., Rawat, S., Mandzhieva, S., Sushkova, S., Shuvaeva, V., Nazarenko, O., Rajput, P., Komariah, Verma, K.K., Singh, A.K., Rao, M., Upadhyay, S.K., Nano-enabled products: Challenges and

opportunities for sustainable agriculture. *Plants (Basel, Switzerland), 10*, 12, 2727, 2021.

40. Raliya, R., Saharan, V., Dimkpa, C., Biswas, P., Nanofertilizer for precision and sustainable agriculture: Current state and future perspectives. *J. Agric. Food Chem.*, *66*, 26, 6487–6503, 2018.
41. Rankic, I., Zelinka, R., Ridoskova, A., Gagic, M., Pelcova, P., Huska, D., Nano/microparticles in conjunction with microalgae extract as novel insecticides against Mealworm beetles, Tenebrio molitor. *Sci. Rep.*, *11*, 1, 17125, 2021.
42. Rawtani, D., Khatri, N., Tyagi, S., Pandey, G., Nanotechnology-based recent approaches for sensing and remediation of pesticides. *J. Environ. Manag.*, *206*, 749–762, 2018.
43. Santamaria A., Historical overview of nanotechnology and nanotoxicology. *Methods in Molecular Biology* (Clifton, N.J.), 926, 1–12, 2012. https://doi.org/10.1007/978-1-62703-002-1_1
44. Scott, N.R., Chen, H., Cui, H., Nanotechnology applications and implications of agrochemicals toward sustainable agriculture and food systems. *J. Agric. Food Chem.*, *66*, 26, 6451–6456, 2018.
45. Shahzad, K. and Manzoor, F., Nanoformulations and their mode of action in insects: A review of biological interactions. *Drug Chem. Toxicol.*, *44*, 1, 1–11, 2021.
46. Singh, A., Tiwari, S., Pandey, J., Lata, C., Singh, I.K., Role of nanoparticles in crop improvement and abiotic stress management. *J. Biotechnol.*, *337*, 57–70, 2021.
47. Singh, H., Sharma, A., Bhardwaj, S.K., Arya, S.K., Bhardwaj, N., Khatri, M., Recent advances in the applications of nano-agrochemicals for sustainable agricultural development. *Environ. Sci. Process Impacts*, *23*, 2, 213–239, 2021.
48. Singh, S., Sangwan, S., Sharma, P., Devi, P., Moond, M., Nanotechnology for sustainable agriculture: An emerging perspective. *J. Nanosci. Nanotechnol.*, *21*, 6, 3453–3465, 2021.
49. Szigeti, Z. and Lehoczki, E., A review of physiological and biochemical aspects of resistance to atrazine and paraquat in Hungarian weeds. *Pest Manag. Sci.*, *59*, 4, 451–458, 2003.
50. Tippannanavar, M., Verma, A., Kumar, R., Gogoi, R., Kundu, A., Patanjali, N., Preparation of nanofungicides based on imidazole drugs and their antifungal evaluation. *J. Agric. Food Chem.*, *68*, 16, 4566–4578, 2020.
51. Veigas, B., Pedrosa, P., Carlos, F.F., Mancio-Silva, L., Grosso, A.R., Fortunato, E., Mota, M.M., Baptista, P.V., One nanoprobe, two pathogens: Gold nanoprobes multiplexing for point-of-care. *J. Nanobiotechnology*, *13*, 48, 2015.
52. Zeng, L., Liu, Y., Pan, J., Liu, X., Formulation and evaluation of norcanthridin nanoemulsions against the Plutella xylostella (Lepidotera: Plutellidae). *BMC Biotechnol.*, *19*, 1, 16, 2019.
53. Zhao, M., Hierarchical assemblies of polypeptoids for rational design of advanced functional nanomaterials. *Biopolymers*, *112*, 9, e23469, 2021.

54. Zhao, X., Cui, H., Wang, Y., Sun, C., Cui, B., Zeng, Z., Development strategies and prospects of nano-based smart pesticide formulation. *J. Agric. Food Chem.*, *66*, 26, 6504–6512, 2018.
55. Zobir, S., Ali, A., Adzmi, F., Sulaiman, M.R., Ahmad, K., A review on nanopesticides for plant protection synthesized using the supramolecular chemistry of layered hydroxide hosts. *Biology*, *10*, 11, 1077, 2021.
56. Thangadurai, D., Adetunji, C.O., Micheal, O.S., Nwanko, W., Kadiri, O., Adetunji, J.B., Nanostructured biomaterials for targeted drug delivery, in: *Biogenic Nanomaterials*, pp. 83–108, Apple Academic Press, 2022.
57. Adetunji, C.O., Olaniyan, O.T., Okeke, N.E., Production of next-generation biodiesel from high yielding strains of microorganisms: Recent advances, in: *Sustainable Green Chemical Processes and their Allied Applications. Nanotechnology in the Life Sciences*, A.A. Inamuddin (Ed.), Springer, Cham, 2020, https://doi.org/10.1007/978-3-030-42284-4_2.
58. Jhala, Y.K., Panpatte, D.G., Adetunji, C.O., Vyas, R.V., Shelat, H.N., Management of biotic and abiotic stress affecting agricultural productivity using beneficial microorganisms isolated from higher altitude agro-ecosystems: A remedy for sustainable agriculture, in: *Microbiological Advancements for Higher Altitude Agro-Ecosystems & Sustainability. Rhizosphere Biology*, R. Goel, R. Soni, D. Suyal (Eds.), Springer, Singapore, 2020, https://doi.org/10.1007/978-981-15-1902-4_7.
59. Adetunji, C.O., Samuel, M.O., Adetunji, J.B., Oluranti, O.I., Corn silk and health benefits, in: *Medical Biotechnology, Biopharmaceutics, Forensic Science and Bioinformatics*, pp. 167–178, CRC Press, 2022.
60. Olaniyan, O.T., Adetunji, C.O., Adetunji, J.B., Oloke, J.K., Okakeye, O.O., Hefft, D.I., Ubi, B.E., Soy-based food products consumed in Africa: A Panacea to mitigate food insecurity and health challenges, in: *Fermentation and Algal Biotechnologies for the Food, Beverage and Other Bioproduct Industries*, pp. 87–103, CRC Press, 2022.
61. Adetunji, C.O. and Oyeyemi, O.T., Antiprotozoal activity of some medicinal plants against Entamoeba histolytica, the causative agent of amoebiasis, in: *Medical Biotechnology, Biopharmaceutics, Forensic Science and Bioinformatics*, pp. 341–358, CRC Press, 2022.
62. Panpatte, D.G., Jhala, Y.K., Adetunji, C.O. (Eds.), *Microbial Rejuvenation of Polluted Environment*. Springer, 2021.
63. Adetunji, C.O., Michael, O.S., Nwankwo, W., Anani, O.A., Adetunji, J.B., layinka, A. S., & Akram, M., Biogenic nanoparticles based drugs derived from medicinal plants: A sustainable Panacea for the treatment of malaria, in: *Green Synthesis in Nanomedicine and Human Health*, pp. 103–122, CRC Press, 2021.

64. Adetunji, J.B., Adetunji, C.O., Olaniyan, O.T., Masajuwa, F.U., Islam, S., Thangadurai, D., ... & Ubi, B.E., Nutritional and health Benefits of nutraceutical beverages derived from cocoa and other caffeine products: A comprehensive review, in: *Fermentation and Algal Biotechnologies for the Food, Beverage and Other Bioproduct Industries*, pp. 105–118, 2022.
65. Adetunji, C.O., Olaniyan, O.T., Adetunji, J.B., Osemwegie, O.O., Ubi, B.E, African mushrooms as functional foods and nutraceuticals, in: *Fermentation and Algal Biotechnologies for the Food, Beverage and Other Bioproduct Industries*, pp. 233–251, 2022.

12

Relevance of Nanomaterials Derived From Medicinal Plants for Marine and Terrestrial Environments: Recent Advances

Charles Oluwaseun Adetunji[1*], Olalekan Akinbo[15], John Tsado Mathew[2], Chukwuebuka Egbuna[3,11], Abel Inobeme[4], Olotu Titilayo[5], Olulope Olufemi Ajayi[6], Wadazani Dauda[7], Shakira Ghazanfar[8], Frank Abimbola Ogundolie[9], Juliana Bunmi Adetunji[10], Babatunde Oluwafemi Adetuyi[13], Shakirat Oloruntoyin Ajenifujah-Solebo[14] and Abdullahi Tunde Aborode[12]

[1]*Applied Microbiology, Biotechnology and Nanotechnology Laboratory, Department of Microbiology, Edo University Iyamho, Auchi, Edo State, Nigeria*
[2]*Department of Chemistry, Ibrahim Badamasi Babangida University Lapai, Niger State, Nigeria*
[3]*Department of Biochemistry, Faculty of Natural Sciences, Chukwuemeka Odumegwu Ojukwu University, Uli Campus, Anambra State, Nigeria; Nutritional Biochemistry and Toxicology Unit, World Bank Africa Centre of Excellence, Centre for Public Health and Toxicological Research (PUTOR), University of Port-Harcourt, Rivers State, Nigeria*
[4]*Department of Chemistry, Edo University Iyamho, Nigeria*
[5]*Department of Microbiology, Adeleke University, Ede Osun State, Nigeria*
[6]*Department of Biochemistry, Edo University Iyamho, Edo State, Nigeria*
[7]*Department of Crop Protection, Federal University Gasua, Nigeria*
[8]*Functional Genomics and Bioinformatics, National Agricultural Research Centre, Islamabad, Pakistan*
[9]*Department of Biotechnology, Baze University Abuja, Nigeria*
[10]*Department of Biochemistry, Osun State University, Osogbo, Nigeria*
[11]*Nutritional Biochemistry and Toxicology Unit, World Bank Africa Centre of Excellence, Centre for Public Health and Toxicological Research (PUTOR), University of Port-Harcourt, Rivers State, Nigeria*
[12]*Healthy Africans Platform, Research and Development, Ibadan, Nigeria*
[13]*Department of Natural Sciences, Biochemistry Unit, Faculty of Pure and Applied Sciences, Precious Cornerstone University, Ibadan, Nigeria*

**Corresponding author*: adetunjicharles@gmail.com; adetunji.charles@edouniversity.edu.ng
ORCID: 0000-0003-3524-6441

Charles Oluwaseun Adetunji and Julius Kola Oloke (eds.) Handbook of Agricultural Biotechnology: Volume I Nanopesticides, (311–326) © 2024 Scrivener Publishing LLC

[14]*National Biotechnology Development Agency Umar Musa Yardua Way, Lugbe, Wuse, Abuja Nigeria*
[15]*AUDA-NEPAD Centre of Excellence in Science, Technology and Innovation, Stellenbosch, South Africa*

Abstract
Over the years, significant attention has been given to nanomaterials derived from plant-based products for several health-enhancing physiological properties like antiviral, antimicrobial, antioxidant, antineoplastic, and anti-inflammatory activity. Many of the bioactive molecules derived from these plants, such as chitosan, carrageenan, polysaccharides, and fucoidan, are known to possess useful therapeutic properties against several metabolic and nonmetabolic diseases, like cancer, wound healing, hypertension, infertility, and several others. The major challenge in the utilization of nanoparticles is the genetic heterogeneity that contributes to the change in the efficacy of nanoparticles, thereby limiting their pharmacological effectiveness. Nanoparticles are also exploited for genome editing utilizing patients' genetic information due to their significant biocompatibility, renewability, and safety. Thus, this chapter offers a general indication of the therapeutic potentials of nanodrugs synthesized using genetically engineered metabolites derived from medicinal plants from the marine and terrestrial environment.

Keywords: Nanomaterials, physiological properties, bioactive molecules, diseases, medicinal plants

12.1 Introduction

Nanoparticles designed with plant products are currently gaining relevance. Plant biomolecules, including carrageenan and chitosan are known for their health-enhancing effects and are suitable for drug delivery [1]. Marine plant-derived polysaccharides have been reported to be effective in treating cancers [2]. Fucoidan, derived from marine plants plays a significant role in wound healing; this is based on its anticoagulant properties.

Nanotechnology has greatly improved the diagnosis and treatment of diseases. Certain challenges are emerging which include genetic heterogeneity of animal species. This has limited treatment in patients because genetics could change the effectiveness of nanoparticles. Therefore, the design of genetically engineered nanodrugs is dependent on patient's information and is aimed at overcoming challenges as they apply to the general population [3].

Cancer is a heterogeneous disease, hence, requires precision mechanisms in the treatment. Mannose is usually employed to target macrophages and dendritic cells. It is evident that nanoparticles coated with dextran, galactose and sialoadhesin can effectively target macrophages. Polymeric hydrogels, as well as a scaffold, can effectively interact with APC, therefore NPs-specific for APC ensures reprogramming APCs [4] and enhancement of the interaction of antigen with the APC, thereby reducing the dose required for therapeutic effects and boosting antigen-dependent treatments. Stimulator of interferon genes (STING) also results in the activation of immune cells with antitumor effects. The downside of cyclic dinucleotides despite their antitumor effects is their instability and polarity which minimizes their uptake by the cells.

tLyp1 peptide is being used to target regulatory T cells, known for its immunosuppressive effects [5]. The effectiveness of NPs for delivering monoclonal antibodies is known. Their ability to deliver small interfering RNAs has also been reported [6]. Nanoparticles are also explored for genome editing. Significantly, efforts have been made in TALEN (transcription activator-like effector nuclease) and CRISPR and zin finger nuclease (ZFN). Studies have shown that lipid and polymer-based nanoparticles are used for nucleic acid delivery. Onpattro, an LNP siRNA drug is used for treating amyloidosis [7].

12.2 Modes of Action of Nanodrugs Synthesized Using Genetically Engineered Metabolite

Marine seaweed polysaccharides have received attention in recent years for their possible use in enhancing health. They are known to possess health-boosting properties including; antiviral, antimicrobial, antioxidant, antineoplastic, and anti-inflammatory properties. Their targeted drug delivery potential is also being explored because of their biocompatibility, renewability, and safety [1, 8–14].

Carrageenan (CG) is an anionic sulphated polysaccharide consisting of α-D-galactose and β-D-galactose or 3, 6-anhydro-α-D-galactose deposits linked to each other via β-1, 4, and α-1, 3 glycosidic bonds in an alternating arrangement. Carrageenan is an important component of red seaweeds such as *Eucheuma denticulatum, Kappaphycus alvarezii* [15]. There are six types of CG depending on sulphate content and position; Theta-CG, lambda-CG, Iota-CG, Mu-CG Kappa-CG, and Nu-CG.

Agar is polymeric sugar-containing agarose and agaropectin primarily found in Gracilariaceae and Gelidiaceae [15]. Its solubility in boiling water, gel formation at temperatures ranging from 32°C to 40°C and its liquefaction at 80°C makes it an emulsifier, lubricant and thickener and hence, its application in the pharmaceutical and food industries [16]. Alginate is a polysaccharide in a linear orientation consisting of β-D-mannuronic acid (M) and α-L-guluronic acid (G) residues in two building successive or alternating blocks [17]. It is primarily obtained from brown seaweeds including; Lessonia Ecklonia, Ascophyllum, Sargassum, etc. in calcium or salt forms [18]. Its biodegradable, biocompatible, and nontoxic properties make it suitable for pharmaceutical use.

Fucoidans is a major constituent of the cell wall of brown algae. They are established in seaweeds like *Ecklonia cava, Ascophyllum nodusum.* It consists of sulphate, as well as L-fucose ester groups alongside other monosaccharides [1]. Laminaran is found in brown seaweeds with higher concentrations in *Saccharina* and *Laminaria* species. It is a major storage carbohydrate. Furcellaran is obtained from *Furcellaria fastigiata*, a red seaweed. It consists of D-galactose, 3, 6-anhydro-D-galactose along with the sulphated types of these sugars. Its antioxidant, apoptotic, and hypolipidemic effects have been reported [16]. Ulvan is a sulphated polyelectrolyte consisting of disaccharides units; rhamnose and xylose, as well as glucuronic acid [19]. Its derivatives are iduronic acid and sulphated rhamnose [20].

Nanoparticles consisting of carrageenan and chitosan showed effective therapeutic and drug delivery effects. Furthermore, no toxicity was observed [19]. Oligocarrageenan modified with polycaprolactone loaded with curcumin showed an efficient delivery of curcumin, with no side effects [1].

Polysaccharides derived from marine plants have shown antineoplastic effects. They are more effective than conventional therapies. They kill cancer cells and avert metastasis via mechanisms including nitric oxide production, depolarization of mitochondrial membrane and cell cycle arrest. The upregulation of TNF-α, interleukins 10 and 6 by sulphated polysaccharides has been reported [2].

In a study, fucoidan modulated the effects of transforming growth factor-beta1 (TGF-β1) on wound repopulation and fibroblast proliferation [21]. Fucoidan has been utilized as an anticoagulant and as an anti-inflammatory agent in the healing of the wound [22].

The role of fucoidan in tissue engineering has been reported. It enhances bone matrix mineral deposition via growing the expression of osteocalcin, alkaline phosphatase and type-1 collagen activity [23]. Fucoidan has as well been established to be efficient in treating osteoarthritis [24] Alginate is also important in bone and cartilage regeneration [25]. Alginate conjugated with chitosan enhanced bone strength and structural stability. It also improved vascularization and enhanced osteogenesis [26].

Marine microalgae are currently receiving attention because they are sources of health-boosting constituents, although a lot of its species are yet to be explored. They are easily cultured also have shorter generation times relative to terrestrial plants [27]. Furthermore, they are of both importance in primary and secondary metabolites. The anticancer potentials of diatoms are currently being studied. This is premised on their wide distribution, range of chemical compounds and abundance [28, 29]. Diatoms are known to produce about 20% of the global oxygen supply and 40% of the ocean's annual organic carbon [30].

Diatoms are rich sources of anticancer compounds including; oxylipins, monoacylglycerides, fucoxanthin, chrysolaminaran polysaccharide, haslenes, stigmasterol, marennine, and fatty alcohol ester [27].

Monacylglycerides: This was derived from *Skeletonema marinoi.* It induced cell death in U-937, as well as HCT-116 cell lines via caspase 3/7 [31]. Fucoxanthin is a major carotenoid found in Bacillariophyta diatoms. Studies have shown its antineoplastic, anti-inflammatory, and antioxidant properties. The antiproliferative could be attributed to the induction of apoptosis by the activation of caspase 3/7. In another study, the induction of apoptosis by fucoxanthin led to DNA fragmentation, reduction in cell sizes, nuclei damage, and chromatin condensation [32]. Crysolaminaran polysaccharides have been explored for their immune-modulatory potentials in cancer immunotherapy, hence a suitable alternative to conventional cancer chemotherapy [33]. Reports have shown its use in treating colon cancer [27].

Oxylipins (OXLs). They are produced by an integration of oxygen through the carbon chains of polyunsaturated fatty acids during hydroxylase/lipoxygenase [34]. Their antiproliferative effects by the induction of extrinsic pathways have been reported [27].

Fatty alcohol ester nonyl 8-acetoxy-6-methylocatanoate (NAMO). This is obtained from *P. tricornutum* Bohlin. Its anticancer effect on promyelocytic leukaemia has been reported. A report showed DNA condensation, increase in pro-apoptotic protein (Bax), as well as cell cycle arrest at sub-G1 phase, caspase, p53 as well as repression of Bcl-x in HL-60-7 cells treated with NAMO [35].

A considerable quantity of adenosine is found in diatoms. The role of adenosine in cellular energy transduction is well known. Its antineoplastic effect includes activation of caspases via the extrinsic as well as intrinsic pathways [36]. It also increases the manufacture of reactive oxygen species in cancerous cells [36]. Marennine: this is derived from *Haslea ostrearia.* Reports have shown its antioxidant and anti-inflammatory effects on lung, kidney and skin cancers, in which considerable induction of apoptosis was observed [27]. Stigmasterol: This is a monohydroxy phytosterol that structurally looks like cholesterol. It lowers cholesterol levels in the plasma hence protecting against cardiovascular diseases. Furthermore, it induced apoptosis via the intrinsic pathway when administered to HepG2 cells [37]. Haslene: also known as Hasla-6(17), 9,13,23-tetraene. It was accounted to induce cell cycle arrest at the G1/S phase [38].

Thalassiosira pseudonana, a genetically modified diatom was produced to enhance IgG-binding on biosilica surface with the view to targeting cancerous cells [39]. In a study, anti-p75NTR loaded GB1 biosilica frustule was prepared which reacted with the antibody for p75 neurotrophin receptor located on the neuroblastoma cell's surface. This suggests site-specific drug delivery potentials.

Bacterial biofilm has been implicated in a lot of pathologies including cystic fibrosis, bacterial endocarditis, otitis media and hospital-acquired infections which do not usually respond to antibiotics [40]. Biofilm-related organisms are implicated in over 60% of human microbial infections [41]. Melia dubia bark extract suppressed hemolysis and biofilm formation of Escherichia coli [42]. Furthermore, biofilm formation was inhibited by the extract of Lagerstroemia speciosa fruit [43]. Green tea inhibited *Streptococcus mutans* growth [44]. The antioxidant, antibacterial and anti-inflammatory potentials of honey have been reported. It inhibits biofilm formation. Reports showed that it inhibited *Enterococcus* spp biofilm formation and maybe therapy for biofilm-associated enterococcal infections [40]. The ability of honey to produce defensin1, an antimicrobial peptide also enhances its antibiofilm potential. Essential oils derived from cinnamon, *cuminum cyminum* and oregano have been found effective against biofilm formation. Cumin oil obtained from Cuminum cyminum has been reported to possess antibiofilm properties. This was observed on *K. pneumonia* [45]. Cinnamon oil inhibited *S mutans* and *Lactobacillus plantarum*-biofilm [46]. Oregano oil inhibited *E. coli* biofilm [40].

12.3 Nanodrugs Synthesized Using Genetically Engineered Metabolites From Plants

Green synthesis technology uses microbes and plants to make nanoparticles that are cost-effective, biologically safe, and environmentally benign.

Plants and microbes have developed the ability to consume as well as build up inorganic metal ions from their surroundings. Both extracellular and intracellular nanoparticles are known to be synthesized by living entities. The ability of a biological system to apply its organic chemistry mechanisms to modify inorganic metallic ions into nanoparticles has opened up a hitherto unknown field of biochemical research. Algae, actinomycetes, cyanobacteria, viruses, bacteria, fungus, yeasts, and plants are examples of living things of both eukaryotic and prokaryotic origin that are involved in nanobiotechnology. The ability of each biological system to provide metallic nanoparticles differs. Furthermore, due to their inherent metabolic processes and enzymatic activities, not all living species can create nanoparticles. These biologically synthesized metallic nanoparticles can be used for a wide range of pharmaceutical formulations, comprising gene delivery or drug, protein detection or pathogen, and tissue engineering. By use of nanotechnology to improve tissue engineering and medication delivery has contributed significantly to biomedical research on pharmaceutical goods and uses. This study addresses both the green manufacturing of nanoparticles and their utilization using different biological systems [47].

Metal nanoparticles are currently being used in a variety of medicinal, pharmacological, and agricultural purposes. Nanobiotechnology, when integrated using green chemistry, has enormous promise for developing unique and essential products that environment, enhance human health, and industry. Green chemistry plays a significant role because of its contribution to novel methods of synthesis of gold as well as silver nanoparticles through plant extracts, which have shown antimicrobial properties among several other remarkable qualities. Biodiversity rich nations must accumulate as well as convert knowledge from the natural resource into techniques, chemicals, schemes, and instruments, all while ensuring that biological variety is used and exploited sustainably [48–50]. As a result, this research discusses the applicable synthesis of green silver along with gold nanoparticles through the plant extracts, as well as their ability to operate as antibacterial compounds in the field of agriculture to combat fungal and bacterial pathogens which could cause plant, foodborne and waterborne infections. Furthermore, this paper provides a brief overview of nanoparticles' role in the treatment of water and the creation of "eco-friendly" nanopesticides, nanoherbicides, and nanofertilizers, as well as the negative impacts of nanoparticle deposition in soils and plants [51].

Through the means of various plant extracts and microbes such as fungi, microalgae, bacteria as well as viruses, a wide number of metallic nanoparticles have been successfully manufactured. Phytopathogens demonstrated considerable antibacterial activity against certain of these metallic nanoparticles. These green-synthesized nanoparticles using plants and microorganisms, as well as their applications in the treatment of plant diseases are presented here. The potential negative impacts of nanostructured materials on plants and beneficial microbial populations connected utilizing plants were also considered. In general, this study highlights the need for more research into the application of synthesized green metallic nanoparticles in plant disease prevention, as well as clarification of the hazards to plants, a plant related to microbial activity, and the ecosystem before they are used in agriculture [52].

Nanotechnology is a branch of research that combines physics, biology, material sciences and chemistry to create unique medicinal nano-sized components for biomedical and pharmaceutical purposes. Different macro–microscopic species, such as plants, fungi, bacteria, microalgae, and seaweeds, carry out biological nanoparticle synthesis. Biosynthesized nanomaterials have proven to be successful in managing endemic diseases while having fewer side effects. Natural chemicals including flavonoids, alkaloids, steroids, saponins, tannins, and other nutritious components abound in plants. Plant materials including stems, leaves, shoots, roots, barks, seeds as well as flowers, are used to make these organic remedies. Several recent investigations have shown that plant extracts can be used as a nonhazardous precursor for the creation of nanomaterials. The plant extract functions as a reducing along with stabilizing agent for the bioreduction process to create unique metallic nanoparticles because it comprises diverse secondary metabolites. In the creation of nanoparticles, nonbiological processes (physical and chemical) are utilized, posing a major risk and high toxicity to living creatures. Furthermore, the biological synthesis of metallic nanoparticles is a low-cost, one-step, as well as environmentally benign process. Cobalt, copper, silver, gold, palladium, zinc oxide, platinum, as well as magnetite are among the greener nanoparticles effectively synthesized using plants. Furthermore, plant-mediated nanomaterials may be employed to take care of a variety of ailments, including cancer, malaria, hepatitis, HIV, and other acute illnesses [53].

Many types of research demonstrate superior pharmacological effects of green generated nanomaterials comparable to those produced through other techniques, in addition to demonstrating the convenience and environmentally friendly approach of synthesizing nanomaterials. This is most likely due to bioactive chemicals being deposited on the nanoparticles' surfaces. Current investigations, on the other hand, have proven biocompatibility and the nanodimension of metal ash (Bhasma) formulations, that have long been solubilized with biological products and utilized in Indian medication for the cure of numerous diseases [54].

Nature functions as a vast "biolaboratory," including algae, plants, yeast, fungi, and other biomolecule-based organisms. These naturally present biomolecules have already been discovered to provide a key role in the production of nanoparticles of various sizes and shapes, paving the way for the development of greener, safer, and environmentally friendly nanoparticle synthesis techniques. The goal of this review is to compare biogenic nanoparticle formation and processes utilizing algae and waste sources (like agrowaste when biomolecules are present). The use of waste products not just lowers synthesis costs, but also decreases the need for harmful chemicals as well as promotes "green synthesis." Also, it covers the analytical components of biomolecule adsorption to nanoparticles, as well as some of the biologically synthesized nanoparticles applications in the catalysis, biosensors as well as biomedical domains [55].

The current study focused on the biosynthesis of AgNPs comprising phenolic chemicals, glycosidic pharmacological macromolecules along structural polymers using *Coriander sativum* (CS). Plant phenolic chemicals, including carotenoids and flavonoids, could be potent antioxidants, attractants, and lignin. AgNPs (Silver nanoparticles) were made extracellularly mostly in the future through combining the reduction and stabilizing actions of the CS leaf extract. UV spectroscopy, scanning electron microscopy (SEM), zeta potential determination, and energy-dispersive X-ray (EDX) analysis was used to characterize as well as validate the production of crystalline nanoparticles in biologically produced CS-AgNPs. These nanoparticles were shown to have substantial antibacterial action against all of the microbial strains tested to varying degrees. In comparison to normal antioxidant activity, the shielding effect of CS-AgNPs on reactive oxygen species exhibited significant antioxidant effectiveness using hydroxyl radicals and superoxide at various doses. The existence of anticancer properties *in vitro* was revealed on the MCF-7 cell line at various concentrations, as evidenced by a reduction in cell effectiveness that was commensurately dependent on the concentration of CS-AgNPs, demonstrating the existence of the pathogen of synthesized nanoparticles on cancer cells [56].

12.4 Nanoparticles Derived From Medicinal Plants Terrestrial Environment

Nanoparticles (NPs) have a small size and a high specific surface area, which confers certain physicochemical qualities such as electrical conductivity, strength, and optical properties. Natural or anthropogenic sources, including designed or incidental/unwanted NPs, can be used to make NPs. Engineered NPs can be used in a range of fields, such as biomedical, electronic, cosmetic, pharmaceutical, environmental, energy, materials research and catalysis, morphology and dimension. More information about the effects of these nanomaterials (NMs) in the environment, especially on animal/human health and plant productivity, is essential because nanotechnology is a creative and innovative development field with accelerating manufacturing. As a result, NP study as an emerging pollutant is a new subject in environmental health. Finally, the impact of NPs on plants at many levels is highlighted, including physiology, genotoxicity and morphology. By examining the data gathered, it is clear that investigation on NPs phytotoxicity is still only getting started, and that more extensive investigations are necessary not just on NPs genotoxicity and cytotoxicity, as well as on the best along with most effective methodologies for determining NPs toxicity [57].

Plant-mediated nanomaterial production is gaining prominence due to its cost-effectiveness and environmental friendliness. Researchers used aqueous extracts of fresh leaves of Lantana Camara and Impatiens balsamina medicinal plants as reducing agents to make silver (Ag) nanoparticles in this investigation. Transmission electron microscopy (TEM) and UV-visible (UV-vis) spectrophotometry were used to establish the nanoparticles synthesis using this approach. After treatment with Ag precursors, the color of the fresh leaf extracts of I. balsamina as well as *L. camara* changed to brownish-yellow and grayish brown, respectively, according to UV-vis spectra and ocular inspection. Furthermore, transmission electron microscopy investigation indicated that AgNO3 solutions formed Ag nanoparticles at all concentrations, with an average size of less than 24 nm. Furthermore, the antibacterial activity of aqueous leaf extracts of *I. balsamina* along with *L. camara* against Gram-positive *Staphylococcus aureus* and Gram-negative *Escherichia coli* bacteria was investigated separately [58].

Because of the abundance of plants and their different metabolites, plant-mediated production of silver nanoparticles is chosen. The diversity and potential of biologically active compounds of AgNPs derived from the synthesis of silver with phytoconstituents of largely historically used Sudanese medicinal and aromatic plants are investigated in this paper. A comprehensive assessment of the green methods of synthesis of silver nanoparticles facilitated by over 45 historically utilized medicinal herbs is presented. Parameters affecting the production of plant-mediated AgNPs, as well as their characterization approaches and biologically active compounds, are outlined and explored [59].

Plant cells are routinely exposed to a variety of harmful environmental conditions affecting agricultural crop yield and plant growth. Rapid advancements in nanotechnology, as well as the indiscriminate release of metal-containing nanoparticles (NPs) into the ecosystem, create a major hazard to environmental receptors, such as plants. Chemical, physical, biological, or hybrid approaches are used to make engineered nanoparticles. Volcanic eruptions, mechanical grinding of earthquake-producing instabilities in the ocean spray, ultrafine cosmic dust and Earth's crust are among natural sources of NPs in the environment. It is crucial to understand the nature of symbioses using NPs to measure their absorption and dissemination and to estimate cytotoxic effects. Proteomic approaches due to mass spectrometry enable for exact detection of low abundance proteins, in-depth analysis of biological signaling networks and protein–protein interactions. Using high-throughput proteomics approaches, the current knowledge of the response of plants to NPs is highlighted in this study [60–62].

References

1. Bilal, M. and Iqbal, H.M.N., Marine seaweed polysaccharides-based engineered cues for the modern biomedical sector. *Mar. Drugs*, 18, 7, 2020.
2. Chen, X., Song, L., Wang, H., Liu, S., Yu, H., Wang, X., Li, P., Partial characterization, the immune modulation and anticancer activities of sulfated polysaccharides from filamentous microalgae Tribonema sp. *Molecules*, 24, 322, 2019.
3. Mitchell, M.J., Billingsley, M.M., Haley, R.M., Wechsler, M.E., Peppas, N.A., Langer, R., Engineering precision nanoparticles for drug delivery. *Nat. Rev. Drug Discov.*, 20, 2, 101–124, 2021.
4. Singh, B.N., Singh, H.B., Singh, A., Singh, B.R., Mishra, A., Nautiyal, C.S., *Lagerstroemia speciosa* fruit extract modulates quorum sensing-controlled virulence factor production and biofilm formation in *Pseudomonas aeruginosa*. *Microbiol.*, 158, Part 2, 529–38, 2012.
5. Ou, W., Thapa, R.K., Jiang, L., Soe, Z.C., Gautam, M., Chang, J.H., Jeong, J.H., Ku, S.K., Choi, H.G., Yong, C.S., Kim, J.O., Regulatory T cell-targeted hybrid nanoparticles combined with immuno-checkpoint blockage for cancer immunotherapy. *J. Control. Release*, 281, 84–96, 2018.
6. Deng, H. and Zhang, Z., The application of nanotechnology in immune checkpoint blockade for cancer treatment. *J. Control. Release*, 290, 28–45, 2018.

7. Riley, R.S., June, C.H., Langer, R., Mitchell, M.J., Delivery technologies for cancer immunotherapy. *Nat. Rev. Drug Discov.*, 18, 175–196, 2019.
8. Jhala, Y.K., Panpatte, D.G., Adetunji, C.O., Vyas, R.V., Shelat, H.N., management of biotic and abiotic stress affecting agricultural productivity using beneficial microorganisms isolated from higher altitude agro-ecosystems: A remedy for sustainable agriculture, in: *Microbiological Advancements for Higher Altitude Agro-Ecosystems & Sustainability*, Rhizosphere Biology, R. Goel, R. Soni, D. Suyal (Eds.), Springer, Singapore, 2020, https://doi.org/10.1007/978-981-15-1902-4_7.
9. Adetunji, C.O., Fawole, O.B., Arowora, K.A., Nwaubani, S.I., Ajayi, E.S., Oloke, J.K., Aina, J.A., Adetunji, J.B., Ajani, A.O., Postharvest quality and safety maintenance of the physical properties of *Daucus carota L.* fruits by Neem oil and Moringa oil treatment: A new edible coatings. *Agrosearch*, 13, 1, 131–141, 2013.
10. Adetunji, C.O., Adetunji, C.O., Michael, O.S., Kadiri, O., Varma, A., Akram, M., Oloke, J.K., Shafique, H., Adetunji, J.B., Jain, A., Bodunrinde, R.E., Ozolua, P., Ubi, B.E., Quinoa: From farm to traditional healing, food application, and phytopharmacology, in: *Biology and Biotechnology of Quinoa*, A. Varma (Ed.), Springer, Singapore, 2021, https://doi.org/10.1007/978-981-16-3832-9_20.
11. Adetunji, C.O., Akande, S.A., Oladipo, A.K., Salawu, R.A., Onyegbula, A.F., Determination of the microbiological quality and proximate composition of fermented cassava food products sold in Ilorin-west local government area, Nigeria. *Ruhuna J. Sci.*, 8, 76–89, 2017. http://doi.org/10.4038/rjs.v8i2.28.
12. Adetunji, C.O., Olaniyan, O.T., Adeyomoye, O., Dare, A., Adeniyi, M.J., Alex, E., Rebezov, M., Garipova, L., Shariati, M.A. *et al.*, eHealth, mHealth, and telemedicine for COVID-19 pandemic, in: *Assessing COVID-19 and Other Pandemics and Epidemics using Computational Modelling and Data Analysis*, S.K. Pani, S. Dash, W.P. dos Santos, S.A. Chan Bukhari, F. Flammini (Eds.), Springer, Cham, 2022, https://doi.org/10.1007/978-3-030-79753-9_10.
13. Adetunji, C.O., Michael, O.S., Varma, A., Oloke, J.K., Oloke, J.K., Akram, M., Bodunrinde, R.E., Imtiaz, A., Adetunji, J.B., Shahzad, K., Jain, A., Ubi, B.E., Majeed, N., Phebean Ozolua, P., Olisaka, F.N., Recent advances in the application of biotechnology for improving the production of secondary metabolites from Quinoa, in: *Biology and Biotechnology of Quinoa*, A. Varma (Ed.), Springer, Singapore, 2021, https://doi.org/10.1007/978-981-16-3832-9_17.
15. Kraan, S., Algal polysaccharides, novel applications and outlook, in: *Carbohydrates—Comprehensive Studies on Glycobiology and Glycotechnology*, C.F. Chang (Ed.), pp. 489–532, InTech, Rijeka, Croatia, 2012.
16. Venugopal, V., Sulfated and non-sulfated polysaccharides from seaweeds and their uses: An overview. *EC Nutr.*, 14, 126–141, 2019.
17. Venkatesan, J., Bhatnagar, I., Manivasagan, P., Kang, K.H., Kim, S.K., Alginate composites for bone tissue engineering: A review. *Int. J. Biol. Macromol.*, 72, 269–281, 2015.

18. Rupérez, P., Gómez-Ordóñez, E., Jiménez-Escrig, A., Biological activity of algal sulfated and nonsulfated polysaccharides, in: *Bioactive Compounds from Marine Foods: Plant and Animal Sources*, pp. 219–247, 2013, Ch11.
19. Cunha, L. and Grenha, A., Sulfated seaweed polysaccharides as multifunctional materials in drug delivery applications. *Mar. Drugs*, 14, E42, 2016.
20. Lahaye, M. and Robic, A., Structure and functional properties of ulvan, a polysaccharide from green seaweeds. *Biomacromolecules*, 8, 1765–1774, 2007.
21. O'Leary, R., Rerek, M., Wood, E.J., Fucoidan modulates the effect of transforming growth factor (TGF)-β1 on fibroblast proliferation and wound repopulation *in vitro* models of dermal wound repair. *Biol. Pharm. Bull.*, 27, 266–270, 2004.
22. Obluchinsksya, E.D., Makarova, M.N., Pozharitskaya, O.N., Shikov, A.N., Effects of ultrasound treatment on the chemical composition and anticoagulant properties of dry fucus extract. *Pharm. Chem. J.*, 49, 183–186, 2015.
23. Cho, Y.S., Jung, W.K., Kim, J.A., Choi, I.W., Kim, S.K., Beneficial effects of fucoidan on osteoblastic MG-63 cell differentiation. *Food Chem.*, 116, 990–994, 2009.
24. Park, S.B., Chun, K.R., Kim, J.K., Suk, K., Jung, Y.M., Lee, W.H., The differential effect of high and low molecular weight fucoidans on the severity of collagen-induced arthritis in mice. *Phytother. Res.*, 24, 1384–1391, 2010.
25. Barralet, J.E., Wang, L., Lawson, M., Triffitt, J.T., Cooper, P.R., Shelton, R.M., Comparison of bone marrow cell growth on 2D and 3D alginate hydrogels. *J. Mater. Sci. Mater. Med.*, 16, 515–519, 2005.
26. Li, Z., Ramay, H.R., Hauch, K.D., Xiao, D., Zhang, M., Chitosan–alginate hybrid scaffolds for bone tissue engineering. *Biomaterials*, 26, 3919–3928, 2005.
27. Hussein, H.A. and Abdullah, M.A., Anticancer compounds derived from marine diatoms. *Mar. Drugs*, 18, 356, 2020.
28. Sansone, C., Braca, A., Ercolesi, E., Romano, G., Palumbo, A., Casotti, R., Francone, M., Ianora, A., Diatom-derived polyunsaturated aldehydes activate cell death in human cancer cell lines but not normal cells. *PLoS One*, 9, e101220, 2014.
29. Delasoie, J., Rossier, J., Haeni, L., Rothen-Rutishauser, B., Zobi, F., Slow-targeted release of a ruthenium anticancer agent from vitamin B12 functionalized marine diatom microalgae. *Dalton Trans.*, 47, 17221–17232, 2018.
30. Dolatabadi, J.E.N. and de la Guardia, M., Applications of diatoms and silica nanotechnology in biosensing, drug and gene delivery, and formation of complex metal nanostructures. *TrAC Trends Anal. Chem.*, 30, 1538–1548, 2011.
31. Miceli, M., Cutignano, A., Conte, M., Ummarino, R., Romanelli, A., Ruvo, M., Leone, M., Mercurio, F.A., Doti, N., Manzo, E., Romano, G., Monoacylglycerides from the diatom *Skeletonema marinoi* induce selective cell death in cancer cells. *Mar. Drugs*, 17, 625, 2019.

32. Moreau, D., Tomasoni, C., Jacquot, C., Kaas, R., Le Guedes, R., Cadoret, J.P., Muller-Feuga, A., Kontiza, I., Vagias, C., Roussis, V., Roussakis, C., Cultivated microalgae and the carotenoid fucoxanthin from *Odontella aurita* as potent anti-proliferative agents in bronchopulmonary and epithelial cell lines. *Environ. Toxicol. Pharmacol.*, 22, 97–103, 2006.
33. Kusaikin, M.I., Ermakova, S.P., Shevchenko, N.M., Isakov, V.V., Gorshkov, A.G., Vereshchagin, A.L., Grachev, M.A., Zvyagintseva, T.N., Structural characteristics and antitumor activity of a new chrysolaminaran from the diatom alga Synedra acus. *Chem. Nat. Compd.*, 46, 1–4, 2010.
34. Yi, Z., *Biotechnological approaches to enhance fucoxanthin production in a model diatom phaeodactylum tricornutum*, Ph.D. Thesis, Faculty of Life and Environmental Sciences, University of Iceland, Reykjavík, Iceland, 2018, November 2018.
35. Samarakoon, K.W., Ko, J.Y., Lee, J.H., Kwon, O.N., Kim, S.W., Jeon, Y.J., Apoptotic anticancer activity of a novel fatty alcohol ester isolated from cultured marine diatom, *Phaeodactylum tricornutum. J. Funct. Foods*, 6, 231–240, 2014.
36. Liu, Y., Wang, J., Wang, W., Zhang, H., Zhang, X., Han, C., The chemical constituents and pharmacological actions of *Cordyceps sinensis. Evidence-Based Complementary and Alternative Medicine (eCAM)*, 2015, 1–15, 2015.
37. Kim, Y.S., Li, X.F., Kang, K.H., Ryu, B., Kim, S.K., Stigmasterol isolated from marine microalgae *Navicula incerta* induces apoptosis in human hepatoma HepG2 cells. *BMB Rep.*, 47, 433–438, 2014.
38. Rowland, S.J., Belt, S.T., Wraige, E.J., Masse, G., Roussakis, C., Robert, J., Effects of temperature on polyunsaturation in cytostatic lipids of *Haslea ostrearia. Phytochemistry*, 56, 597–602, 2001.
39. Delalat, B., Sheppard, V.C., Rasi Ghaemi, S., Rao, S., Prestidge, C.A., McPhee, G., Rogers, M.L., Donoghue, J.F., Pillay, V., Johns, T.G., Targeted drug delivery using genetically engineered diatom biosilica. *Nat. Commun.*, 6, 8791, 2015.
40. Sadekuzzaman, M., Yang, S., Mizan, M.F.R., Ha, S.D., Current and recent advanced strategies for combating biofilms. *Compr. Rev. Food Sci. Food Saf.*, 14, 491–509, 2015.
41. Shunmugaperumal, T., Biofilm irradication and prevention, a pharmaceuticals approach to medical device infection, in: *Introduction and Overview of Biofilm*, John Wiley & Sons, Inc, Hoboken, NJ, USA, 2010, published July 22, 2010.
42. Ravichandiran, V., Shanmugam, K., Anupama, K., Thomas, S., Princy, A., Structure-based virtual screening for plant-derived SdiA-selective ligands as potential antivirulent agents against uropathogenic *Escherichia coli. Eur. J. Med. Chem.*, 48, 200–205, 2012.
43. Singh, A. and Peppas, N.A., Hydrogels and scaffolds for immunomodulation. *Adv. Mater.*, 26, 6530–6541, 2014.

44. Faraz, N., Islam, Z.U., Rehman, R., Sehrish, Antibiofilm forming activity of naturally occurring compound. *Biomedica*, 28, 171- 175, 2012.
45. Safoura, D., Morteza, S., Mohsen, B., Effect of cumin (*Cuminum cyminum*) seed essential oil biofilm formation and plasmid integrity of *Klebsiella pneumoniae. Pharmacogn. Mag.*, 6, 57–61, 2010.
46. Filoche, S.K., Soma, K., Sissons, C.H., Antimicrobial effects of essential oil in combination with chlorhexidine digluconate. *Oral. Microbiol. Immunol.*, 20, 221–5, 2005.
47. Zhang, D., Ma, X., Gu, Y., Huang, H., Zhang, G., Green synthesis of metallic nanoparticles and their potential applications to treat cancer. *Front. Chem.*, *8*, 799, 2020.
48. Adetunji, C.O., Michael, O.S., Nwankwo, W., Anani, O.A., Adetunji, J.B., Olayinka, A.S., Akram, M., Based drugs derived from medicinal plants, in: *Green Synthesis in Nanomedicine and Human Health*, p. 103, 2021.
49. Adetunji, C.O. and Oyeyemi, O.T., Antiprotozoal activity of some medicinal plants against Entamoeba histolytica, the causative agent of amoebiasis, in: *Medical Biotechnology, Biopharmaceutics, Forensic Science and Bioinformatics*, pp. 341–358, CRC Press, 2022.
50. Adetunji, C.O., Olaniyan, O.T., Adeyomoye, O., Dare, A., Adeniyi, M.J., Alex, E., Rebezov, M., Koriagina, N., Shariati, M.A., Diverse techniques applied for effective diagnosis of COVID-19, in: *Assessing COVID-19 and Other Pandemics and Epidemics using Computational Modelling and Data Analysis*, S.K. Pani, S. Dash, W.P. dos Santos, S.A. Chan Bukhari, F. Flammini (Eds.), Springer, Cham, 2022, https://doi.org/10.1007/978-3-030-79753-9_3.
51. Castillo-Henríquez, L., Alfaro-Aguilar, K., Ugalde-Álvarez., J., Vega-Fernández, L., Montes, V.G., Vega-Baudrit, J.R., Green synthesis of gold and silver nanoparticles from plant extracts and their possible applications as antimicrobial agents in the agricultural area. *Nanomaterials, 10, 9, 1763*, 2020.
52. Ali, M.A., Ahmed, T., Wu, W., Hossain, A., Hafeez, R., Islam, M., Md., M., Wang, Y., An, Q., Sun, G., Li, B., Advancements in plant and microbe-based synthesis of metallic nanoparticles and their antimicrobial activity against plant pathogens. *Nanomaterials, 10, 6, 1146*, 2020.
53. Kuppusamy, P., Yusoff, M.M., Maniam, G.P., Govindan, N., Biosynthesis of metallic nanoparticles using plant derivatives and their new avenues in pharmacological applications - an updated report. *Saudi Pharm. J. : SPJ: Off. Publ. Saudi Pharm. Soc.*, *24*, 4, 473–484, 2016. https://doi.org/10.1016/j.jsps.2014.11.013.
54. Rajendran, K.S. and Gregory, F., Prospective application of nanoparticles green synthesized using medicinal plant extracts as novel nanomedicines. *Nanotechnol. Sci. Appl.*, *14*, 179–195, 2021.
55. Sharma, D., Kanchi, S., Bisetty, K., Biogenic synthesis of nanoparticles: A review. *Arabian J. Chem.*, 12, 8, 3576–3600, 2015.

56. Alsubki, R., Tabassum, H., Abudawood, M., Rabaan, A.A., Alsobaie, S.F., Ansar, S., Green synthesis, characterization, enhanced functionality and biological evaluation of silver nanoparticles based on Coriander sativum. *Saudi J. Biol. Sci.*, *28*, *4*, 2102–2108, 2021.
57. Remédios, C., Rosário, F., Bastos, V., Environmental nanoparticles interactions with plants: Morphological, physiological, and genotoxic aspects. *J. Bot.*, 2012, 1–8, 2012.
58. Aritonang, H.F., Koleangan, H., Wuntu, A.D., Synthesis of silver nanoparticles using aqueous extract of medicinal plants' (Impatiens balsamina and Lantana camara) fresh leaves and analysis of antimicrobial activity. *Int. J. Microbiol.*, *21*, *1–8*, 2019.
59. Ahmed, R.H. and Mustafa, D.E., Green synthesis of silver nanoparticles mediated by traditionally used medicinal plants in Sudan. *Int. Nano Lett.*, 10, 1–14, 2020. https://doi.org/10.1007/s40089-019-00291-9.
60. Hossain, Z., Yasmeen, F., Komatsu, S., Nanoparticles: Synthesis, morphophysiological effects, and proteomic responses of crop plants. *Int. J. Mol. Sci.*, *21*, *9*, 3056, 2020.
61. Adetunji, C.O., Olaniyan, O.T., Adeyomoye, O., Dare, A., Adeniyi, M.J., Alex, E., Rebezov, M., Isabekova, O., Shariati, M.A., Smart sensing for COVID-19 pandemic, in: *Assessing COVID-19 and Other Pandemics and Epidemics Using Computational Modelling and Data Analysis*, S.K. Pani, S. Dash, W.P. dos Santos, S.A. Chan Bukhari, F. Flammini (Eds.), Springer, Cham, 2022, https://doi.org/10.1007/978-3-030-79753-9_9.
62. Adetunji, C.O., Olaniyan, O.T., Adeyomoye, O., Dare, A., Adeniyi, M.J., Alex, E., Rebezov, M., Petukhova, E., Shariati, M.A., Internet of health things (IoHT) for COVID-19, in: *Assessing COVID-19 and Other Pandemics and Epidemics Using Computational Modelling and Data Analysis*, S.K. Pani, S. Dash, W.P. dos Santos, S.A. Chan Bukhari, F. Flammini (Eds.), Springer, Cham, 2022, https://doi.org/10.1007/978-3-030-79753-9_5.

13

Biological Activities of Nanomaterials From Biogenic Source for the Treatment of Diseases and Its Role in Regenerative and Tissue Engineering

Charles Oluwaseun Adetunji[1*], Olugbemi Tope Olaniyan[2], Chukwuebuka Egbuna[3,4], Olulope Olufemi Ajayi[5], Wadazani Dauda[6], Juliana Bunmi Adetunji[7], Shakira Ghazanfar[8], Frank Abimbola Ogundolie[9], John Tsado Mathew[10], Abel Inobeme[11], Olotu Titilayo[12], Abdullahi Tunde Aborode[13], Babatunde Oluwafemi Adetuyi[14], Shakirat Oloruntoyin Ajenifujah-Solebo[15] and Olalekan Akinbo[16]

[1]*Applied Microbiology, Biotechnology and Nanotechnology Laboratory, Department of Microbiology, Edo University Iyamho, Auchi, Edo State, Nigeria*
[2]*Laboratory for Reproductive Biology and Developmental Programming, Department of Physiology, Rhema University, Aba, Nigeria*
[3]*Department of Biochemistry, Faculty of Natural Sciences, Chukwuemeka Odumegwu Ojukwu University, Uli Campus, Anambra State, Nigeria*
[4]*Nutritional Biochemistry and Toxicology Unit, World Bank Africa Centre of Excellence, Centre for Public Health and Toxicological Research (PUTOR), University of Port-Harcourt, Rivers State, Nigeria*
[5]*Department of Biochemistry, Edo University Iyamho, Edo State, Nigeria*
[6]*Department of Crop Protection, Federal University Gasua, Nigeria*
[7]*Department of Biochemistry, Osun State University, Osogbo, Nigeria*
[8]*Functional Genomics and Bioinformatics, National Agricultural Research Centre, Islamabad, Pakistan*
[9]*Department of Biotechnology, Baze University Abuja, Nigeria*
[10]*Department of Chemistry, Ibrahim Badamasi Babangida University Lapai, Niger State, Nigeria*
[11]*Department of Chemistry, Edo University Iyamho, Nigeria*
[12]*Department of Microbiology, Adeleke University, Ede Osun State, Nigeria*
[13]*Healthy Africans Platform, Research and Development, Ibadan, Nigeria*

**Corresponding author*: adetunjicharles@gmail.com; adetunji.charles@edouniversity.edu.ng
http://orcid.org/0000-0003-3524-6441

Charles Oluwaseun Adetunji and Julius Kola Oloke (eds.) Handbook of Agricultural Biotechnology: Volume I Nanopesticides, (327–346) © 2024 Scrivener Publishing LLC

[14]Department of Natural Sciences, Biochemistry Unit, Faculty of Pure and Applied Sciences, Precious Cornerstone University, Ibadan, Nigeria
[15]National Biotechnology Development Agency Umar Musa Yardua Way, Lugbe P.M.B., Wuse, Abuja, Nigeria
[16]AUDA-NEPAD Centre of Excellence in Science, Technology and Innovation, Stellenbosch, South Africa

Abstract
Stem cell and regenerative medicine have provided a significant opportunity in tissue engineering, enhancement of the extracellular matrix formation, improved uptake of nutrients, and elimination of metabolic wastes and prompt degeneration of scaffold materials. Many natural and synthetic polymers are currently being utilized generally as regenerative materials in tissue engineering and medicine owing to their bioavailability, biocompatibility, nontoxic, mechanical properties, ease of timely degradation, ease of adaptable porosity, and physiological properties. Some of them include chitin, chitosan, fibroin, elastin, gelatin, collagen, fibrin, while synthetic polymers are polyglycolic acid (PGA), polylactic acid, polylactic coglycolic acid, and polyhydroxyalkanoate. This book chapter highlights several applications of nanotechnology in stem cell, regenerative medicine, tissue engineering, and treatment of different diseases.

***Keywords*:** Tissue engineering, stem cell, polymers, physiological properties, nanotechnology, antimicrobial

13.1 Introduction

Tissue engineering is the process of repairing or recreating live tissue or organs. It entails the knowledge of materials chemistry, biomaterial engineering, cell biology, immunology, physiology and clinical evaluations. The high demand for tissue or organ transplants necessitates rigorous research into the revitalization of cells [1]. Rejuvenation of cells is dependent on the kind and source of cells, materials utilized, scaffold design, the intracellular and extracellular milieu for tissue/organ formation [1].

Generally, advent tissue engineering in stem cell physiology heralded a new dimension in regenerative medicine. Scaffolds are necessary constituents for tissue regeneration in three-dimensional (3D) cells which imitate live tissues. They confer structural strength to tissue to be regenerated, enhance extracellular matrix (ECM) formation, enhance uptake of nutrients and removal of wastes and prompt degeneration of scaffolding materials.

Synthetic and natural polymers are used for preparing 3D scaffoldings materials. Natural bio/polymers because of their biocompatibility used include; chitin, chitosan, fibroin, elastin, gelatin, collagen, fibrin, while synthetic polymers are polyglycolic acid, polylactic coglycolic acid, polylactic acid, and polyhydroxyalkanoate are used because of their ease of timely degradation, mechanical properties and ease of adaptable porosity [2].

The effectiveness of nanoparticles containing polymers, such as PLA, PLGA, and polycaprolactone, in determining the fate of stem cells has been reported. Carbon-based nanoparticles are utilized in stem cell regenerative medicine due to their strength, biocompatibility, bioavailability and mechanical stability [3, 4].

Graphene, the basic structure of graphite, consisting of a single layer of carbons have been used in constructing scaffolds. It consists of reduced graphene oxide, graphene oxide, graphene quantum dots, monolayer graphene, graphene nanosheets, etc. The distinct features of graphene are; enormous surface area, huge intrinsic mobility, ideal thermal conductivity, great optical transmittance and outstanding electrical conductivity [5–9]. Additionally, graphene nanomaterials are biocompatible making them the perfect choice in stem cell tissue engineering among a range of uses [1]. The incorporation of graphene components has tremendously improved cytocompatibility of biomaterials via osteointegration, stem cell separation into diverse cellular structures, enhanced development and growth, diverging of neurons, hence, increasing vascular endothelial growth factor for restoration of cardiac muscles [1].

Graphene-based 3D scaffolds showed tremendous ability in ensuring electrical stimulation in an extracellular milieu in cardiac tissue engineering. It thus provided ECM for culturing cardiac cells [10]. Another report showed that rat's embryonic stem cells cultured on graphene 3D scaffoldings prepared from PCL nanofibers differentiated into cardiomyocytes [11]. A report also indicated that graphene-based materials improved bone formation as they exhibited proosteodifferentiation ability on scaffold materials via simple surface modification [12, 13].

The combination of RGO with nanohydroxyapatite (nHA) and bone mineral led to a self-assembled biomimetic 3D scaffolding appropriate for bone and another tissue engineering. RGO-nHA–centered biomimetic 3D scaffolding exhibited improved cell multiplication, outstanding osteogenic ability and increased activity of alkaline phosphatase in comparison with RGO. RGO-nHA scaffolding also displayed novel collagen, osteoid, and bone matrix tissue in another study [1].

The graphene-based 3D scaffold has been utilized in skin tissue engineering. Scaffold made of GO, PLGA and collagen regenerated skin tissues for wound healing. Furthermore, GO and collagen also enhanced skin regeneration and also affects dermal fibroblast's proliferation and migration [6]. A 3D scaffold prepared from graphene foam (GF) with mesenchymal bone marrow stem cells (MSC) enhanced skin wound healing going by its high retention of water ability. The vascular endothelial growth factor and fibroblast growth factor were regulated by the MSC which led to neovascularization [14]. Transforming growth factor-beta 1(TGF-β1), TGF-β3, and alpha-smooth muscle actin were responsible for the observed antiscarring effect.

The liver is an important organ of the body known for its secretory, biosynthetic, detoxifying, metabolic, storage and excretory functions. Liver failure caused by alcoholism, drug abuse, viral infections, and metabolic diseases is the cause of many deaths, globally etc. Despite therapies developed for liver diseases, transplantation appears to be the only curative therapy. The unavailability of the liver donor is a major limitation to this therapy. Hence, tissue and stem cell engineering seems to be a viable substitute in which nanotechnology performs a vital role.

The production of decellularized scaffolds from the liver cells with a 3D bioprinting scheme as well as nanobased 3D scaffoldings that could enhance liver cell internal environment are promising in hepatocytes and tissue engineering. Furthermore, genetic modulation that is specific for micro-RNAs and other low molecular weight molecules could enhance the rate of hepatic differentiation [15].

13.2 General Overview

Regenerative medicine and tissue engineering have been profoundly impacted by advancements in nanoparticle (NP) development, production and regulation over nanoscale structures. It is possible to improve engineered tissues and organs by using nanoparticles with low toxicity, differing properties, targeted and stimulus-reaction distribution, and detailed regulation of behavior performance through external stimulation like electromagnetic fields) thanks to their unique properties [16].

Biological events must be controlled with high spatial and temporal precision during functional tissue and organ replacements as well as be monitored in real time. To exercise control over the engineered tissues, content delivery (growth factors, chemokines, inhibitors, and cytokines) and contrast mediators delivered in a precise fashion are essential.

The need for growth factors to be delivered, imaging contrast to be maintained, and scaffold properties to be controlled led to the use of NP-based structures in tissue and stem cell engineering scaffoldings. The production of nanoparticles and materials like polymers, ceramics, metals, and their diverse constituents can be used, depending on the application [17].

Transplanting tissues and organs has many drawbacks including donor shortages, immunosuppression requirements and low success rates (rejection of transplants). In regenerative medicine and tissue engineering, there is an increased call for regenerative medicine and tissue engineering solutions or results. Artificial tissue and organs are being developed that copy the native tissue and organs through the incorporation of physiological, material, and engineering, both as implantable systems and as miniature organ models [18, 19].

To achieve cellular adhesion with the right mechanical strength, monitor cellular activity, and deliver bioactive agents, it is necessary to construct a 3D scaffold for cells that respond to the extracellular matrix [20].

It is possible to manipulate the properties of scaffolds by using nanoparticles (NPs), like alteration their mechanical power and offering an organized release of bioactive compounds [21–25]. In addition, bioactive constituents (growth factors, steroids, inhibitors, cytokines, drugs, genes) have disadvantages and limitations like low solubility, very short circulation half-life and unstable bioactivity. These characteristics have made NPs an attractive candidate for the delivery of bioactive agents and monitoring of their effects [21–24].

Regenerative medicine and tissue engineering have evolved from simple and traditional approaches towards complex and efficient ones as a result of the impact of nanotechnology. Along with nanoparticles, other materials of nanoscale technology like nanofibers and nanopatterned planes are utilized to manipulate cellular behavior.

By using concurrent imaging systems and therapeutic, implanting new bioparticles with higher spatiotemporal regulation within scaffoldings, modifying various bioactive constituents, namely growth factors to control stem cell fates and organ morphogenesis, and altering mechanical properties, we will be able to direct stem cell fates, morphogenesis, and other outcomes. As part of regenerative medicine and tissue engineering, NPs are applied to minimize toxicity and increase biocompatibility by delivering them to specific tissues [21, 26].

Tissue and stem cell engineering is a field that combines the knowledge of biology, materials science, engineering, physiology and microbiology with a focus on restoring, maintaining, repairing, rebuilding, replacing, and enhancing the physiology of tissues using bioreplacements [27]. The field of transplantable tissues has seen tremendous progress in recent decades, and new implantable tissues are now being developed, which are now being utilized in humans like skin, bone, cartilage or are in clinical trials like bladders, blood vessels and soft tissues [27].

In contrast, many tissue engineering approaches depend on the principle that, under appropriately controlled circumstances in the bioreactor, cells recruited or seeded onto 3D biocompatible scaffoldings can reconvene into useful, native-like structures. On a macroscopic level, initial synthetic scaffoldings were aimed to facilitate cell growth, but they only achieved moderate success. As a result, scaffolds should also consist of a tissue-specific milieu to preserve and control cell activity to recapitulate proper tissue functionality [28].

A normal mesh of structured nanofibers surrounds cells in tissues, which are encapsulated within the extracellular matrix (ECM) [29]. Nanoarchitecture is important for cell support, as well as for regulating cell behavior through communications between the intracellular and extracellular milieu. Additionally, the extracellular matrix allows for the storage, release, and activation of a wide variety of physiological factors, as well as assisting cell-to-cell and cell-to-soluble factor interfaces [30].

To regenerate tissues successfully, it is therefore imperative to design biomaterials that closely resemble ECM in terms of both complexity and functionality. Nevertheless, advances in nanotechnology have enabled the creation of biomimetic microenvironments at the nanoscale, which can be used to mimic native ECMs [30]. Nanotopographic surfaces and nanofeatured scaffolds have been created using these technologies, and drugs can be encapsulated and controlled spatiotemporally (e.g. growth factors) via these technologies. By directing cellular behaviors, such as cell adhesion and gene expression, these nanodevices enable cellular behavior control.

Additionally to the benefits discussed above, nanotechnology should also make significant contributions to creating new strategies for regenerating tissues like cell sheet engineering and to overcoming additional significant impediments to tissue and stem cell engineering—including the characterization and development of novel materials. The development of microtechnologies and nanotechnologies for high throughput assays for screening biomaterials based on cells is one example [31].

Although there are many challenges, it is clear that the attainment of functional, synthetic tissues and organs would greatly impact a positive regenerative approach in medical science. A stimulating prospect in the future is the concept of "organs-on-a-chip" which, within the next few years, would substitute for life-consuming and expensive animal experiments that are currently utilized for drug design and development through investigation of nanoparticle structures used to deliver drugs [32].

Nanotechnology has been described as the driving force with the capacity to propel and change many approaches in the field of medical science. The influence of nanotechnology on stem cell, regenerative medicine and drug delivery has assisted and increased the effectiveness of existing therapeutics thereby creating completely novel therapeutic targets.

Nanotechnology has opened up new approaches in tissue and stem cell engineering that can excite the restoration of intricate tissue structures. With the improvement of health care regulations regarding nanotechnology and the continued investment and influx of energy, nanomedicine will not only improve traditional treatments but may also help people globally and individually by addressing the shortcomings of traditional medicine [33].

The *in vivo* development of engineered tissues are generally known to require a cascade of event or sequence like the host's cell immunology, circulatory capacity with cell components of the embedded material [34]. The characteristics or activities of the scaffoldings with the presence of immobilized bioactive constituents in the scaffolding could assist in achieving exact maturation regulation.

Several suggestions have been made concerning the future of tissue and stem cell engineering in providing solutions to many health complications and challenges, such as complex constituents, that could be utilized for better regulation, monitoring, integration, and safety of engineered products.

In nanoscale drug delivery methods, the focus should be on the production of cutting-edge drug delivery methods as well as the evaluation of general cytotoxic effects and immune reactions of the systems. This research may offer an improved understanding of the bioavailability and biocompatibility of a wide range of nanoscale drug delivery methods and guide future research output in enhancing success rates with fewer cost approaches. To this end, different kinds of nanoparticles from diverse materials (ceramics, metals, magnetic/paramagnetic and polymers) deliver a reliable toolkit for the functionalization of tissue and stem engineered synthetic organs.

13.3 Application of Nanotechnology in Tissue and Stem Cell Engineering

Tissue and stem cell engineering has been described as one of the best suitable techniques for regenerating numerous organs and tissues, such as the cornea. Furthermore, one of several key problems that limit the regeneration of injured corneal tissues seems to be the failure of biomaterial scaffoldings to effectively implement into the ecosystem of nearby structures. As a result, controlling cellular and molecular pathways is critical for long-term survival and successful graft integration. Signaling pathways, as well as cell behavior, can be influenced by interactions among cells, biomaterial surfaces, and their surroundings. For the present coordination, understanding, and targeted management of molecular cell–transplant interactions in the treatment of corneal lesions, nanotechnology seems to be a useful tool. As a result, through regulating and modulating host graft interaction patterns towards appropriate integration as well as various nanotechnological methods, the continuing functionality of the transplant was to bring novel clarifications to issues concerning corneal inflammatory processes [35].

Tissue and stem cell engineering is a rapidly growing scientific discipline that uses cell or cell combinations with bioparticles or bioactive constituents to repair, treat, create, or replace cells, tissues, and organs. It aids the production of constituents that closely look like the body's native structure. Current medicines have been revolutionized, owing to tissue and stem cell engineering, thereby improving the quality of life for millions of individuals or patients. Tissue and stem cell engineering is a field that bridges materials science, technology, biology and medical science. In traditional tissue engineering, cells have always been placed on biomimetic scaffolds to adhesive surfaces, after which cells accumulate with their protein to create each other more biocompatible. However, failure to vascularize correctly, reduced mechanical power of engineered cells, lack of functional cells, nutrient limitation and lack of immune-compatibility with the host, are all common problems in tissue and stem cell engineering [36].

Tissue engineering has progressed from a fledgling scientific discipline susceptible to establishing proof of concept and creating diverse functioning tissues using various methodologies to a mature scientific discipline capable of creating major selling points for clinical use. Tissue-engineered skin is among the first commercially available products based on the techniques of tissue engineering. Tissue engineering had also sparked tremendous interest as a unique method for exploring cells along with systems biology and creating innovative therapeutics utilizing tissues generated in 3D environments, in addition to clinical uses [37].

This section presents a summary of recent advances in micro and nanotechnologies and their applications in tissue and stem cells engineering. Nanotechnologies and micro as both the methods of production and also as analytical techniques have become increasingly recognized as powerful tools for designing advanced TE strategies. Many such techniques could be used to create scaffolds to improve functionality that would function as active agents in tissue regeneration rather than just substrates for cellular adhesion. Furthermore, these techniques could be utilized to investigate but also regulate cellular microenvironmental phenomena. The primary technologies under/developed development as well as their various possible applications [38].

Tissue engineering (TE) is a multidisciplinary system that incorporates material science, medical biology along with engineering to produce biological replacements for organ and tissue repair, enhancement and replacement. Current TE approaches confront challenges, such as a lack of suitable nanomaterials, poor tissue regeneration, and a lack of tools for capturing optimal physiological structures, but also insufficient and unstable growth factor manufacturing to drive cell response and communication. In addition, challenges, such as biomolecular detection, as well as biosensors, and the failure to regulate cellular physiology with their several activities (mechanical, biological, and electrochemical), all contribute to the recent limits in this sector. Nanomaterials are still at the cutting edge of nanotechnology, along with their size-dependent features have already revealed potential in addressing several challenges that TE faces currently. Despite significant advancements in the utilization of nanotechnology throughout the last twenty years, the full promise of nanotechnology uses in TE challenges is still to be exploited. This study gives extensive information on a wide range of uses of different kinds of nanoparticles in stem cell technologies, as well as the hurdles that must be solved for nanoparticles to attain their maximum potential [39].

Tissue engineering (TE) is a wide and broad science discipline that covers several fields with combined principles of material science, bioengineering, cell biology, anatomy and physiology to repair tissues. Nanotechnology allows for the creation of novel materials, which could lead to extra tissue and stem cell engineering uses. Therefore, the use of nanomaterials and nanotechnology promise to emulate nature in a biomimetic manner. The recent spectrum of nanotechnology application potential is mostly in tissue and stem cell engineering of vascular, muscle, and bone implants using arrangements on nanofibrous materials [40].

Tissue engineering solutions that combine engineering ideas and procedures with biological sciences to create implanted tissues have just been recognized as among the most promising technology for restoring damaged organs and tissues and tissues due to organ and tissue scarcity. Nanobiomaterials including nanotubes, nanofibers, nanoparticles, along some nanofabricated equipment can contribute to tissue formation and cell growth because native organs or tissues are constituted of proteins at the cells and nanoscale directly interact with nanomaterials extracellular matrices (ECM) [41].

Nanotechnology is a fast-evolving area in science with an extensive range of uses, particularly in medicine. Tissue engineering, medication delivery, tissue regeneration, and regenerative medicine all benefit from it. The Cas9/CRISPR system is essential in tissue engineering because it is a revolutionary genome-editing technique utilized for tissue and stem cell engineering and focuses on stem cell architecture development, cell differentiation, immune reaction circumvention, as well as infection simulation techniques. Furthermore, tissue and stem cell engineering, research costs, regenerative medicine therapeutic development are very expensive, thus, these processes must be made more cost-effective. As a result, the enhanced methodology of nanoscale materials in CRISPR and regenerative medicine would undoubtedly transform the concept of disease therapy, treatment, and detection [42].

Nanoparticles are created to combat limitations in novel ocular medication delivery by improving drug permeation at the specified location and prolonging drug levels by several internal parts of administering drugs under much lower dosages even without protracted toxicity than conventional eye drops provide. Nanoparticles provide high specificity and multifunctionality DNA, which could result in improved new therapeutic improved efficacy. Fibrin is a powerful and flexible scaffold for tissue engineering uses that molecular biology and combines cells with components and mechanical engineering to substitute tissues and organs that have been damaged. Fibrin is an essential part of homeostasis and has been utilized thoroughly in tissue-engineered as a natural polymers scaffold. As a result, fibrin is a flexible biopolymer with a lot of applications in tissue regeneration as well as wound healing. As evidenced by studies, nanotechnology is altering molecular engineering on a far smaller scale [43].

13.4 Biochemical and Specific Modes of Action Involved in the Application of Nanodrugs for the Management of Diseases

13.4.1 Nanodrugs and Bacterial Infections

Pathogenic bacteria have lived with humans for ages. The development of antibiotics and vaccines has helped by reducing mortality and morbidity associated with these pathogenic microorganisms [44]. However, there are instances of drug resistance, which is currently prevalent. Multidrug-resistant pathogens are a huge challenge to treat with the current treatment protocols and in certain cases can be untreatable. A typical example is methicillin-resistant *Staphylococcus aureus* [45]. To worsen the matter, the pharmaceutical industry appears to have reduced effort in producing novel antibiotics, probably due to reduced profit on their investment [44].

Nanomedicine appears to be altering the world of infectious diseases by bringing a ray of hope in the treatment of multidrug-resistant pathogens. Nanomaterials by their properties perform significant functions in the diagnosis and treatment of infectious microbes. Some of these nanomaterials target multidrug-resistant bacteria, some are also efficient drug carriers, hence, reducing adverse drug reactions that characterize conventional antibiotics. The incorporation of nanomaterials in medical devices has also been reported to prevent microbial adhesion and infectivity. Furthermore, vaccine adjuvants made of nanomaterials could provoke immune responses targeted at the microbial infection [44].

Bacterial resistance is attributed to gene mutation or via horizontal gene transfer by conjugation, transformation or transduction [46]. Bacterial resistance to antibiotics is via mechanisms involving the production of the competitive inhibitor to antimicrobial drugs, modification of substrate to which antibiotics bind, reduced influx and elevated efflux of drug from microbes and microbial formation of biofilms as seen in Chlamydia spp. and Mycobacterium leper [47, 48].

Nanomedicine could effectively manage drug resistance and disallows further resistance [49–52]. This is via targeted drug delivery which increased bioavailability of drugs, specificity for multiple pathways, and inhibition of biofilm formation [44]. The bactericidal properties of some metals and metal oxides are well documented. There is evidence that it is hard for microbes to acquire resistance to them. This is because of their multiple mechanisms of action [53–55].

Silver nanoparticles have gained more popularity and are recognized for destroying gram-negative and positive bacteria pathogens. They have been utilized in treating drug-resistant bacteria including *Pseudomonas aeruginosa* [56]. Therefore, the plausible mode of action of silver nanoparticles includes; direct interference with microbial DNA, enzymes and cell membrane. Silver nanoparticles also enhance the generation of oxidative stress markers [57]. The antimicrobial effects of silver nanoparticles are influenced by oxidation state, hydrophobicity, surface roughness as well as surface functionalization [57, 58]. Additionally, tellurium and bismuth nanoparticles have also proven to be effective antibiotics [44].

Some metal oxide nanomaterials have also been studied for their antimicrobial effects. These include copper oxide, titanium oxide, zinc oxide, aluminium oxide etc. Metal oxide nanoparticles are effective antimicrobials by their ability to induce the photocatalytic generation of oxidative stress markers, inhibit the synthesis of DNA, reduce bacterial membrane integrity, disrupts energy production and respiratory enzyme activity [20, 59, 60].

The antimicrobial effects of carbon-based nanomaterials have been reported. Examples are single-walled carbon nanotubes, fullerene and multiwalled carbon nanotubes. They are known to disrupt microbial cell membranes by direct contact or by photothermal effect [61, 62]. Furthermore, they can also induce oxidative stress which comprises microbial membrane integrity [50].

Cationic antimicrobial peptides (CAP) are effective against a broad spectrum of microorganisms and MDR bacteria [63]. They act by damaging the negatively charged membranes of the microorganisms. Synthetic cationic polymers are currently been utilized because they are of improved enzymatic stability and are cheap [64]. Cationic polymers using phosphonium functional groups and quaternary ammonium do imitate CAP. They disrupt microbial membranes, hence, hindering microbial progression [50].

Reports showed the general effectiveness of chitosan nanoparticles against both gram-negative and positive bacteria pathogens [65, 66]. Furthermore, nanoparticle-containing chitosan and alginate exerted a bactericidal effect on *Propionibacterium acnes* via a mechanism involving inhibition of production of acne-induced cytokine [67].

13.4.2 Nanodrugs and Cardiovascular Diseases

A variety of nanomaterials are utilized for the management of prolonged diseases. Cardiovascular diseases (CVDs) are a global health challenge with a high mortality rate. Nanomedicine has proven to be effective in the management of CVDs because they enhance drug stability, improve drug bioavailability, reduce the enzymatic breakdown of drugs [68].

RNA interference (RNAi) has been described as a gene-specific silencing tool that presents double-stranded RNA into cells. These double-stranded RNAs consists of sense and antisense RNAs that match the mRNA sequence into cells, thereby inhibiting the expression of cell-specific genes [68]. The role of RNAi in the management of CVDs is currently being explored.

Furthermore, small interfering RNA (siRNA) also appears promising in the treatment of CVDs. In a reported study, chitosan nanoparticles conjugated with siRNA designed against the appearance of PDGF-B mRNA showed significant inhibition of PDGF-B mRNA and PCNA expression in the intima of rabbit's arterial wall [69]. Stearic acid altered carboxymethyl chitosan nanopolymers conjugated to a homing peptide for distribution to hypertrophied heart muscle led to regression of cardiac hypertrophy [70].

13.4.3 Nanodrugs and Type 2 Diabetes Mellitus

Oxidative stress has been associated with the etiology of chronic diseases like type 2 diabetes (T2DM) mellitus. Chronic hyperglycemia enhances the assembly of progressive glycation end products (AGEs), the additional production of which results in oxidative stress. Furthermore, telomere shortening that is associated with oxidative stress has been implicated in the pathogenesis of T2DM [71]. Certain metallic nanoparticles (MNPs) have been revealed to possess enzyme-like antioxidant features which are also named nanoantioxidants. This is shown by their ability to mop up free radicals. Examples are cerium oxide, copper oxide, manganese oxide vanadium pentoxide. The antioxidant effect of these MNPs could be attributed to their catalytic and redox properties, electronic configuration high surface to volume ratio and multioxidation states [45].

The modulation of antioxidant pathways including Nrf2 by cerium oxide is another plausible mechanism [72]. Diabetic foot ulcer (DFU) is a complication of Diabetes mellitus that results from neuropathy, peripheral arterial diseases. The sustained elevated blood glucose increases the actions of sorbitol dehydrogenase and aldose reductase catalyze the alteration of glucose to fructose and sorbitol, the buildup of which leads to reduced production of myoinositol in the nerve [73, 74].

Nanotechnology proves to be effective in the management of DFU. This entails the incorporation of peptides, nucleic acids, antibiotics, and antioxidants into nanoparticles for delivery to target sites. The application of zinc NP in managing diabetes mellitus could be due to being a cofactor to over 300 enzymes. Its hypoglycemic effects could be attributed to its ability to inhibit glucose absorption, as well as facilitate glucose absorption through the skeletal muscles and fat tissues [75]. Nanofibers enhance wound healing by their peculiar features including; enhanced porosity, capability to encapsulate nanoparticles with other constituents, increased surface area/volume ratio [76, 77].

13.5 Conclusion

This chapter has provided significant information on the application of nanotechnology in stem cells, regenerative science and tissue engineering. Nanotechnology, this is a fast-evolving area with an extensive range of applications, particularly in biochemistry, physiology, anatomy, and pharmacology for tissue and stem cell engineering, tissue regeneration, drug delivery, wound healing, tissue replacement, and regenerative medicine. Through the utilization of molecular biology and material science, diverse creations of scaffolds materials can be generated to improve functionality and provide active support in tissue regeneration and cellular adhesion.

References

1. Bai, R.G., Muthoosamy, K., Manickam, S., Hilal-Alnaqbi, A., Graphene-based 3D scaffolds in tissue engineering: Fabrication, applications, and futurescope in liver tissue engineering. *Int. J. Nanomed.*, 14, 5753–5783, 2019.
2. Dhandayuthapani, B., Yoshida, Y., Maekawa, T., Kumar, D.S., Polymeric scaffolds in tissue engineering application: A review. *Int. J. Polym. Sci.*, 2011, 290602, 2011.
3. Binulal, N., Deepthy, M., Selvamurugan, N. *et al.*, Role of nanofibrous poly (caprolactone) scaffolds in human mesenchymal stem cell attachment and spreading for *in vitro* bone tissue engineering – response to osteogenic regulators. *Tissue Eng. Part A.*, 16, 2, 393–404, 2010.
4. Zhao, C., Tan, A., Pastorin, G., Ho, H.K., Nanomaterial scaffolds for stem cell proliferation and differentiation in tissue engineering. *Biotechnol. Adv.*, 31, 5, 654–668, 2013.
5. Bolotin, K.I., Sikes, K., Jiang, Z. *et al.*, Ultrahigh electron mobility in suspended graphene. *Solid State Commun.*, 146, 9, 351–355, 2008.
6. Lee, C., Wei, X., Kysar, J.W., Hone, J., Measurement of the elastic properties and intrinsic strength of monolayer graphene. *Science.*, 321, 5887, 385–388, 2008.
7. Morozov, S., Novoselov, K., Katsnelson, M. *et al.*, Giant intrinsic carrier mobilities in graphene and its bilayer. *Phys. Rev. Lett.*, 100, 1, 016602, 2008.
8. Balandin, A.A., Ghosh, S., Bao, W. *et al.*, Superior thermal conductivity of single-layer graphene. *Nano Lett.*, 8, 3, 902–907, 2008.
9. Cai, W., Zhu, Y., Li, X., Piner, R.D., Ruoff, R.S., Large area few-layer graphene/graphite films as transparent thin conducting electrodes. *Appl. Phys. Lett.*, 95, 12, 123115, 2009.
10. Ameri, S., Singh, P., D'Angelo, R., Stoppel, W., Black, L., Sonkusale, S., Three dimensional graphene scaffold for cardiac tissue engineering and *in-situ*

electrical recording. Paper presented at: *2016 IEEE 38th Annual International Conference of the Engineering in Medicine and Biology Society (EMBC)*, Orlando, FL, USA, 2016.
11. Aphale, A., Hitscherich, P., Narula, R. *et al.*, Biocompatibility of PCLgraphene nanostructured scaffolds with mouse embryonic stem cell-derived cardiomyocytes. *Materials*, 8, 12, 2015.
12. Gu, M., Liu, Y., Chen, T. *et al.*, Is graphene a promising nano-material for promoting surface modification of implants or scaffold materials in bone tissue engineering? *Tissue Eng. Part B Rev.*, 20, 5, 477–491, 2014.
13. Shadjou, N. and Hasanzadeh, M., Graphene and its nanostructure derivatives for use in bone tissue engineering: Recent advances. *J. Biomed. Mater. Res. B Appl. Biomater.*, 104, 1250–1275, 2016.
14. Li, Z., Wang, H., Yang, B., Sun, Y., Huo, R., Three-dimensional graphene foams loaded with bone marrow-derived mesenchymal stem cells promote skin wound healing with reduced scarring. *Mater. Sci. Eng. C.*, 57, 181–188, 2015.
15. Hosseini, V., Maroufi, N.F., Saghati, S., Asadi, N., Darabi, M., Ahmad, S.N.S., Hosseinkhani, H., Rahbarghazi, R., Current progress in hepatic tissue regeneration by tissue engineering. *J. Transl. Med.*, 17, 383, 2019. https://doi.org/10.1186/s12967-019-02137-6.
16. Amoli-Diva, M., Sadighi-Bonabi, R., Pourghazi, K., Switchable on/off drug release from gold nanoparticles-grafted dual light- and temperature-responsive hydrogel for controlled drug delivery. *Mater. Sci. Eng. C.*, 76, 242–248, 2017.
17. Gaharwar, A.K., Peppas, N.A., Khademhosseini, A., Nanocomposite hydrogels for biomedical applications. *Biotechnol. Bioeng.*, 111, 441–453, 2014.
18. Dvir, T., Timko, B.P., Kohane, D.S., Langer, R., Nanotechnological strategies for engineering complex tissues. *Nat. Nano.*, 6, 13–22, 2011.
19. Zorlutuna, P., Vrana, N.E., Khademhosseini, A., The expanding world of tissue engineering: The building blocks and new applications of tissue-engineered constructs. *IEEE Rev. Biomed. Eng.*, 6, 47–62, 2013.
20. Ataol, S., Tezcaner, A., Duygulu, O., Keskin, D., Machin, N.E., Synthesis and characterization of nanosized calcium phosphates by flame spray pyrolysis, and their effect on osteogenic differentiation of stem cells. *J. Nanopart. Res.*, 17, 95, 2015.
21. Park, J., Wrzesinski, S.H., Stern, E., Look, M., Criscione, J., Ragheb, R. *et al.*, Combination delivery of TGF-beta inhibitor and IL-2 by nanoscale liposomal polymeric gels enhances tumour immunotherapy. *Nat. Mater.*, 11, 895–905, 2012.
22. Pérez, R.A., Won, J.E., Knowles, J.C., Kim, H.W., Naturally and synthetic smart composite biomaterials for tissue regeneration. *Adv. Drug Deliv. Rev.*, 65, 471–496, 2013.

23. Cheng, C.J., Tietjen, G.T., Saucier-Sawyer, J.K., Saltzman, W.M., A holistic approach to targeting disease with polymeric nanoparticles. *Nat. Rev. Drug Discovery*, 14, 239–247, 2015.
24. Bahal, R., Ali McNeer, N., Quijano, E., Liu, Y., Sulkowski, P., Turchick, A. *et al.*, *In vivo* correction of anaemia in b-thalassemic mice by gPNAmediated gene editing with nanoparticle delivery. *Nat. Commun.*, 7, 13304, 2016.
25. Mi, P., Kokuryo, D., Cabral, H., Wu, H., Terada, Y., Saga, T. *et al.*, A pH-activatable nanoparticle with signal-amplification capabilities for noninvasive imaging of tumour malignancy. *Nat. Nanotechnol.*, 11, 724–730, 2016.
26. Gao, G., Schilling, A.F., Yonezawa, T., Wang, J., Dai, G., Cui, X., Bioactive nanoparticles stimulate bone tissue formation in bioprinted three-dimensional scaffold and human mesenchymal stem cells. *Biotechnol. J.*, 9, 1304–1311, 2014.
27. Herranz-Blanco, B., Ginestar, E., Zhang, H., Hirvonen, J., Santos, H.A., Microfluidics platform for glass capillaries and its application in droplet and nanoparticle fabrication. *Int. J. Pharm.*, 516, 100–105, 2017.
28. Marino, A., Arai, S., Hou, Y., Degl'Innocenti, A., Cappello, V., Mazzolai, B. *et al.*, Gold nanoshell-mediated remote myotube activation. *ACS Nano.*, 11, 2494–2508, 2017.
29. Hidouri, M., Dorozhkin, S.V., Albeladi, N., Thermal behavior, sintering and mechanical characterization of multiple ion-substituted hydroxyapatite bioceramics. *J. Inorg. Organomet. Polym. Mater.*, 29, 87–100, 2018.
30. Medeiros, S.F., Santos, A.M., Fessi, H., Elaissari, A., Stimuli-responsive magnetic particles for biomedical applications. *Int. J. Pharm.*, 403, 139–161, 2011.
31. Kunath, S., Panagiotopoulou, M., Maximilien, J., Marchyk, N., Sanger, J., Haupt, K., Cell and tissue imaging with molecularly imprinted polymers as plastic antibody mimics. *Adv. Healthc. Mater.*, 4, 1322–1326, 2015.
32. Labala, S., Jose, A., Chawla, S.R., Khan, M.S., Bhatnagar, S., Kulkarni, O.P. *et al.*, Effective melanoma cancer suppression by iontophoretic co-delivery of STAT3 siRNA and imatinib using gold nanoparticles. *Int. J. Pharm.*, 525, 407–417, 2017.
33. Panikkanvalappil, S.R., Hooshmand, N., El-Sayed, M.A., Intracellular assembly of nuclear-targeted gold nanosphere enables selective plasmonic photothermal therapy of cancer by shifting their absorption wavelength toward the near-infrared region. *Bioconjug. Chem.*, 28, 2452–2460, 2017.
34. Pankongadisak, P., Ruktanonchai, U.R., Supaphol, P., Suwantong, O., Gelatin scaffolds functionalized by silver nanoparticle-containing calcium alginate beads for wound care applications. *Polym. Adv. Technol.*, 28, 849–858, 2017.
35. Mijanović, O., Pylaev, T., Nikitkina, A., Artyukhova, M., Branković, A., Peshkova, M., Bikmulina, P., Turk, B., Bolevich, S., Avetisov, S., Tissue engineering meets nanotechnology: Molecular mechanism modulations in cornea regeneration. *Micromachines*, 12, 1336, 2021. https://doi.org/10.3390/mi12111336.

36. Danie, K.J., Ranjan, S., Dasgupta, N., Saha, P., Nanotechnology for tissue engineering: Need, techniques and applications. *J. Pharm. Res.*, 7, 2, 200–204, 2013.
37. Hasan, A., Tissue engineering for artificial organs (Regenerative Medicine, Smart Diagnostics and Personalized Medicine), in: *Introduction to Tissue Engineering*, pp. 1–34, 2017.
38. Pallua, N. and Suscheck, C.V., *Tissue engineering, micro- and nanotechnology in tissue engineering*, pp. 3–29, Springer Berlin Heidelberg, 2011.
39. Hasan, A., Morshed, M., Memic, A., Hassan, S., Webster, T., Marei, H., Nanoparticles in tissue engineering: Applications, challenges and prospects. *Int. J. Nanomed.*, 13, 5637–5655, 2018.
40. Bigdeli, A.K., Lyer, S., Detsch, R., Boccaccini, A.R., Beier, J.P., Kneser, U., Horch, R.E., Arkudas, A., Nanotechnologies in tissue engineering. *Nanotechnol. Rev.*, 2, 4, 411–425, 2013.
41. Yu, X., Wang, J., Du, Y., Ma, Z., He, W., Nanotechnology for tissue engineering applications. *J. Nanomater.*, 20, 1–2, 2011.
42. Ali, Q., Sabeen, M., Arif, M., Muhammad, N.H., Said, S., Role of modern technologies in tissue engineering. *Arch. Neurosci.*, 7, 1, e90394, 2020.
43. Patel, P., Patel, D., Patel, S., Dewi, M.H., Athiyah, U., Pathak, Y., Ocular tissue engineering and regeneration: Nano biomaterial applications. *Adv. Nanosci. Nanotechnol.*, 3, 2, 1–3, 2019.
44. Zhu, X., Radovic-Moreno, A.F., Wu, J., Langer, R., Shi, J., Nanomedicine in the management of microbial infection - overview and perspectives. *Nano Today*, 9, 4, 478–498, 2014.
45. Limbago, B.M., Kallen, A.J., Zhu, W., Eggers, P., McDougal, L.K., Albrecht, V.S., Report of the 13th vancomycin-resistant Staphylococcus aureus isolate from the United States. *J. Clin. Microbiol.*, 52, 998–1002, 2014.
46. Jayaraman, R., Antibiotic resistance: An overview of mechanisms and a paradigm shift. *Curr. Sci.*, 96, 11, 1475–1484, 2009.
47. Pelgrift, R.Y. and Friedman, A.J., Nanotechnology as a therapeutic tool to combat microbial resistance. *Adv. Drug Deliv. Rev.*, 65, 1803–1815, 2013.
48. Ray, K., Marteyn, B., Sansonetti, P.J., Tang, C.M., Life on the inside: The intracellular lifestyle of cytosolic bacteria. *Nat. Rev. Microbiol.*, 7, 5, 333–340, 2009.
49. Adetunji, C.O., Fawole, O.B., Arowora, K.A., Nwaubani, S.I., Ajayi, E.S., Oloke, J.K., Aina, J.A., Adetunji, J.B., Ajani, A.O., Postharvest quality and safety maintenance of the physical properties of Daucus *carota L.* fruits by Neem oil and Moringa oil treatment: A new edible coatings. *Agrosearch*, 13, 1, 131–141, 2013.
50. Adetunji, C.O., Adetunji, C.O., Michael, O.S., Kadiri, O., Varma, A., Akram, M., Oloke, J.K., Shafique, H., Adetunji, J.B., Jain, A., Bodunrinde, R.E., Ozolua, P., Ubi, B.E., Varma, A., Quinoa: From farm to traditional healing, food application, and phytopharmacology, in: *Biology and Biotechnology of Quinoa*, Springer, Singapore, 2021, https://doi.org/10.1007/978-981-16-3832-9_20.

51. Adetunji, C.O., Akande, S.A., Oladipo, A.K., Salawu, R.A., Onyegbula, A.F., Determination of the microbiological quality and proximate composition of fermented cassava food products sold in Ilorin-west local government area, Nigeria. *Ruhuna J. Sci.*, 8, 76–89, 2017. http://doi.org/10.4038/rjs.v8i2.28
52. Adetunji, C.O., Olaniyan, O.T., Adeyomoye, O., Dare, A., Adeniyi, M.J., Alex, E., Rebezov, R., Garipova, L., Shariati, M.A. *et al.*, eHealth, mHealth, and telemedicine for COVID-19 pandemic, in: *Assessing COVID-19 and Other Pandemics and Epidemics using Computational Modelling and Data Analysis*, S.K. Pani, S. Dash, W.P. dos Santos, S.A. Chan Bukhari, F. Flammini (Eds.), Springer, Cham, 2022, https://doi.org/10.1007/978-3-030-79753-9_10.
53. Adetunji, C.O., Michael, O.S., Varma, A., Oloke, J.K., Kadiri, O., Akram, M., Bodunrinde, R.E., Imtiaz, A., Adetunji, J.B., Shahzad, K., Jain, A., Ubi, B.E., Majeed, N., Ozolua, P., Olisaka, F.N., Recent advances in the application of biotechnology for improving the production of secondary metabolites from Quinoa, in: *Biology and Biotechnology of Quinoa*, A. Varma (Ed.), Springer, Singapore, 2021, https://doi.org/10.1007/978-981-16-3832-9_17.
54. Adetunji, C.O., Michael, O.S., Nwankwo, W., Ukhurebor, K.E., Anani, O.A., Oloke, J.K., Varma, A., Kadiri, O., Jain, A., Adetunji, J.B., Quinoa, the next biotech plant: Food security and environmental and health hot spots, in: *Biology and Biotechnology of Quinoa*, A. Varma (Ed.), Springer, Singapore, 2021, https://doi.org/10.1007/978-981-16-3832-9_19.
55. Adetunji, C.O., Olaniyan, O.T., Okeke, N.E., Production of next-generation biodiesel from high yielding strains of microorganisms: Recent advances, in: *Sustainable Green Chemical Processes and their Allied Applications*, Nanotechnology in the Life Sciences, A.A. Inamuddin (Ed.), Springer, Cham, 2020, https://doi.org/10.1007/978-3-030-42284-4_2.
56. Lara, H.H., Ayala-Núnez, N.V., Turrent, L.D.C., II, Padilla, C.R., Bactericidal effect of silver nanoparticles against multidrug-resistant bacteria. *World J. Microbiol. Biotechnol.*, 26, 4, 615–621, 2010.
57. Eckhardt, S., Brunetto, P.S., Gagnon, J., Priebe, M., Giese, B., Fromm, K.M., Nanobio silver: Its interactions with peptides and bacteria, and its uses in medicine. *Chem. Rev.*, 113, 4708–4754, 2013.
58. Li, Q., Mahendra, S., Lyon, D.Y., Brunet, L., Liga, M.V., Li, D., Alvarez, P.J., Antimicrobial nanomaterials for water disinfection and microbial control: Potential applications and implications. *Water Res.*, 42, 4591–4602, 2008.
59. Huang, Z., Zheng, X., Yan, D., Yin, G., Liao, X., Kang, Y., Yao, Y., Huang, D., Hao, B., Toxicological effect of ZnO nanoparticles based on bacteria. *Langmuir*, 24, 8, 4140–4144, 2008.
60. Huh, A.J. and Kwon, Y.J.J., "Nanoantibiotics": A new paradigm for treating infectious diseases using nanomaterials in the antibiotics resistant era. *J. Control. Release*, 156, 2, 128–145, 2011.
61. Kang, S., Herzberg, M., Rodrigues, D.F., Elimelech, M., Antibacterial effects of carbon nanotubes: Size does matter! *Langmuir*, 24, 13, 6409–6413, 2008.

62. Kotagiri, N., Lee, J.S., Kim, J.-W., Selective pathogen targeting and macrophage evading carbon nanotubes through dextran sulfate coating and PEGylation for photothermal theranostics. *J. Biomed. Nanotechnol.*, 9, 6, 1008–1016, 2013.
63. Hancock, R.E. and Sahl, H.-G., Antimicrobial and host-defense peptides as new anti-infective therapeutic strategies. *Nat. Biotechnol.*, 24, 12, 1551–1557, 2006.
64. Engler, A.C., Wiradharma, N., Ong, Z.Y., Coady, D.J., Hedrick, J.L., Yang, Y.-Y., Emerging trends in macromolecular antimicrobials to fight multi-drug-resistant infections. *Nano Today*, 7, 201–222, 2012.
65. Rabea, E.I., Badawy, M.E.-T., Stevens, C.V., Smagghe, G., Steurbaut, W., Chitosan as antimicrobial agent: Applications and mode of action. *Biomacromolecules.*, 4, 1457–1465, 2003.
66. Kong, M., Chen, X.G., Xing, K., Park, H.J., Antimicrobial properties of chitosan and mode of action: A state of the art review. *Int. J. Food Microbiol.*, 144, 1, 51–63, 2010.
67. Friedman, A.J., Phan, J., Schairer, D.O., Champer, J., Qin, M., Pirouz, A., Blecher-Paz, K., Oren, A., Liu, P.T., Modlin, R.L., Kim, J.J., Antimicrobial and anti-inflammatory activity of chitosan–alginate nanoparticles: Targeted therapy for cutaneous pathogens. *J. Invest. Dermatol.*, 133, 5, 1231–1239, 2013.
68. Deng, Y., Zhang, X., Shen, H., He, Q., Wu, Z., Liao, W., Yuan, M., Application of the nano-drug delivery system in treatment of cardiovascular diseases. *Front. Bioeng. Biotechnol.*, 7, 489, 2020.
69. Xia, H., Jun, J., Wen-ping, L., Yi-feng, P., Xiao-ling, C., Chitosan nanoparticle carrying small interfering RNA to platelet-derived growth factor B mRNA inhibits proliferation of smooth muscle cells in rabbit injured arteries. *Vascular*, 21, 301–306, 2013.
70. Rana, S., Datta, K., Reddy, T.L., Chatterjee, E., Sen, P., Pal-Bhadra, M., Bhadra, U., Pramanik, A., Pramanik, P., Chawla-Sarkar, M., Sarkar, S., A spatio-temporal cardiomyocyte targeted vector system for efficient delivery of therapeutic payloads to regress cardiac hypertrophy abating bystander effect. *J. Control. Release*, 200, 167–178, 2015.
71. Lushchak, O., Zayachkivska, A., Vaiserman, A., Metallic nanoantioxidants as potential therapeutics for Type 2 diabetes: A hypothetical background and translational perspectives. *Oxid. Med. Cell. Longev.*, 2018, Article ID 3407375, 9, 2018. https://doi.org/10.1155/2018/3407375.
72. Akhtar, M.J., Ahamed, M., Alhadlaq, H.A., Alshamsan, A., Mechanism of ROS scavenging and antioxidant signaling by redox metallic and fullerene nanomaterials: Potential implications in ROS associated degenerative disorders. *Biochim. Biophys. Acta Gen. Subj.*, 1861, 4, 802–813, 2017.
73. Tang, W.H., Martin, K.A., Hwa, J., Aldose reductase, oxidative stress, and diabetic mellitus. *Front. Pharmacol.*, 3, 87, 2012.

74. Ezhilarasu, H., Vishalli, D., Dheen, S.T., Bay, B.-H., Srinivasan, D.K., Nanoparticle-based therapeutic approach for diabetic wound healing. *Nanomaterials*, 10, 1234, 2020.
75. El-Gharbawy, R.M., Emara, A.M., Abu-Risha, S.E., Zinc oxide nanoparticles and a standard antidiabetic drug restore the function and structure of beta cells in Type-2 diabetes. *BioMed. Pharm.*, 84, 810–820, 2016.
76. Shan, X., Liu, C., Li, F., Ouyang, C., Gao, Q., Zheng, K., Nanoparticles vs. nanofibers: A comparison of two drug delivery systems on assessing drug release performance *in vitro*. *Des. Monomers Polym.*, 18, 678–689, 2015.
77. Ezhilarasu, H., Sadiq, A., Ratheesh, G., Sridhar, S., Ramakrishna, S., Ab Rahim, M.H., Yusoff, M.M., Jose, R., Reddy, V.J., Functionalized core/shell nanofibers for the differentiation of mesenchymal stem cells for vascular tissue engineering. *Nanomedicine*, 14, 201–214, 2019.

14

Application of Plant-Based Nanobiopesticides for Mitigation of Several Abiotic Stress

Babatunde Oluwafemi Adetuyi* **and Oluwakemi Semiloore Omowumi**

Department of Natural Sciences, Faculty of Pure and Applied Sciences, Precious Cornerstone University, Ibadan, Nigeria

Abstract

When a human is under stress, they are less able to function normally, it stunts their potential for natural progression and decreases their ability to reproduce. With respect to farming, stress is the factor that reduces crop productivity and causes biomass to be lost. Differences in resource competition, predation, and allelopathic effects among organisms constitute biotic stress, which contrasts with abiotic stress. These two types of stress can be broadly categorized based on the factors that cause them. Abiotic stress is the adverse effect that nonliving elements have on living things in a particular habitat. Drought, flood, heat injury, radiation injury, cold injury, and other abiotic stresses are examples. Although all of these pressures are unavoidable, they can be reduced to some amount in order to boost crops' potential for productivity. Successful crop production in adverse conditions can be achieved through a variety of management strategies, including breeding, biotechnology, and agronomic approaches. All methods based on the usage of plant-based nanobiopesticides were covered in this chapter.

Keywords: Biotechnology, abiotic stress, nanoparticle, agronomy

14.1 Introduction

Abiotic and biotic stress on the biosphere brought about by environmental change influences their creation and efficiency either straightforwardly or

**Corresponding author*: badetuyi@gmail.com

Charles Oluwaseun Adetunji and Julius Kola Oloke (eds.) Handbook of Agricultural Biotechnology: Volume I Nanopesticides, (347–400) © 2024 Scrivener Publishing LLC

by implication. Hans Selye originally presented pressure as bad ecological limitations for plants in 1936 (Selye, 1936). As far as physical science, stress is the power per unit region that outcomes in layered change on the item to which it is applied. Stress limits a singular's capacity to work typically, which thusly reduces their hereditary limit with respect to development, advancement, and proliferation (Levitt, 1980). Stress with regard to horticulture is an element that brings down crop efficiency and obliterates biomass (Grime, 1979). Levitt characterizes pressure as any natural condition that could be destructive to living things (Levitt, 1980). By recognizing eustress and di-stress, Lichtenthaler extended plant pressure to incorporate the plant's recovery stage (Lichtenthaller, 1988 and 1996). Abiotic stress alludes to the collaboration between a creature and its actual climate, while biotic pressure is the aftereffect of cooperations between organic entities for assets, predation, and allelopathic impacts. These two kinds of pressure can be comprehensively ordered in view of the elements that cause them. Abiotic factors confine crop choice, advance boundless creation, and cause total harvest disappointment. For plants and animals, "abiotic stress" describes any adverse influence from an inanimate source (Ahmadi and Bakar, 2001).

For the nonliving variable to have a major impact on the performance of the organic creature as a whole or any members of its physiology, it must alter the climate in a way that goes beyond its normal range of variation. Abiotic stress factors, otherwise called stressors, are normally happening, much of the time undetectable, and lifeless variables like serious daylight, temperature, or wind that might hurt the plants and creatures in the space impacted (Ahmadi, 1998). This differentiations with biotic pressure, which would incorporate living aggravations like growths or hurtful bugs. Abiotic stress cannot actually be forestalled. Creatures are impacted by abiotic stress, however plants are outstandingly, while possibly not altogether, reliant upon natural factors, making them particularly obliged. The most impeding component influencing the development and yield of harvests overall is abiotic stress (Ali, 2013). The combination of many abiotic stresses has been proven in studies to be particularly dangerous (Allen, 1997; James-Okoro, 2021). Abiotic stress comes in many forms. The most regular stressors are the least demanding for individuals to perceive, yet there are various other, more subtle abiotic stress components that persistently affect living spaces (Anjum *et al.*, 2011). The most widely recognized stressors are as per the following: Strong winds, temperature extremes, Drought, Flood, rapidly spreading fires, Cold, Heat (Anjum *et al.*, 2011). Minor stressors typically have a less impact. Examples of unfavorable edaphic circumstances include excessive rock and pH levels, excessive radiation, soil compaction

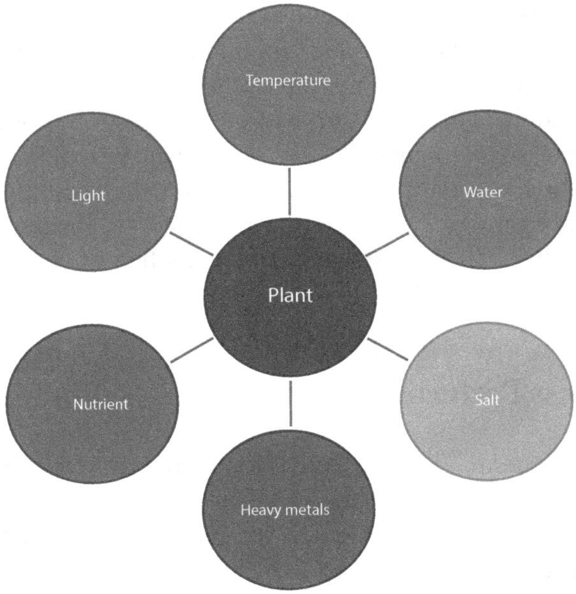

Figure 14.1 Different abiotic factors that influence plants.

and contamination, and the need for rapid rehydration during seed germination (Anjum *et al.*, 2011) as seen in Figure 14.1.

Different biotic and abiotic stresses on crop plants have generally become because of various anthropogenic and normal exercises, which adversely affects plants' hereditary capacity for development and monetary creation (Prasad, 2011). At different harvest development stages, abiotic stress, for example, dampness stress, temperature stress, saltiness stress, and so on, impacts the presence of different biotic pressure, like infections, bugs, and weeds (Mitler, 2006; Prasa, 2011). Abiotic stress can change the physiology of plants and their safeguard components, transforming a minor nuisance into a possible peril (Duveiller, 2007; Scherm and Coakley, 2003). Various kinds of collaborations between different pressure components exist, and Mittler and partners' "stress framework" may assist with bettering comprehend what these communications mean for plant improvement and efficiency (Mittler, 2006; Suzuki *et al.*, 2014).

14.2 Stress Speculations

Stress is the variable that seriously affects plants and has adverse results. This element might be biotic, like weeds, microorganisms, bugs, and so on, or abiotic, like environment and soil. For instance, an adjustment of air temperature could become upsetting in no time, yet an absence of soil minerals can require a long time to become distressing. Plant endurance, development, and yield, as well as the essential ingestion process, like photosynthetic rate, supplement collection, and development rate, are utilized to quantify the impacts of pressure. Stress can be flexible, where plants recuperate once the pressure factor is eliminated, or plastic, where plants disfigure and the change is extremely durable when the pressure factor is available.

14.3 Stress Patterns

In light of its general, physical, and organic applications, stress has various definitions. In any case, stress is characterized as far as horticulture as horrible impacts that bring down the biomass and efficiency of harvest plants. Figures 14.2 and 14.3 portray the organic meanings of the biotic and abiotic stressors. Stress can inflict any kind of damage that isn't promptly clear, for example, changes in catalyst action and layer structure, ending of development and advancement, plant harm aesthetic alterations in microbes, including as corruption, chlorosis, and staining, hinder and adjust genetic structures that are exclusive to individual species (Cramer, 2011; Dreselhaus and Huckeloven, 2018) as shown in Figure 14.2.

- Draft: This pressure results from a mix of states of being that make it challenging for plants to get to water, which dials back their development and improvement.
- Flood: The pressure got on by an excess of wetness the root zone forestalls the trading of gases, principally oxygen,

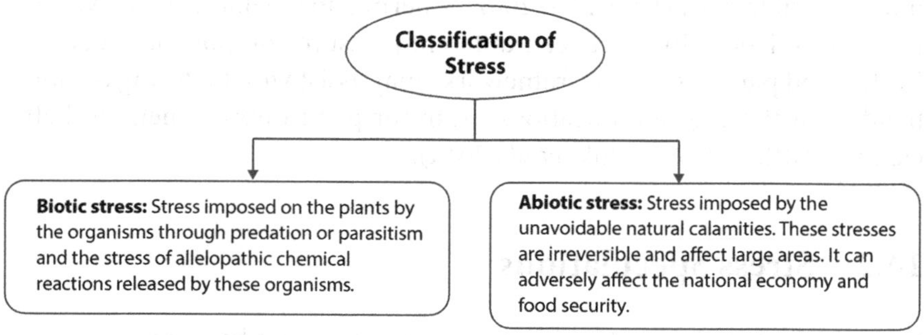

Figure 14.2 Categorization of stresses.

between the dirt climate and plant roots, prompting hypoxic conditions.

- Heat injury: Stress welcomed on by temperatures that are higher than whatever a plant can take, commonly over 40°C.
- Freezing/ice injury: Stress welcomed on by beneath frigid temperatures, i.e., 0°C or 0°C to 10°C, when most of effectively developing plants are annihilated or injured.

Rice, maize, sorghum, and other warm-season crops are especially powerless against harm from chilling injury, which is pressure welcomed on by low temperatures of 10°C to 15°C or less however more noteworthy than 0°C as shown in Figure 14.3.

- Radiation harm: Stress welcomed on by electromagnetic radiations, for example, UV radiation that makes proteins change or debase or IR radiation that causes heat injury in plants.
- Salt pressure: Stress welcomed on by a gathering of an excess of solvent salt in the root zone, which changes mineral supplement retention or increments osmotic potential.

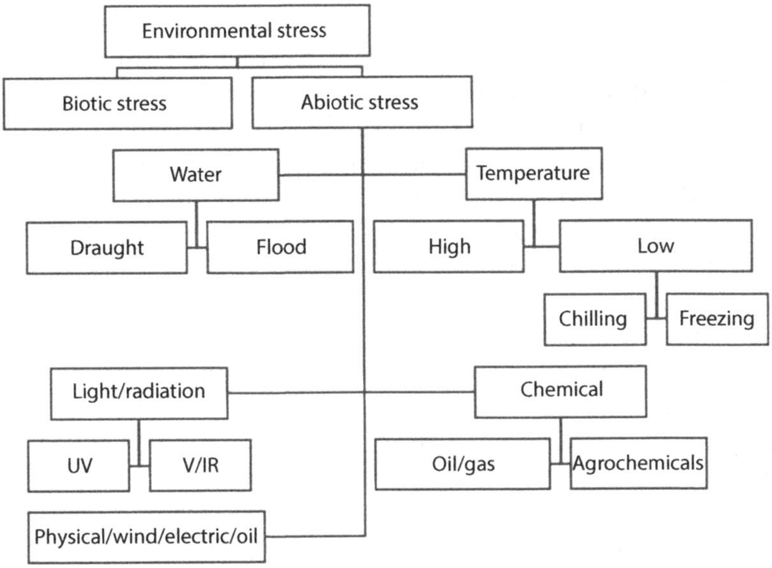

Figure 14.3 Types of pressure.

14.4 Natural Stress

The effect of biotic burdens is similarly essential to abiotic stresses, however there is a tiny contrast between the two; biotic anxieties adversely affect crop plants, yet this effect can be diminished fundamentally by being more wary while cultivating, though abiotic stresses can't be straightforwardly controlled. Crop plants answer shifted natural burdens in various ways by adjusting and changing their reaction systems (Bartels and Salamini, 2001). A complicated climate is comprised of microbial networks that work with the plants to work as holobionts as opposed to as detached genotypes and can assist with growing new courses for protection from risky microbial diseases. Plant species that are consistently presented to these burdens change their quality examples to adjust to these upsetting circumstances. Benzoxazinoids are an example of an auxiliary metabolite in grass that protects the plant from biotic stresses (BXs). Niculaes 2016's study demonstrated the emergence of BXs and other novel organic capacities.

14.5 Organic Stress

Abiotic stresses are those that happen normally yet out of the blue and in a roundabout way affect horticultural plants because of the actual climate. This unfavorably affects crop yield, feasibility, absolute biomass, and creation quality (Grover, 2001; Calanca, 2017). Since food is perhaps of the main component in keeping up with human existence, farming and agrarian land have generally been safeguarded. Be that as it may, to oblige the quickly expanding human populace, rural regions are being changed over into nonarable terrains. Indifferent human way of behaving has essentially expanded the worldwide temperature and lost the planning of the environment. Plants that are creating under pressure have grown preferable survival strategies and flexibility over those that are developing under great and regular circumstances. The previous' pressure responses are fundamentally more multifaceted. Earth's air temperature has expanded altogether since the twentieth 100 years, and plant efficiency keeps on confronting difficulties that make it trying to address the world's issues. Because of its standard openness, it has become critical (Crame, 2011; Kuma, 2018). These covering abiotic provokes make the plants answer in extremely powerful and complex ways. Because of natural stressors like environmental change, the plant local area is continually taken advantage of. The improvement of methods and advancements has essentially diminished how much these double dealings. In the ongoing circumstance, battling abiotic stressors is a steady errand. Despite the fact that advances in atomic and physiological science have expanded how we might interpret pressure resilience, research is as yet expected to give light on the system and quality design.

14.6 Natural Stress

Abiotic stresses are those that happen normally however startlingly and by implication affect rural plants because of the actual climate. This unfavorably affects crop yield, practicality, all out biomass, and creation quality (Grover, 2001; Calaca, 2017). Since food is quite possibly of the main component in keeping up with human existence, farming and agrarian land have customarily been secured. Be that as it may, to oblige the quickly expanding human populace, farming regions are being changed over into nonarable grounds. Uninterested human way of behaving has altogether expanded the worldwide temperature and lost the planning of the environment. Plants that are creating under pressure have grown preferred survival strategies and flexibility over those that are developing under great and average circumstances. The previous' pressure responses are altogether more mind boggling. Earth's air temperature has expanded altogether since the twentieth hundred years, and plant efficiency keeps on confronting difficulties that make it trying to address the world's issues. Because of its normal openness, it has become critical (Crame, 2011; Kuma, 2018). These covering abiotic provokes make the plants answer in extremely powerful and complex ways. Because of natural stressors like environmental change, the plant local area is continually taken advantage of. The advancement of strategies and innovations has fundamentally diminished how much these double dealings. In the ongoing circumstance, battling abiotic stressors is a consistent undertaking. Despite the fact that advances in atomic and physiological science have expanded how we might interpret pressure resilience, research is as yet expected to give light on the system and quality design.

14.7 Thermodynamic Pressure

The more noteworthy temperatures and radiation in heat and humidities thwart plant development and advancement. High temperatures stunt plant development by causing rankling, burn from the sun, and leaf staining (Vollenweider and Gunthardt-Goerge, 2005). Temperature stress is characterized as pressure that confines development, digestion, and yield potential because of openness to temperatures that are either underneath or over the warm edge for the best biochemical, physiological, and morphological turn of events (Greaves, 1996). As per their capacity to bear low, medium, and high temperatures, plants are classified as psychrophiles, mesophiles, and thermophiles (Levitt, 1980). The seriousness of intensity fatigue relies upon the length, stage, and power of the pressure (Fahad, 2016b). Diminished spikelets, florets per plant, and seed set in sorghum are impacts of expanded heat pressure (Prasad 2006; Faha, 2016b). Furthermore, it diminishes nodal length, biomass aggregation, and early leaf senescence in sugarcane, as well as adversely affecting net digestion rate and development in maize (Asraf and Hafez, 2004; Wahid, 2007).

For rice, wheat, and maize, heat pressure influences yield by half, 31%, and 42%, separately (Bala, 2011). Furthermore, it brings down quality since less protein, sugars, and oil are delivered (Wilhel, 1999; Mastri, 2002). Because of diminished water and food conveyance, mass and amount of roots radically decline with the seriousness of intensity stress development (Wahid, 2007; Hal, 2012). It is shown that intensity stress causes passing when there is an absence of water (Machado and Paulsen, 2001). It represses a compound engaged with the production of chlorophyll, which causes chlorophyll a and b to corrupt all the more rapidly in mature leaves (Dutta, 2009; Karim, 1999). Also, oxidative damage was associated with this breakdown. The ideal temperature for photosynthetic movement is expected to ascend with expanded CO2 focus in the air, notwithstanding every one of the physiological and biochemical impacts of expanded heat pressure. When presented to high temperatures, the harm interaction advances from immediate, reversible strain — breath happening more oftentimes than photosynthesis — through backhanded strain, like draft, or direct strain, like starvation (Levitt, 1980; Bita and Gerats, 2013) The ideal plant types for better resistance to warm pressure ought to have better photosynthetic rate, layer thermo-steadiness, and natural product setting and organic product setting under high temperature. Heat pressure lenient plants display negligible harm to photosynthetic organs, cycle, and expansion in biosynthesis of defensive mixtures (Nagarajan *et al.*, 2010). By making heat-lenient cultivars through customary rearing (Ehlers and Hall, 1998), involving QTLs for grain filling and leaf senescence, heat resistance during the conceptive stage, and remain green properties (Farooq, 2011), as well as applying these attributes to the improvement of transgenic creatures, heat pressure can be made due (Bohnert *et al.*, 2006). Premolding plants (Morales *et al.*, 2003), preplanting or early planting (Tikhomirova, 1985), applying calcium (Jiang and Huang, 2001), applying glycine and betain (Wahid and Shabir, 2005), and utilizing spermidine are a few additional strategies for overseeing heat pressure (Murkowski, 2001). Shrinking, blanching, cooking, corruption, and plant demise are side effects of low temperature stress (Levitt, 1980; Witt and Barfield, 1982). An expected 15% of arable land will encounter freezing pressure (Dudal, 1976). moistness stress The world's 28% of land is excessively dry for farming (Kramer and Bouyer, 1995). The assessed yearly yield misfortune in the jungles because of draft is near 17% (Edmeades *et al.*, 1992). Diminished physiology, plant development, and propagation are the results of expanded draft brought about by the situation of an evolving environment (Barnabas, 2008). As per Daryanto, (2016), draft caused yield decreases of 21% and 40% in wheat and maize, separately. Dry spell expands happening and diminishes water accessibility to establish roots, which will in general tip the water balance off course and adversely influences development, supplement and water relations, photosynthesis, acclimatize apportioning, and yield (Anjum, 2011; Farooq, 2009). Plant species answer diversely to dry spell pressure in light of their formative stages and other development boundaries (Demirvska, 2010). High temperature stress changes plant cell division and catalyst movement, as well as the developing season and conveyance of rural harvests (Smetenko, 1997; Watcman, 2005). Dry season represses seed germination and seedling development, as well as early seedling development, length of the hypocotyl, root and shoot dry weight, and vegetative development in field crops like pea, hay, and rice (Farooq, 2009; Manikavel, 2006). Dry season pressure brings about unfortunate yield improvement that is principally brought about by diminished cell bloat, mitosis, and cell prolongation (Taz and Zeiger, 2006; Hussain, 2008). Decreased bloat limits leaf development, which hinders photosynthesis and eases back plant development (Rucker, 1995). Early anthesis is brought about by draft pressure that happens preceding it, while early grain filling in oats is brought about by pressure that happens after it (Estrada-Campuzan, 2008). The key element lessening cereal result is a decline in the action of the proteins that manage grain filling, for example, sucrose synthase, starch synthase, starch expanding catalyst, and adenosine diphosphate glucose pyrophosphorylase (Taz and Zeger, 2006; Ahmad and Baker, 2001). The primary drivers of yield decrease attributable to dry spell pressure are diminishes in photosynthetic rate (Bota, 2004), change in acclimatize dividing (Farooq, 2009b), and diminishes in banner leaf development (Rucker, 1995). Dry season pressure has been connected to yield misfortunes of 53% to 92%, 57%, and 63% to 87% in rice, wheat, and maize, separately (Lafit, 2007; Ball, 2011). Dry spell pressure can be overseen through customary rearing of pressure open minded plants utilizing cultivars that have great agronomic execution,

QTL planning followed by MAS technique, transgenic approach, and prompted pressure resistance, or preparing (Ashra, 2010; Farooq, 2008).

14.8 Stress on Heavy Metals

Weighty metals (53 sorts) incorporate cadmium (Cd), lead (Pb), arsenic (As), silver (Ag), and others that are by and large hurtful at even low focuses (Sharma and Dietz, 2006; Rascio and Navari-Izzo, 2011). As per a few examinations Dembitsky (2003), Igwe and Abia (2006), Ali *et al.*, (2013), the primary drivers of weighty metal pollution in soil are human exercises like mining, purifying, concentrated farming practices, fuel creation, electroplating, and so on. Other potential causes incorporate regular cycles like soil disintegration, unreasonable enduring of rocks and minerals, and volcanic emission. A few weighty metals, known as unnecessary weighty metals, are known to physiologically affect plants, including arsenic (As), lead (Pb), cadmium (Cd), mercury (Hg), and selenium (Se), while fundamental weighty metals, like cobalt (Co), copper (Cu), manganese (Mn), zinc (Zn), iron (Fe), molybdenum (Mo), and nickel, are known to physiologically affect plants (Ni). At the point when Zn is kept from the get-go, it becomes phytotoxic, limits development by thickening the roots, diminishes cell division and stretching, and forestalls the creation of radicles (Rout and Das, 2009).

Superoxide dismutase (SOD), ascorbate peroxidase, glutathione reductase (GR), monodehydroxy ascorbate reductase, and catalase are only a couple of the safeguard systems that plants have set up to safeguard them against ROS (CATs). Ascorbate, tocopherol, carotenoids, glutathione, and other nonenzymatic cell reinforcements, notwithstanding these enzymatic cell reinforcements, are imperative in keeping ROS from harming biomolecules. Grass gives the principal line of guard against oxygen free extremists by changing over O2 into H2O2 and safeguards plants from different burdens among these cancer prevention agents (Banister *et al.*, 1987; Mittler and Zilinskas, 1994; Allen *et al.*, 1997).

14.9 Plant Response to Abiotic Stress

To adjust to the natural circumstances, establishes that are regularly presented to pressures from different biotic and abiotic sources have fostered a few profoundly evolved frameworks. Various plants have many resistance and reaction qualities that they adjust to. A large number of the issues in sorting out the systems of resilience and protection from these burdens through genomic and proteomic examination have been settled by the development and improvement of biotechnologies and omics advances. The revelation of new qualities and the comprehension of their genomic capability are made conceivable by exceptionally successful omics innovations. Because of their continuous openness to pressures at successive spans, plants have developed an assortment of development examples and survival techniques as well as modifying the manner in which they respond to pressure. Specific protein products have been developed by plants in response to environmental stresses via a variety of transcriptional and translational changes (Cushman and Bohnert, 2000). Different species and even genotypes react differently to these viruses. Gardening plant varieties with increased resilience have been divided into four categories. Mannitol, glycine, proline, and betaine are all examples of osmolytes; catalase, ascorbate, and peroxidase are all examples of cellular reinforcements; LEA proteins, chaperones, liquid catalyst proteins, heat shock proteins, and protein kinase are all examples of stress-initiated proteins; and executing factors such as DREB1/CBF, AP2/ERF, DREB2, NAC (Taya, 2004; Sangam, 2005). As a pressure response mitigating transformation, plants change the qualities that outcome in the development of specific results. Flagging variables, which contain signal transduction proteins, and transcriptional factors, which work to keep up with the particle homeostasis and trustworthiness of the phone, are the two classes of utilitarian qualities that give pressure obstruction in plants. Utilitarian protein, be that as it may, is a further vital component notwithstanding these resistance related boundaries. The making of rummaging proteins, cell reinforcement compounds, development chemicals like abscisic acid (ABA), different pressure incited proteins, and protectants is controlled by useful proteins.

High throughput omics innovation has worked on how we might interpret the as of late advanced reaction pathways and systems that the plants have adjusted. How we might interpret how qualities are managed has extended therefore, which controls how certain utilitarian proteins are integrated. Genomic analysis (which profiles coding and noncoding qualities and the statement of their regulatory roles), proteomic analysis (which reveals the proteins' regulatory pathways and primary structures), metabolomic analysis (which reveals the atoms and molecules associated with metabolic activities), and transcriptomic analysis (which profiles gene expression) are all examples of omics tools. Furthermore, bioinformatics became famous by giving subtleties on the declaration of the opposition quality and its similitudes to different species. It is feasible to look into the plants that have been focused on account of the assortment of all the data information utilizing different omics innovations. Primary and framework science can likewise be utilized to analyze factors that influence how different component proteins are organized in light of natural pressure. Various examinations have been directed with individuals' interests over dry spell, intensity, and flooding as a main priority. As indicated by studies, these three burdens altogether affect plants and hurt the development and formative stages. Nonetheless, safeguarding layer breakdown is a key need. Plants answer physiologically to keep up with their pace of happening, photosynthesis, breath, and osmotic balance. Heat pressure adversely influences the conceptive organs and other leaf parts, which makes it challenging for plants to duplicate and may represent up to 40% of creation misfortunes. Nitric oxide (NO), receptive oxygen species (ROS), heat pressure factors (HSF), as well as various other mediocre proteins and flagging components, are completely created by plants in light of intensity stress (Nadeem *et al.*, 2018). On the opposite side, dry spell causes osmotic irregularity and dials back photosynthesis. As indicated by studies, it is important to assess all types of wild plants and to limit the atomic methodologies that can be utilized to explain the fundamental cycles of pressure reactions. Most specialists presently advocate the utilization of transgenic plants to get through testing natural circumstances and limit yield misfortune. By diminishing happening and breath rates, plants could help their capacity to get by under dry season pressure. Some C3 plants, similar to Arabidopsis sp., are intrinsically greater at using water than C4 plants, albeit this quality is not without its drawbacks. It diminishes the take-up of carbon dioxide, which influences photosynthesis and, subsequently, brings down farming result. These plants' development and advancement are additionally stopped at times, which brings down their result. Using QTL examination, which supports distinguishing the special lenient qualities, recognizing the pressure resilience genes is conceivable. Understanding their cycles for making new, altered rural plants might be conceivable with the utilization of the information.

14.10 Plant Abiotic Stress Tolerance Mechanisms

Abiotic stress adversely affects plant development and efficiency wherever over the world. The genuine yields would have to match the plants' likely yields in the event that natural tensions did not occur. While the typical plant yield in many harvests is only 20% to 30% of what it very well may be, Plants have fostered a large number of resistance systems at the full scale and miniature levels because of their sessile nature and the predominance of abiotic stress. When exposed to an abiotic stress, they regularly go through an assortment of morphological, physiological, and subatomic changes that trigger a fountain of obstruction systems at various levels (Grift, 2011; Sinclar and Muhow, 2001). This section examines the fundamental abiotic pressures that influence plants, takes a gander at how they can adjust to them, and summarizes plant tolerance to biotic and abiotic stresses.

14.10.1 Adaptability in Morphology Abiotic Stress Tolerance

A surge in water pressure In order to continue its life cycle, the plant might adapt to or migrate away from its hostile surroundings. To create seeds and make use of the water resources that would not be accessible at the start of the dry spell period, the main system pushes the plant to bypass the vegetative stage and move right into the regeneration stage. Therefore, many plant species, especially those that are compelled to mature early in desert circumstances, choose to escape while they are fully reproducing before the beginning of lethal dryness (Fahad *et al.*, 2017; Farooq *et al.*, 2012). However, some species have adapted by changing their genetic make-up to better withstand dry spells. Modifications are being made primarily to ensure a sufficient supply of water for future development. One way they do this is by enhancing the root system's ability to take in water, and the other is by decreasing the rate at which water is lost through vaporized soil (Blum, 2005). The biomass production of many species drops significantly during the dry season, with large differences between their above- and below-ground parts. Furthermore, a decrease in vegetative biomass is not necessarily associated with a lack of development or an increase in the root foundation's biomass. Root/shoot ratio shifts in rice, as reported by Xu (2015), were brought on by an increase in the severity of shoot disasters rather than a rise in root biomass. Increases in above-ground biomass were mirrored by increases in underground biomass in dry season adaptable chickpea cultivars, suggesting that this variety strategy to retain water and nutrients through massive underground root growth is effective (Rahbarian, 2011). To deal with the dry season issue, modifications must be made to the root design. Most commonly, this means expanding the root system in all directions (in terms of both the number of major roots and the number of fine roots) and speedily reaching the soil surface. Most importantly, plants hit by a dry spell need their deep roots to expand further so that they can better access water. Their dedication has been shown to be beneficial in a range of crops that are vulnerable to the dry season, including as rice (Manschadi, 2010; Wasson, 2012), maize (Hussain, 2019), and wheat (Krishnamurthy, 1999). Chickpea under dry season pressure benefited from an increase in water and mineral supplement retention due to a root length thickness of up to 30 cm in the root's capacity zone (Sadok and Sinclair, 2011).

Then again, a huge root foundation shouldn't necessarily be seen well on the grounds that the more material is distributed for its upkeep, the bigger the root will be, which can add to a decrease in over-the-ground biomass (Braley, 2009; Li, 2010). Dry spell pushed crop plants go through huge primary changes to their leaves, remembering a decrease for the size of the leaf, which is a critical wellspring of water misfortune. By limiting water misfortune through happening, this change is perceived to be a fruitful procedure for conquering water deficiencies, yet it punishes efficiency by having an unfortunate pace of carbon ingestion (Li, 2006; Sinclair and Muchow, 2001). Moreover, expanding fingernail skin thickness, stomatal opposition, stomatal size, and to wrap things up, shifting or moving leaves, are different strategies to decrease light assimilation and water use when under dry spell pressure (Sinclir and Muhow, 2001). To balance the limitations welcomed on by flooding, especially the general decrease in CO2 trade and CO2 obsession, an assortment of morphophysical variations in plants have been made to make due under flood pressure. There are various productive strategies, one of which is the improvement of new roots with gas-occupied spaces (aerenchyma) because of cell partition or customized cell passing. This attribute is available in many plant species and helps establishes that are lowered in water, similar to rice, to give oxygen to their organs. At the point when rice is lowered submerged, oxygen move is actually forestalled, yet this transformation effectively expanded oxygen transport in rice (Cardos, 2013; Coler, 2002; Drew, 1979). The root foundation's capacity to oxidize the rhizosphere and discharge oxygen into the anaerobic soil diminishes the gamble of injury from diminished poisons like Fe2+ (Comer, 2002). Aerenchyma development in certain plants, similar to maize, can be animated by a decline in oxygen and an ascent in ethylene, what begins the course of modified cell demise in the root cortex (Drew, 1979). Moreover, less-extended fundamental roots, enormous root distances across, and profound roots are qualities of root design that elevates resilience to overwhelmed plants (Aguilar, 2003, 2008; Visse, 1997). Sena Gomes and Kozloski (1980) saw in *Eucalyptus* spp. that there seems, by all accounts, to be a positive relationship between dry season lenient and flood open minded, albeit in woody species the transformations are somewhat unique. In maize, the improvement of aerenchyma in the external leaves is conceivable in the event that the shoot base is lowered in water (Masukura, 2000). How many aerenchyma were delivered in flood-pushed wheat depended on the genotype's level of resilience, with higher-resilience occurring in safe genotypes that had more aerenchyma. Clearly, in light of the fact that the root foundation and ethereal parts are interconnected basically, the flood additionally influences the photosynthesis mechanical assembly by exhausting oxygen in the root zone and confining the capacity of roots to retain water. This outcomes in shrinking leaves and a dry spell like condition, which thusly debilitates the capacity to acclimatize carbon. Furthermore, N shortage is oftentimes present during flood pressure, yet establishes attempt to move N from more seasoned passes on to youthful passes on to keep filling to settle this issue, to some degree briefly (Drew and Sisworo, 1977). At the point when harvests are likely to flooding, leaves can frequently go through modifications that decrease light interference by moving lower (epinasty). This conduct has been seen in tomato leaves (Elsworth and Freebarn, 1969).

14.10.2 Outrageous Temperature

Crop plant bottle resilience is the capacity to endure heat pressure, develop, create, and produce a financially sensible yield. Up-and-comer crop plants might be morphologically ready to balance out their inward temperatures just beneath the irreversible place where harms are terminal under heat pressure. Such a system can be accomplished by making changes like those depicted before for transformation to dry season pressure, for example, leaf rolling, changing the direction of the leaves to mirror daylight and lessen light retention, controlling stomatal way of behaving, and advancing the fingernail skin and wax layer on the outer layer of the passes on to increment stomatal opposition (Grif, 2011). Moreover, different characteristics of open minded crop cultivars incorporate the capacity to keep up with photosynthetic action, scattering of the decent carbon, and assimilation of fundamental minerals (Cui, 2006). Moving banner leaves lessens temperature and, thus, happening in wheat crops, expanding the yield of photosynthetic movement (Perkons, 2014). In this way, many plant species, essentially those that are early developing obliged in desert conditions, decide to escape while having full proliferation before the appearance of deadly dryness (Fahad, 2017; Faroq, 2012). Different species, nonetheless, have settled on the choice to adjust their underlying cosmetics to be more dry season lenient to stay away from lack of hydration. The significant objective of these changes is to keep a high water status with the goal that development and improvement might proceed. They either decrease the pace of water misfortune from aeronautical partitions or increment the root foundation's ability to retain water (Blum, 2005). The contribution of break like strategies in plants has been noted. These species regularly show a propensity to develop and deliver seeds right on time before the beginning of the intensity stress period, which undeniably concurs with a dry spell nearby (Trachse, 2009). A few plants will quite often make more slender sharp edge leaves with diminished stress and prostrate development propensities to work on the radiation, use effectiveness, and water asset preservation (Richards, 1996). Advancing happening is a typical practice in plants to bring down temperature and keep away from heat pressure harm. Utilizing this technique requires a profound root foundation to support the water use, yet it has been demonstrated to find true success as Chauhan *et al.*, (2009) noticed that wheat genotypes with lower temperature shelters showed more noteworthy resilience to warm pressure than genotypes with higher temperature coverings. The upsides of tetraploid wheat were likewise answered to have a more prominent yield (Bahar, 2008). At the point when the plant is focused on by intensity or dry spell, major areas of strength for a framework with stretching yet profound principal roots can retain water and mineral necessary supplements. We looked for important data or distributed research demonstrating explicit morphological transformation in crop plants under low temperature; however, we could not find any. This might be on the grounds that yields' virus stress obstruction systems fundamentally depend on physiological and subatomic methodologies as a handy solution to endure the poor temperature.

14.10.3 Saltiness

Saltiness has various unfavorable consequences for the wellness of plants. The root foundation and flying segments of plants have gone through an assortment of morphophysical adaption techniques during development. The changes in morphological elements in this section primarily concern salt-resilient plant species that are not thought of as harvest plants because these variations at this level have typically but not always been seen in plant species possessed in saline waters or regions with high salt content and dry soils. Most of halophytes and glycophytes are monocotyledon families, while they can sporadically be dicotyledonous. Halophytes are plants that can oppose salt, while glycophytes, for instance, can persevere through a low measure of saltiness (Rozema 1991; Sharma *et al.*, 2016). The underground root growth engineering of species that have adjusted to dry spell pressure is more like saltiness since it is joined by an outrageous water shortfall in the dirt. These species have roots that are profound, broad, have a high length thickness and measurement, as well as having roots with a higher root weight. In grasses presented to saltiness, expanded root stress was seen (Seregin and Ivanov, 2001). Contrasting salt-open minded plant species like *Puccinellia distans* and *Leptochloa fusca* L. with salt-bigoted *Pennisetum divisum* L. uncovered that the last option species' biomass was altogether decreased. Salt obstruction is linked to maintaining biomass biosynthesis rate (Ashraf and Yasmin, 1997). *P. vulgaris* and *Gossypium hirsutum* were two glycophyte species whose proportion of the mesophyll area was profoundly reduced open minded species like *Atriplex patula* L. might have the option to safeguard the photosynthesis contraption against salt-incited terminal impacts by expanding, keeping up with, or in any event, expanding how much mesophyll per leaf region. As indicated by Longstreth and Nobel (1979), the most common technique utilized by halophytes to deliver salt particles is the improvement of salt organs. As a transformation component, the thickness of these organs escalated in Zoysia species, working on salt resilience, which continued after the organs were taken out (Marcum, 1998). One more way that some plant species have adjusted to get through saltiness in dry season stricken conditions is to store water from restricted assets in thick, corona delicious passes on to keep up with the water status. In drylands with high saltiness, xerophytes' unmistakably thickly covered leaves with pubescence can help halophytes in decreasing water misfortune and supporting development and advancement (Marcum, 1998). One more way that some plant species have adjusted to get through saltiness in dry season stricken conditions is to store water from restricted assets in thick, radiance delicious passes on to keep up with the water status.

In drylands with high saltiness, xerophytes' unmistakably thickly covered leaves with pubescence can help halophytes in lessening water misfortune and supporting development and improvement (Marcum, 1998).

14.10.4 Heavy Metals

In spite of the fact that plants have fostered an extensive variety of protection components against weighty metals, the majority of these safeguards are biochemical and subatomic in nature. As of late, nonetheless, has examination into how plants have changed their morphology to endure the destructive impacts of weighty metals on their leaves or roots (Shahid *et al.*, 2017). Nonetheless, the meaning of explicit plant changes and organs that were connected to weighty metal resilience were examined, For instance, heavy metal particles can enter and completely penetrate the cuticular structure of a leaf (Vu, 2013). Maize immunized with Glomus sp. forms a symbiotic connection with the mycorrhizal parasite, which, as is common knowledge, aids host plant roots by limiting metal ion mobility and providing host plants with critical nutrients for soil formation. Seedlings that were immunized showed significant improvement over their nonimmunized counterparts, there was an extensive diminishing in the convergence of weighty metal particles (Kaldorf *et al.*, 1999). With regards to the impact of the advantageous association with mycorrhiza on the decrease in the transportation of weighty metal particles to flying parts, various cycles have been proposed (Hall, 2002). It's conceivable that a combination of these instruments is having an effect on everything.

14.11 Biotechnical Techniques to Reduce Plant Abiotic Stress

The improvement of resistance in plants for battling off horrible natural circumstances, especially under the continually changing climatic circumstances, is a critical issue for plant biotechnology in the present day. As indicated by Figure 14.4, abiotic stressors adversely affect plant development and improvement and fundamentally lower crop yield (Jalil and Ansari, 2020a). To increment plant pressure opposition and guarantee plant endurance and farming efficiency, different techniques have been created (Chikkaputtaiah and Marwein, 2019). Conventional reproducing techniques utilize genetic contrasts that as of now exist in different germplasms or that are made by intergeneric and interspecific hybridization, actuated transformation, and substantial change. Regardless of their viability, these strategies have various disadvantages since they simply prompted the making of few new pressure safe sorts that can get by in the field (Lin *et al.*, 2013). Moreover, conventional techniques are hampered by the low hereditary variety of plants that can live in testing natural circumstances, the intricacy of stress obstruction characteristics, and powerful determination processes (Lin *et al.*, 2010). Late endeavors to work on plants' capacity to endure abiotic stressors have gained critical headway. Ansari *et al.*, (2011), stress responsive qualities (Ma *et al.*, 2018), and proteins engaged with the age of metabolites are subject to hereditary articulation that influences pressure the board as well as flagging systems in hereditary designing strategies for plants. The techniques utilized in plant hereditary designing to increment stress resilience incorporate Figure 14.5 meddling at different reactions, levels of responses, administrative parts, record elements, and cancer prevention agents to invigorate endogenous frameworks (Jacob *et al.*, 2017). Moreover, there have been extensive progressions in the use

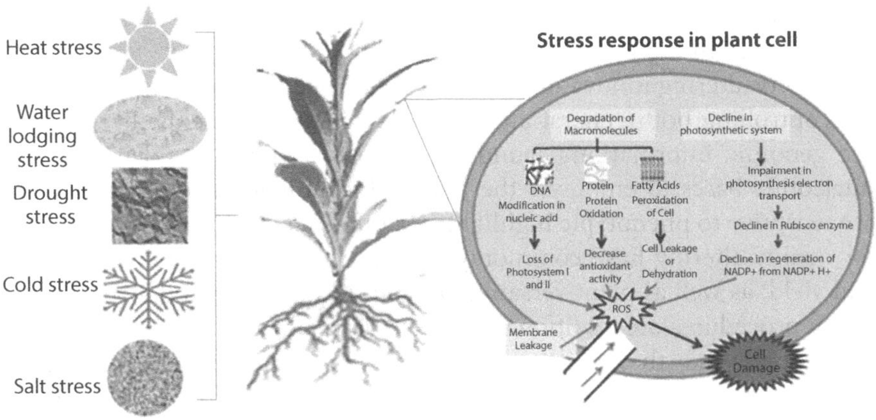

Figure 14.4 A simplified depiction of the cellular responses of plants to abiotic stress. Abiotic stress causes a plant to produce an abundance of reactive oxygen species (ROS), which in turn leads to lipid peroxidation, enzyme inactivation, and DNA damage, all of which ultimately result in cell death.

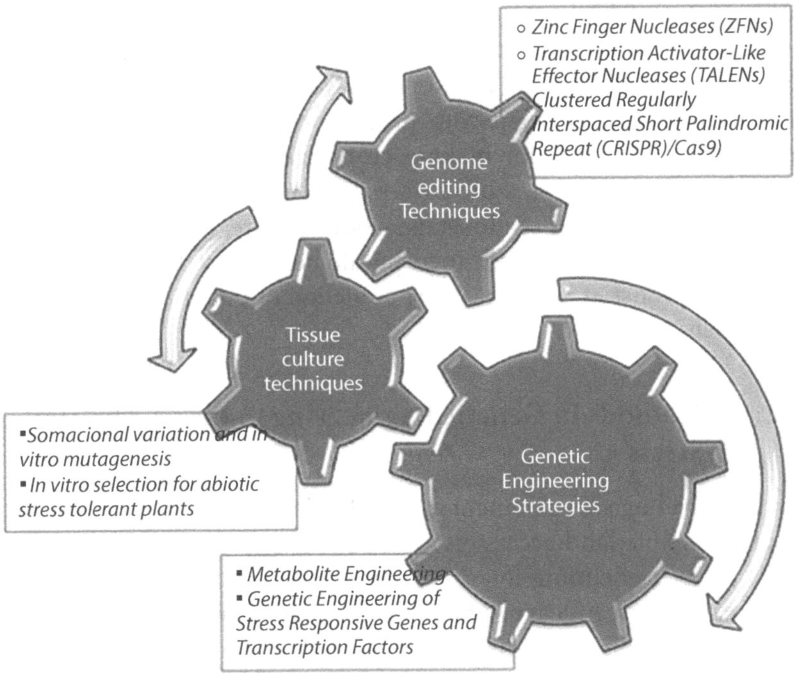

Figure 14.5 Biotech methods for protecting plants from environmental hazards.

of atomic strategies, which might expand plants' capacity to endure abiotic stress as seen in Figure 14.4.

Quality altering innovations have as of late opened up another strategy for controlling both general and specialty digestion systems. It changed the genomic information's found successions, zeroing in solely on few nucleotide bases to empower the practical examination of qualities that are receptive to phenotypic highlights (Mahfouz *et al.*, 2014). Instruments for quality altering have commonly been used to increment crop quality and yield, as well as plant versatility to abiotic stressors (Abdallah, 2016). These procedures are additionally viewed as nonhereditarily changed on the grounds that they abstain from controlling extended groupings and adjusting plants with unfamiliar qualities (Abdallah, 2016). These biotechnological methods enabled the production of various metabolites and bioparticles, which in turn regulated transcriptional and posttranscriptional processes, increased plant adaptability, and improved our ability to understand the key components involved in pressure resistance. The improvement of powerful strategies for hereditary designing and quality altering in plants, as well as examinations of tissue-explicit advertisers and quality characterization, have all profited from propels in biotechnological advancements (Chikkaputtaiah and Marwein, 2019). This biotechnology strategy gives the chance to work on healthful quality, crop yield, and stress resistance. The advancement of chemical flagging pathways, cell associations, and crosstalk between pathways, as well as the examination of plant atomic components engaged with protection from abiotic pressure, can work on the techniques for making crop-delivering, stress-lenient plant assortments. To foster pressure safe plants or to improve abiotic stress resilience in plants, this section talks about progresses in biotechnological procedures. These methods incorporate hereditary designing, genome altering, and, most as of late, atomic guideline of qualities for supporting yield efficiency as shown in Figure 14.5.

14.11.1 Methods in Genetic Engineering to Resist Abiotic Stress

Because of the huge development in worldwide populace and the restricted measure of arable land accessible, as well as continuous natural changes, there is an overwhelming interest in expanded farming yield. To resolve this issue, various endeavors have been made to propel biotechnology and rearing techniques for making pressure safe harvests with enormous yields and worked on dietary benefit. Then again, it keeps on being the farming business' significant test (Joshi *et al.*, 2020). Because of the intricacy of abiotic stress opposition qualities, conventional rearing strategies have had almost no achievement. Conversely, present day biotechnological approaches, like hereditary designing techniques, stand out of specialists since they include the presentation of new unfamiliar qualities or modification of endogenous pressure related quality articulations to work on plants' capacity to endure pressure. It was fundamental to comprehend the subatomic components engaged with the transmission of stress signs to cell compartments to actuate different reactions, which required the recognizable proof of huge pressure related qualities and pathways (Sanchez *et al.*, 2011). A few exploration have utilized forefront sequencing and utilitarian genomics strategies to give light on the subatomic component of qualities that answer stressors (Chikkaputtaiah *et al.*, 2017).

14.11.2 Metabolite Engineering to Increase Resistance to Abiotic Stress

Ammonium mixtures like betaine, polyols, sugars, and a couple of amino acids (glycine, proline) that collected in plants under horrible natural circumstances are osmoprotectants (Ashraf and Harris, 2004). At the point when cells are under pressure, these osmoprotectants deal with the osmotic change by balancing out proteins and films (Zandalinas *et al.*, 2018). Using metabolic designing procedures to mimic the combination of osmoprotectants in such plants to build their resilience to different abiotic stress circumstances is legitimate on the grounds that assorted harvests blend very little osmoprotectants (McNeil *et al.*, 1999). A review of the literature reveals that salt and drought stress cause *Ailanthus altissima* (Miller) to increase its 1-pyrroline-5-carboxylate synthetase activity. Proline is collected in various plants under abiotic conditions (Filippou *et al.*, 2014). Furthermore, *Zea mays* plant under pressure showed significant proline amassing to oppose abiotic stress reactions (Huang *et al.*, 2018). Stress-safe hereditarily designed crops have been created utilizing the aggregation of osmoprotactants that occurs during osmotic change under troublesome ecological circumstances (Per *et al.*, 2018). Moreover, by moving the betA quality for glycine betaine biosynthesis, saltiness stress opposition in *Triticum aestivum* was illustrated (He *et al.*, 2010). Because of the level of GB1 quality articulation, *Zea mays* and Glycine max plants have higher glycine betaine 1 substance when contrasted with control plants (Castiglioni, 2018). Because of dry spell and cold stressors, hereditarily designed *Oryza sativa* plants with the bacterial choline dehydrogenase quality showed elevated degrees of glycine betaine development. Similarly, *Triticum aestivum* plants with the *Vignaa coniti* folia quality of pyrroline T carboxylate synthetase (P5CS) have demonstrated resistance to dry season. Furthermore, *O. sativa* seedlings conveying the P5CS quality show expanded proline levels (Sawahel and Hassan, 2002). Plants genetically modified to overexpress the ornithine delta-aminotransferase quality (OsOAT) of *Oryza sativa* had higher levels of the amino acid ornithine and the proline needed to withstand a variety of stresses (You *et al.*, 2012). Under dry season and salt pressure, transgenic *O. sativa*'s yield potential was expanded by improving trehalose biosynthesis (Joshi *et al.*, 2020). During antagonistic circumstances, the GABA shunt course controls GABA digestion. GABA likewise plays a critical capability as an osmolyte, lessening the harmful impacts on plants (Jalil, 2019; Jali and Asari, 2020). Isoprenoids are terpenes that assume a part with all due respect systems against abiotic stressors and oxidizing ecological circumstances. Two particular biosynthetic courses are utilized by plants to deliver isoprenoids. Traditional reproductive tactics have failed miserably because of the complexity of abiotic stress opposing features. Conversely, scientists have zeroed in on cutting edge biotechnological approaches like hereditary designing strategies, which include presenting new unfamiliar qualities or adjusting endogenous pressure related quality articulations to work on plants' capacity to endure pressure. Pressure resistance ability in plants for combating abiotic challenges was developed by the heritable alteration of these pathway qualities that enable isoprenoids overexpression that responds against abiotic stress by altering ROS age with direct responses, circuitous responses with ROS. Development of transgenic plants with resistance to stress necessitated fundamental data on the atomic components (Jali and Asari, 2020c).

14.11.3 Stress-Responsive Qualities and Record Variables Can Be Hereditarily Adjusted

Hereditary engineering is used to increase stress resistance in plants by presenting various pressure administrative qualities such as atomic escorts, qualities encoded by heat shock proteins (HSPs), histone guidance, helicase and microRNA modification, and different record factors (TFs). Various cell processes, including protein collapsing, transport, and debasement, are completed by atomic chaperones, and it has been shown that a wide range of proteins display chaperone-like capabilities. Then again, different chaperones are pressure reaction proteins and are known as HSPs (Wang *et al.*, 2004). Moreover, during heat pressure, Glycine max Hsp90 quality overexpression in A. thaliana brought about diminished pressure initiated reactions and improved yield and phynotypic highlights (Xu *et al.*, 2013). Moreover, rice seedlings designed by hereditary designing to overexpress OsHSP18.6 showed improved resistance to abiotic stressors. Rice seedlings adjusted by the exogenous articulation of the HSP70 quality showed upgraded protection from salt pressure (Hoang *et al.*, 2015). Moreover, EaHSP70-initiated sugarcane plants show resistance to dry season and salt pressure (Augustine *et al.*, 2015). In an alternate examination, the *Z. mays* heat-prompted HsfA2 quality (record factor), Overexpression of ZmHsf04 in A. thaliana, which shown resistance to intensity and salt pressure (Jiang *et al.*, 2017). Histones were adjusted in the N terminal district by methylation, acetylation, phosphorylation, and ubiquitination; these changes either enacted or hushed qualities that answer pressure (Asensi-Fabado, 2017; Kim, 2015). Hyperacetylation of histones H3K9, H3K14ac, and H3K27ac is responsible for the OsDREBb quality in *O. sativa*, which provides enhanced resistance to salt and cold stress (Scott *et al.*, 2014). Histone modification in the MYB, bZIP, and AP2/DREB TFs of G. max provides additional defense against salt pressure (Ci *et al.*, 2015). To *et al.* (2011) report that the histone deacetylase quality HDA6 is crucial for plants to adapt to salt conditions and survive in cold temperatures (Chen and Wu, 2010). In addition, HD2 proteins are more resistant to highly salty environments (Luo *et al.*, 2012). Helicases have been proven in a number of studies to have an important role in mitigating the effects of potentially dangerous environmental factors (Macovei *et al.*, 2012; Tuteja *et al.*, 2012; Liu and Imai, 2018; Raikwar *et al.*, 2015). Reports suggest that transgenic *Nicotiana tabacum* with pea PDH45 quality overexpression is resistant to salt pressure (Sanan-Mishra *et al.*, 2005). The *O. sativa* ATP-restriction protein (OsABP) also shown a response to abiotic stresses (Macovei *et al.*, 2012). In addition, enhancing cancer-prevention drug action and facilitating the declaration of stress features, transgenic stew's overexpression of the pea DNA helicase PDH45 helped reduce various abiotic stress situations (Shivakumara *et al.*, 2017). Further, in comparison to wild kind plants, which exhibit early foliar senescence in response to various stressors, Nicotiana plants engineered through genetic engineering to overexpress the pea p68 quality accrue low Na1 and high K1 levels (Tuteja *et al.*, 2014).

Another investigation discovered that groundnut establishes that communicated the PgeIF4A quality had expanded versatility to dry season and saltiness stress (Rao *et al.*, 2017). Portraying these objectives might be useful in clarifying the administrative job of miRNA in light of unfavorable ecological circumstances in light of the fact that microRNAs are noncoding

RNAs that adversely manage the posttranscriptional articulations of stress-related qualities and can impact articulation give transformation benefits (Chen *et al.*, 2017). The improvement of stress open-minded transgenic plants can be worked with by the portrayal of these RNAs utilizing genomic and proteomic procedures (Shriram *et al.*, 2016). Furthermore, *O. sativa*'s miR529a-intervened overexpression of the qualities for OsSPL2 and OsSPL14 exhibited upgraded resistance to oxidative pressure (Yue *et al.*, 2017). Furthermore, Cicer arietinum overexpressing miR408 showed further developed resilience to dry spell pressure (Hajyzadeh *et al.*, 2015). It has been noticed that transgenic A. thaliana that overexpresses soybean miRNA172c and MIR394a shows upgraded resilience to saltiness and drying out pressure.

In light of chilling pressure, the overexpressed *O. sativa* seedlings communicating miR319a and miR319b showed resilience (Yang *et al.*, 2013). Moreover, it has been found that A. thaliana, which overexpresses the CBF quality while stifling ESK1 by means of siRNA quality articulation, displays strength to dry season pressure (Xu *et al.*, 2014).

14.11.4 Devices for Gene Editing to Increase Plant Stress Resistance

Ways to deal with genome altering utilizing nuclease-explicit arrangements have become compelling instruments for research on hereditary qualities and plant advancement. The accessibility of first rate genome altering strategies makes the way for the possibility of upgrading plant quality for viable and designated development of plant attributes, the increased resistance to abiotic stresses. The development of engineered nucleases represents a double-stranded break that will alter cellular biology by providing a new way to introduce desirable traits into the human genome. Genome editing with ZFNs (Kim, 1996) and TALENs (Christian, 2010) is now more commonplace because to the development of clustered regularly interspaced short palindromic repeats (CRISPR/Cas) frameworks (Jinek, 2012), which provide the versatility and ease of targeted genome editing.

14.11.5 Zinc Finger Nucleases

Very deliberate nucleases, known as zinc finger nucleases (ZFNs), are generated when the Cys2-His2 zinc finger (ZF) domain is introduced (Kim *et al.*, 1996; Pabo *et al.*, 2001; Palpant and Dudzinski, 2013). Although the initial ZF emphasis successfully fused certain animal and plant species

(Gaj, 2013). The focus of ZFNs is on the ineffective endogenous traits in various plant species (Townsend, 2009). The use of base replacement and site-specific mutagenesis is crucial for plant genetic engineering (Osakabe, 2010). Therefore, ZFNs are a major technique used in various yields to increase resistance to abiotic stress. An ABA-INSENSITIVE4 from the AP2/ERF family, which is involved in abiotic stress response, was the target of an intentional ZFN framing with HSPs (Osakabe, 2010). In Arabidopsis, ZFNs are intended to target specific regions of DNA and trigger corresponding addition and cancellation alterations (dePater, 2009; Hou, 2014).

With the guide of specific protein activator-like effectors (TALE) proteins, record activator-like effector nucleases (TALENs) produce and manage the designated DNA (Jankele and Svoboda, 2014). A segment of proteins is devoted to the elements of the filling site as a quality interpretation instrument, DNA restricting, and cell flagging (Schornack, 2006). In this way, TALE proteins can impede DNA, and after a couple of years, scientists had the possibility of deciphering the unique code of the designated DNA design that incorporates TALE proteins (Boch, 2009). TALENs are an effective tool for promoting site-specific mutations in a variety of organisms (Hou, 2014; hfouz, 2011). *A. thaliana*, *Nicotiana tabacum*, and *Oryza sativa* are only a few examples of the many plants that can benefit from TALENs (Curtin *et al.*, 2012; Shan *et al.*, 2013). On top of that, *O. sativa* has undergone a number of changes because of TALEN (Zhang, 2016). In addition, TALEN's application to potatoes has revealed the system's exposed adaptability in terms of handling and storage (Clasen, 2016).

Additional options for hereditary adjustment using designated features for specific plant highlights become available with the development of genome altering procedures like Clustered Regularly Interspaced Short Palindromic Repeat (CRISPR)/Cas9. Due of its true prospective benefits over other genome changing technologies like ZFN and TALEN, scientists have made astounding assumptions regarding CRISPR/CRISPR-related protein-9 (Cas9) (Mao, 2013). CRISPR-Cas9 is based on the ability of RNA-DNA to generate a double strand, rather than on specific nucleases that target proteins. The advantages of CRISPR Cas9 include the ease of multiplexing zeroing in on variations in different hereditary articulations and the efficacy of presenting alterations via the detailing of Cas9 and guide RNA, despite the fact that these nucleases are specific to the objective framework (Zafar, 2019). In order to activate (CRISPR or CRISPRa execution) or inactivate (CRISPR impedance or CRISPRi) quality reconciliation notwithstanding mutagenesis, this technique couples a synthetically dynamic Cas9 with a transcriptional activator or repressor (Bortesi and Fischer, 2015). Covers the exhibit of abiotic stress hushing, nonetheless, are deficient for such exploration. This procedure disrupted the hereditary change of tomatoes that supported plant strength to warm pressure by coordinating pressure qualities like SlAGAMOUS-LIKE 6 (SlAGL6) (Klap, 2017). Subsequently, a few plants have successfully utilized CRISPR-Cas9 to zero in on different qualities in a solitary living structure (Char, 2017). Consequently, by zeroing in on a couple of pressure related qualities in the valuable however essential harvest yields, this genome-altering strategy can possibly make establishes that are more lenient to push. Moreover, using CRISPR, which is much of the time used to make hereditary adjustments, open minded qualities can be overexpressed. The use of this method for genome altering has as of late expanded as more exhaustive genomic studies are being led to raise crop quality (Rodrguez-Leal, 2017). It very well may be an elective strategy to developing traditional plants that depend on finding plant networks with adequate hereditary assortment to give advantageous characteristics to plants. This procedure could bring about clever allelic varieties in plants, making it conceivable to recognize novel alleles connected to a specific positive aggregate in the aide RNA grouping (Eid, 2018). Subsequently, this instrument was vital in the advancement of stress-safe cultivars.

14.12 An Approach for Future Applications of Nanomaterials in Combating Plant Stress

It has been observed that harvest plants' capacity to create significant returns relies to a great extent upon the nature of the dirt, the accessibility of water, and various other ecological variables, including light, temperature, etc. (vanderLaan, 2017). Any progressions to the previously mentioned ecological factors might affect plant creation by diminishing result (He, 2017). Notwithstanding, climatic changes and human contribution prompted a sharp fall in soil richness, which thusly diminished crop yield (Akhtar, 2015). Furthermore, the utilization of out of date frameworks for water system, compost, and different parts of farming has made this issue decline (vanderLaan, 2017). Using state of the art water system and preparing innovations while likewise distinguishing plant sustenance that is less perilous to food crops is the answer for this issue (Bargaz, 2018). Moreover, the advancement of new plant species that can endure biotic and abiotic stressors is required (Rejeb, 2014). To the extent that abiotic factors go, normally created nanoparticles are a typical part of organic frameworks that have a great many underlying varieties and natural capabilities (Hedayati, 2016).

Nanotechnology is a quickly developing part of science that can assist with further developing these pressure factors through various means, for example, the making of compelling cell reinforcement frameworks and composts that are both more successful and less harming (Zuverza-Mena, 2017). Plant pressure causes that are both abiotic and biotic in nature have been accounted for by various examinations (Calanca, 2017). Abiotic stressors, which incorporate saltiness, flooding, dry season, UV radiation, freezing, and chilling as well as different circumstances, make rural creation endure hurt on a worldwide scale (Wani, 2016). Various specialists have archived plant abiotic stress, which incorporates extra burdens welcomed on by ominous ecological factors such temperature swings, extreme UV beams, dry spell, freezing, salt, weighty metal fixations, and hypoxia (Waqas, 2019). Moreover, late examinations on the cell, subcellular, and atomic degrees of plant abiotic stress have been distributed (Iqbal, 2020). There is a dire requirement for transformation and adjustment of plant abiotic stress factors because of the probability of worldwide environment variety and various abiotic challenges (Raza, 2019). Accordingly, there should be broad exploration on plant development and formative stages because of different impacts of abiotic stressors at the morphophysiological, biochemical, and subatomic levels (Wani, 2016). Safeguarding plant development and improvement as well as various plant frameworks in light of different abiotic

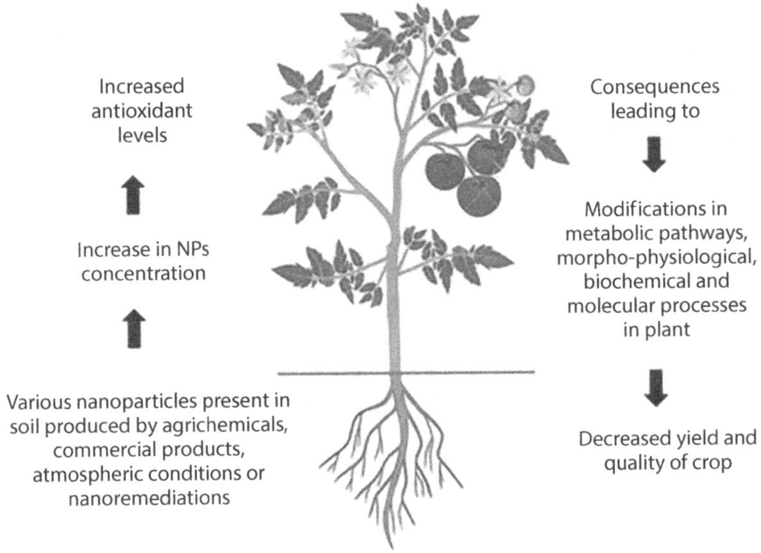

Figure 14.6 Probable sources of nanoparticles and its effect on plant with respect to its yield and productivity.

difficulties would be of excellent significance (Zhu, 2016). In spite of the way that the utilization of nanomaterials is creating, significant enhancements should be made to the laid out approaches for how plants answer abiotic stressors (Iqbal, 2020) as seen in Figure 14.6.

Nanomaterials have previously been utilized in an extensive variety of modern, clinical, and farming applications because of their remarkable physiognomies (Jeevanandam, 2018). It has been noticed that the utilization of nanomaterials in horticulture has expanded and will keep on doing as such. Models incorporate soil nanoreclaimants, nanopesticides (Chipa and Josh, 2016), nanofertilizers (Ahmed, 2019), and nanosensors for nanocultivating (Ylmaz, 2017), soil and water nanoremediators (Floris, 2017), as well as (Paul, 2018). Using such nanomaterials may assist with limiting the utilization of different agrochemicals, for example, manures and pesticides, which might lessen ecological contamination and backing economical agribusiness (Shang, 2019). A few investigations have been delivered giving more data about the association between plant pressure and nanomaterials (Khan, 2017). The clearest aftereffect of abiotic stress on plants is oxidative pressure, which is ordinarily their most memorable line of safeguard (Serin and Whit, 2016). Nanomaterials might assist pushed plants with further developing their guard framework, which incorporates enzymatic cell reinforcements like catalase (CAT), superoxide dismutase (SOD), and peroxidase (POX), because of oxidative pressure (Patra *et al.*, 2016).

Nonetheless, when used in higher focuses, these nanomaterials may cause oxidative pressure in plants, which is welcomed on by the making of receptive oxygen or nitrogen species (ROS or RNS), which can hurt cells or their organelles (Iqbal, 2018). In this manner, it is important to explore the communications among plants and nanomaterials under assorted morphophysiological, biochemical, and atomic levels. The objective of this section is to all the more likely comprehend how nanoparticles collaborate with plants and how plants answer pressure to empower the utilization of nanoparticles in maintainable agribusiness strategies later on.

14.12.1 Take-Up, Translocation, and Biological Impacts of Plants

A wary, however, exact assessment of collaborations among plants and nanoparticles is expected to create nanoparticle nanotechnology's suggestions for plant science (Shang, 2019). It could give knowledge into how nanoparticles are taken up by plants, moved around within them, and in the long run gathered. Furthermore, it offers an appraisal of the likely effect of contention on plant development and improvement (Behzadi, 2017). The take-up of nanoparticles by plants is flighty as a result of various boundaries, including net charge, surface functionalization, substance content, shape, and size of the nanoparticles (Gupta and Xie, 2018). What's more, various different variables, for example, application techniques, associations with ecological components like soil type, water openness, and soil microorganisms, are additionally to fault for how nanoparticles influence plants (Remedios, 2012). In any case, it was noticed that different plant species' life structures, morphology, and physiology collaborate with nanoparticles in various ways (Remedios, 2012). Past examinations guaranteed that basic microscopy observing and location strategies made it more obvious how metal particles and metal oxide nanoparticles cooperate in plants (Plascencia-Villa, 2012). A restricted measure of data is accessible about the kinds of nanoparticles, explicitly in regards to their shape, size, charge, and covering with regards to their cooperations with plants (Garca-Gómez, 2018) as seen in Figure 14.7.

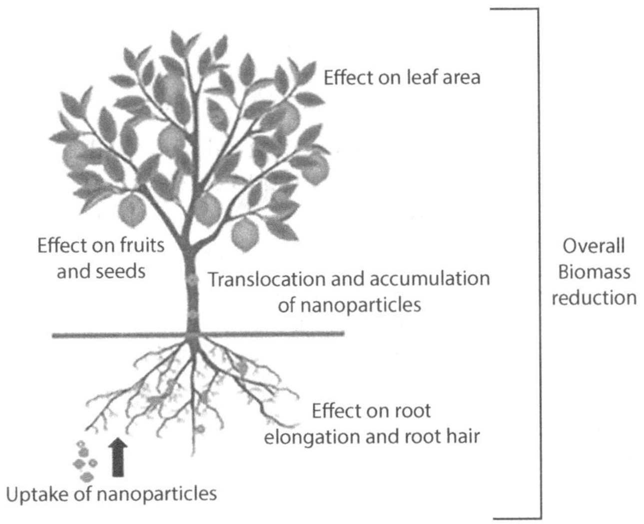

Figure 14.7 Nanoparticles uptake, translocation and accumulation in plants and its biological effect (Yen and Chen, 2019).

14.13 Nanobiopesticides

14.13.1 How Do Nanobiopesticides Work?

Nanobiopesticides can be made in two distinct ways. Naturally delivered dynamic pesticide synthetic substances or mixtures are joined with nanoparticles and a reasonable polymer to make nanobiopesticides in the principal class. In the next iteration, the active pesticide component depletes a metallic salt to produce metallic NPs, which are subsequently homogenized into a suitable polymer. Benefits such as slow and controlled delivery, improved mobility, and biodegradability of the pesticide ingredient and polymer motivated the development of APC with nanoparticles and blends in with viable vectors such as polymers, micelles, nanofibers, liposomes, and nanospheres. In addition, a wide variety of polymers can be used to create even more distinguishable nanocomposites. Extensive research into the usage of nanobiopesticide would reduce the amount of synthetics sprayed on crops by ensuring the precise delivery of active pesticide ingredients to kill insects for plant security (Ragaei and Sabry, 2014).

14.13.2 Advantages of Nanobiopesticides

Due to their nanosize, high surface region/volume proportion, soundness, upgraded viability, better dissolvability, versatility, and low poisonousness, nanobiopesticides are ideal (Sason, 2007). Albeit a few changes are made to a poisonous synthetic or pesticide utilized in nanodetails to forestall nanopesticide release in air, water, and soil, the tox ins might collect in the pecking order, raising serious ecological and well-being concerns. Nanopesticide particles are coordinated into conveyance materials as biopolymers, micelles, or composites to hold and work on the insecticidal worth while likewise limiting natural development to forestall such events. A natural trigger, like temperature, stickiness, wetness, or light, causes the arrival of these nanobiopesticides. Dirt nanotubes (halloysite), planned as pesticide transporters, are delivered by the natural trigger to cooperate better with bugs, like how biopolymers do. Like nanopesticides, nanofertilizers are encased in a specific transporter to permit controlled discharge for explicit impacts (Batiha, 2021; Awoyelu, 2021; Manjunatha, 2016). For the nanobiopesticides to show more desirable characteristics including firmness, vulnerability, crystallinity, warm steadiness, solvency, and biodegradability, they are supported into biopolymers (Adetuyi, 2015, 2021, 2022; Ogunlana, 2020, 2021; Bordes, 2009). One more advantage of utilizing nanomaterials is that they didn't seem to affect soil microorganisms

(Shah and Belozerova, 2009). Moreover, the phototoxicity of the powdered Ag nanoparticles was smothered by covering them with the nanocovering poly vinyl pyrrole, which is biocompatible (Lu, 2010; Farombi, 2021). The chance of utilizing nanomaterials in plant germination and development is expanded by examining such biocompatible covers to turn around the harmfulness of nanomaterials. One more opportunities for natural cultivating, as per Oancea, (2009), is the controlled arrival of dynamic plant development triggers and different mixtures encased in nano composites shaped of layered twofold hydroxides (anionic muds). Nanoemulsions, nanotypifies, nanoholders, and nanoconfines have as of late been portrayed as a few nanopesticide conveyance procedures with different functionalities for plant security (Adetuyi, 2022; Adeleke, 2022; Ogunlana, 2021). To make the nanostructure of nanobiopesticides, the current section utilizes electrospinning methods to fathom green nano biopesticide details (NBP). Also, the aggregation of NBP in plants, bugs, and other living things is analyzed, as well as the capability of NBP *in vitro* and *in vivo.* At last, themes connected with quality, wellbeing, human wellbeing, and morals are covered.

14.13.3 Delivery of Nanobiopesticide

Utilization of NPs for the conveyance of pesticides and biopesticides in the field faces various challenges, including numerous natural aggravations, the size of the areas covered by the splash, and eventually cost viability. Nanomaterial-based arrangements are expected to involve low-volume, high-esteem applications rather than the run of the mill splashing system, which includes soaking the whole harvest with the compound for effortlessness of organization and a huge volume, low-esteem readiness. A designated conveyance methodology in light of understanding the existence cycle and conduct of the infection or irritation is essential for these controlled nanoparticle conveyance frameworks (Manjunatha, 2016; Adetuyi, 2022). Figure 14.8 delineates how nanobiopesticides are frequently applied to plants to stifle bothers. These nanopesticides may contain both natural (like polymers) and inorganic (like metal oxides) constituents, and they might do as such in various structures, including micelles, nanofibers, nanogels, and particles (Ragaei and Sabry, 2014). Plants are treated with nanobiopesticides to shield them from bugs and bugs as well as to control them. However, we do not yet completely comprehend how these nanobiopesticides influence the two plants and bugs. Deciding the poisonousness of biopesticide definitions at nontoxic-limit dosages is essential as shown in Figure 14.8.

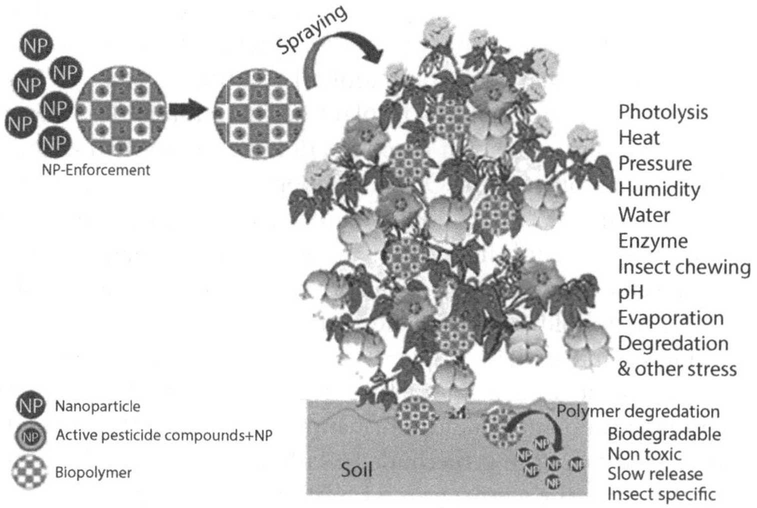

Figure 14.8 Application of nanobiopesticides delivery (Alvarez-Paimo, 2017).

14.13.4 The Potential Impacts of Nanobiopesticides

Through embodiment, the biopesticides could be mixed and combined with specific polymers for proficient activity and the sluggish delivery component. The breaking of the biopolymer, (for example, chitosan) and the deferred arrival of nanoparticles and pesticides, (for example, biocide) could happen once the light levels or other determined conditions, like temperature, water, dampness, or pH, are reached. This specific situation allows the chitosan polymer to vanish or corrupt, delivering the dynamic nanobiopesticides high up, soil, and vegetation. These nanoformulations are regularly tracked down in different pesticides and have the advantage of working on the evident solvency of an ineffectively solvent dynamic part, delivering the dynamic fixing gradually and explicitly, or potentially safeguarding the dynamic fixing from untimely corruption (Ragai and Saby 2014; Douza, 2018). While the polymeric particles are innocuous and disintegrate subsequent to conveying the nanoparticles, the nanobiopesticides target bugs or different nuisances that obliterate plants (Adetuyi, 2022).

14.13.5 Gathering of Nanoparticles in Plants

Despite the fact that the impact and gathering in plants might be pivotal, NPs regularly enter through the cell wall, leaf surfaces, or stomata or are taken up by the root from the dirt. These NPs interface by an unidentified strategy, incidentally acting to help and upgrade plant and seed development. NPs, be that as it may, can stifle plant and seed development clandestinely (Didunyemi, 2019, 2020). NPs might affect cell cytoplasm or mitochondria after entering the plant through leaf surfaces, tie the ribosome, plant proteins, and catalysts, and definitely disturb their normal capability to bring about cell demise. Moreover, they communicate with a few cell capabilities, changing phytohormones, metabolites, photosynthesis, transportation, and digestion, which brings about apoptosis. NP restricting modifies how cells work, causing oxidative pressure that causes proteins, lipids, and nucleic acids to oxidize. This influences pressure related qualities and lifts the blend of cell reinforcements for ROS movement (Rastogi, 2017; Nazir, 2022; Muniz, 2022). Accordingly, the NBP items should be regular and nonharmful, and a definitive sum that collects in plants should be painstakingly analyzed. As well as being naturally useful for plants and the climate, pesticides in nanoform or nanomaterial should likewise work specifically to ruin pests or insects. It is essential to completely explore the NBP's belongings and gathering in the collections of pests as well as their objective regions (on and inside the pests) where they connect or tie and the activities they achieve.

14.14 Conclusion

The mitigation of various abiotic stress in plants can be done in many ways, one of the ways is the utilization of plant-based nanobiopesticides on plants.

References

Abd-Elrahman, S.H. and Mostafa, M.A.M., Applications of nanotechnology in agriculture: An overview. *Egypt J. Soil Sci.*, 55, 2, 197214, 2015.

Abiri, R., Shaharuddin, N.A., Maziah, M., Role of ethylene and the APETALA 2/ ethylene response factor superfamily in rice under various abiotic and biotic stress conditions. *Env. Exp. Bot.*, 134, 33–44, 2017.

Abobatta, W.F., Nanotechnology application in agriculture. *Acta Sci. Agric.*, 2, 6, 99–102, 2018.

Adetuyi, B.O., Adebayo, P.F., Olajide, P.A., Atanda, O.O., Oloke, J.K., Involvement of free radicals in the ageing of cutaneous membrane. *World News Nat. Sci.*, 43, 11–37, 2022.

Adetuyi, B.O., Adebisi, O.A., Adetuyi, O.A., Ogunlana, O.O., Toloyai, T.Y., Egbuna, C., Uche, C.J., Khan, J., Adumanya, O.C.U., Patrick-Iwuanyanwu, K.C., Ficus exasperata attenuates acetaminophen-induced hepatic damage via NF-κB signaling mechanism in experimental rat model. *BioMed. Res. Int.*, 2022, 1, Article number 6032511, 2022.

Adetuyi, B.O., Adebisi, O.A., Awoyelu, E.H., Adetuyi, O.A., Ogunlana, O.O., Phytochemical and toxicological effect of ethanol extract of *Heliotropium indicum on* liver of male albino rats. *Lett. Appl. NanoBiosci. (LIANB)*, 10, 2, 2085–2095, 2020. https://doi.org/10.33263/LIANBS102.20852095.

Adetuyi, B.O. and Farombi, E.O., 6 Gingerol, an active constituent of ginger, attenuates lipopolysaccharide-induced oxido-inflammation, cognitive deficits, neuroplasticity and amyloidogenesis in rat. *J. Food Biochem.*, 45, 4, e13660, 2021. https://doi.org/10.1111/jfbc.13660.

Adetuyi, B.O., Dairo, J.O., Didunyemi, M.O., Anti-hyperglycemic potency of *Jatropha Gossypiifolia in* alloxan induced diabetes. *Biochem. Pharmacol. (Los Angel)*, 4, 5, 193, 2015.

Adetuyi, B.O., Odine, G.O., Olajide, P.A., Adetuyi, O.A., Atanda, O.O., Oloke, J.K., Nutraceuticals: Role in metabolic disease, prevention and treatment. *World News Nat. Sci.*, 42, 1–27, 2022.

Adetuyi, B.O., Ogundipe, A.E., Ogunlana, O.O., Egbuna, C., Estella, O.U., Mishra, A.P., Akram, M., Achar, R.R., Banana peel as a source of nutraceuticals, Food and agricultural byproducts as important source of valuable nutraceuticals, 1st ed., p. 400, Springer, Berlin, 2022.

Adetuyi, B.O., Okeowo, T.O., Adetuyi, O.A., Adebisi, O.A., Ogunlana, O.O., Oretade, O.J., Marraiki, N., Beshbishy, A.M., Welson, N.N., Batiha, G.E.S., Ganoderma lucidum from red mushroom attenuates formaldehyde-induced liver damage in experimental male rat model. *Biology*, 9, 10, 313, 2020. https://doi.org/10.3390/biology9100313.

Adetuyi, B.O., Olajide, P.A., Adetuyi, O.A., Oloke, J.K., Preventive phytochemicals of cancer as speed breakers in inflammatory signaling. *Res. J. Life Sci. Bioinform. Pharm. Chem. Sci.*, 8, 1, 30–61, 2022.

Adetuyi, B.O., Olajide, P.A., Awoyelu, E.H., Adetuyi, O.A., Adebisi, O.A., Oloke, J.K., Epidemiology and therapeutic measure for COVID-19, A review. *Afr. J. Reprod. Health (Special Edition COVID-19)*, 24, 2, 142, 2020. https://www.ajrh.info/index.php/ajrh/article/view/2300.

Adetuyi, B.O., Oluwole, E.O., Dairo, J.O., Biochemical effects of shea butter and groundnut oils on white albino rats. *Int. J. Chem. Chem. Processes (IJCC)*, 1, 8, 1–17, 2015.

Adetuyi, B.O., Oluwole, E.O., Dairo, J.O., Chemoprotective potential of ethanol extract of *Ganoderma Lucidum* on liver and kidney parameters in *Plasmodium Beghei*-induced mice. *Int. J. Chem. Chem. Processes (IJCC)*, 1, 8, 29–36, 2015.

Adetuyi, B.O., Omolabi, F.K., Olajide, P.A., Oloke, J.K., Pharmacological, biochemical and therapeutic potential of milk thistle (Silymarin): A review. *World News Nat. Sci.*, 37, 75–91, 2021.

Adetuyi, B.O., Toloyai, P.Y., Ojugbeli, E.T., Oyebanjo, O.T., Adetuyi, O.A., Uche, C.Z., Olisah, M.C., Adumanya, O.C.U., Jude, C., Chikwendu, J.K., Akram, M., Awuchi, C.G., Egbuna, C., Neurorestorative roles of microgliosis and astrogliosis in neuroinflammation and neurodegeneration. *Scicom J. Med. Appl. Med. Sci.*, 1, 1, 1–5, 2021.

Afsharinejad, A., Davy, A., Jennings, B., Brennan, C., Performance analysis of plant monitoring nanosensor networks at THz frequencies. *IEEE Internet Things J.*, 3, 5969, 2016.

Ahmad, A., Ghouri, M., Khan, Z., *et al.*, Nanotechnology: Transformation of agriculture and food security. *Am. J. Biotech. Biosci.*, 3, 19, 2020.

Ahmad, P. and Prasad, M.N.V., *Abiotic stress responses in plants: Metabolism, productivity and sustainability*, Springer, New York, 2012.

Ahmadi, A. and Baker, D.A., The effect of water stress on the activities of key regulatory enzymes of the sucrose to starch pathway in wheat. *Plant Growth Regul.*, 35, 81–91, 2001.

Ahmadi, A.A., Effect of postanthesis water stress on yield regulating processes in wheat *(Triticum aestivum L.)*, Ph.D. Thesis, University of London, Wye College, Wye, Ashford, U.K., 1998.

Ahmed, S., Hassan, A.A., Salama, D.M. *et al.*, Green synthesis of nanofertilizers and their application as a foliar for Cucurbita pepo L. *J. Nano.*, 2019, 1–11, 2019.

Akbari, G.A., Morteza, E., Moaveni, P. *et al.*, Pigments apparatus and anthocyanins reactions of borage to irrigation, methylalchol and titanium dioxide. *Int. J. Biosci.*, 4, 7, 192–208, 2014.

Ali, H., Khan, E., Sajad., M.A., Phytoremediation of heavy metals-concepts and applications. *Chemosphere*, 91, 91, 869–881, 2013.

Allen, R.D., Webb, R.P., Schake, S.A., Use of transgenic plants to study antioxidant defenses. *Free Rad. Biol. Med.*, 23, 473–479, 1997.

Almutairi, Z.M., Influence of silver nanoparticles on the salt resistance of tomato (Solanum lycopersicum L.) during germination. *Int. J. Agric. Biol.*, 18, 2, 449–457, 2016.

Alvarez-Paino, M., Muñoz-Bonilla, A., Fernández-García, M., Antimicrobial polymers in the nano-world. *Nano.*, 7, 48, 2017. https://doi.org/10.3390/nano7020048.

Amini, S., Ghadiri, H., Chen, C., Marschner, P., Salt-affected soils, reclamation, carbon dynamics, and biochar: A review. *J. Soils Sediments*, 16, 939–953, 2016.

Anand, R. and Bhagat, M., Silver nanoparticles (AgNPs): As nanopesticides and nanofertilizers. *MOJ Biol. Med.*, 4, 1, 1920, 2019.

Anjum, S.A., Xie, X.-Y., Wang, L.-C., Saleem, M.F., Man, C., Lei, W., Morphological, physiological and biochemical responses of plants to drought stress. *Afr. J. Agric. Res.*, 6, 2026–2032, 2011.

Arizona State University, 2018. Designing nanoparticles. Ask a biologist. https://askabiologist.asu.edu/building-right-nanoparticles (accessed June 5, 2023).

Asensi-Fabado, M.A., Amtmann, A., Perrella, G., Plant responses to abiotic stress: The chromatin context of transcriptional regulation. *Biochim. Biophys. Acta (BBA) - Gene Reg. Mechan.*, 1860, 1, 106–122, 2017.

Ashraf, M., Ozturk, M., Athar, H., Salinity and water stress: Improving crop efficiency, Springer, Netherlands, 2009.

Ashraf, M., Breeding for salinity tolerance in plants. *Crit. Rev. Plant Sci.*, 13, 17–42, 1994.

Ashraf, M. and Wu, L., Breeding for salinity tolerance in plants. *Crit. Rev. Plant Sci.*, 13, 1, 17–42, 1994.

Aslam, Z., Khattak, J.Z.K., Ahmed, M., Drought tolerance in cereal grain crops under changing climate, in: *Quantification of Climate Variability, Adaptation and Mitigation for Agricultural Sustainability*, A. Ahmed and C.O. Stockle (Eds.), 18209, Springer, Cham, 2017.

ASTM E2456-06 (2006). Terminology Relating to Nanotechnology. West Conshohocken, PA: ASTM International, United State.

Atef, M.M.S. and Behle, R.W., Comparing formulations for a mixed-microbial biopesticide with Bacillus thuringiensis var. kurstaki and Beauveria bassiana blastospores. *J. Arch. Phytopathol. Plant Prot.*, 50, 15–16, 2017.

Awoyelu, E.H., Oladipo, E.K., Adetuyi, B.O., Senbadejo, T.Y., Oyawoye, O.M., Oloke, J.K., Phyloevolutionary analysis of SARS-CoV-2 in Nigeria. *New Microbes New Infect.*, 36, 100717, 2020. https://doi.org/10.1016/j.nmni.2020.100717.

BaCgli, M.A., Moraes, J.C., Carvalho, G.A., Ecole, C.C., Gonçalves-Gervásio, R.D., Effect of sodium silicate application on the resistance of wheat plants to the green-aphids Schizaphisgraminum (Rond.) (Hemiptera: Aphididae). *Neotrop. Entomol.*, 32, 659–663, 2003.

Banerjee, J. and Kole, C., Plant nanotechnology: An overview on concepts, strategies and tools, in: *Plant Nanotechnology*, C. Kole *et al.* (Eds.), p. 114, Springer, Cham, 2016, https://doi.org/10.1007/9783-319-42154-4_1.

Bargaz, A., Lyamlouli, K., Chtouki, M. *et al.*, Soil microbial resources for impsroving fertilizers efficiency in an integrated plant nutrient management system. *Front. Microbiol.*, 9, 1606, 2018.

Bartels, D. and Salamini, F., Desiccation tolerance in the resurrection plant Craterstigma plantagineum. A contribution to the study of drought tolerance at the molecular level. *Plant Physiol.*, 127, 4, 1346–1353, 2001.

Batiha, G.E.B., Awad, D.A., Abdelazeem, M.A., Nyamota, R., Wahed, M.I., Shah, M.A., Amin, M.N., Adetuyi, B.O., Hetta, H.F., Cruz-Marins, N., Koirala, N., Ghosh, A., Sabatier, J., Diary-derived and egg white proteins in enhancing immune system against COVID-19. *Front. Nutr. (Nutritional Immunology)*, 8, 629440, 2021.

Baybordi, A., Effect of zinc, iron, manganese and copper on wheat quality under salt stress conditions. *J. Water Soil*, 140, 150–170, 2005.

Behzadi, S., Serpooshan, V., Tao, W. *et al.*, Cellular uptake of nanoparticles: Journey inside the cell. *Chem. Soc. Rev.*, 46, 14, 4218–4244, 2017.

Benelli, G., Plant-mediated biosynthesis of nanoparticles as an emerging tool against mosquitoes of medical and veterinary importance: A review. *J. Parasitol. Res.*, 115, 23–34, 2016.

Benelli, G., Mode of action of nanoparticles against insects. *Environ. Sci. Pollut. Res.*, 25, 13, 12329–12341, 2018.

Benfey, P.N., Ren, L., Chua, N.H., Tissue-specific expression from CaMV 35S enhancer subdomains in early stages of plant development. *EMBO J*, 9, 1677–1684, 1990.

Bergeson, L.L., Nanosilver: US EPA's pesticide office considers how best to proceed. *Environ. Qual. Manag.*, 19, 3, 79–85, 2010a.

Bergeson, L.L., Nanosilver pesticide products: What does the future hold? *Environ. Qual. Manag.*, 19, 4, 73–82, 2010b.

Bioinicia, S.L., Innovative material application, in: *Electrospinning Process Bioinicia*, 2017, https://bioinicia.com/electrospinning-equipment-fiber-fabrication/electrospinning-process-bioinicia-2/.

Bhatnagar-Mathur, P., Devi, M.J., Reddy, D.S. *et al.*, Stress-inducible expression of at DREB1A in transgenic peanut (Arachis hypogaea L.) increases transpiration efficiency under water-limiting conditions. *Plant Cell Rep.*, 26, 2071–2082, 2008.

Bisotto-de-Oliveira, R., De Jorgea, B.C., Roggiaa, I., Sant'Anab, J., Pereira, C.V., Nanofibers as a vehicle for the synthetic attactant TRIMEDLURE to be used for ceratitis capitata Wied: (Diptera, Tethritidae) capture. *J. Res. Updat. Polym Sci.*, 3, 40–47, 2014.

Bogue, R., Nanosensors: A review of recent research. *Sensor Review - SENS Rev.*, 29, 310–315, 2009.

Bogutska, K.I., Sklyarov, Y.P., Prylutskyy, Y.I., Zinc and zinc nanoparticles: Biological role and application in biomedicine. *Ukr. Bioorg. Acata*, 1, 9–16, 2013.

Bordes, P., Pollet, E., Avérous, L., Nano-biocomposites: Biodegradable polyester/nanoclay systems. *Prog. Polym. Sci.*, 34, 125–155, 2009.

Bouwmeester, H., Dekkers, S., Noordam, M.Y., Hagens, W.I., Bulder, A.S., de Heer, C., ten Voorde, SECGS, Wijnhoven, W.P., Marvin, H.J.P., Sips, A.J.A.M., Review of health safety aspects of nanotechnologies in food production. *Regul. Toxicol. Pharmacol.*, 53, 1, 52–62, 2009.

Boyer, J.S., Plant productivity and environment. *Science*, 218, 4571, 443448, 1982.

Bray, E.A., Bailey-Serres, J., Weretilnyk, E., Responses to abiotic stresses, in: *Biochemistry and Molecular Biology of Plants*, B.B. Buchanan, W. Gruissem, R.L. Jones (Eds.), American Society of Plant Physiologists, pp. 1158–1203, 2000.

Broman, K.W. and Speed, T.P., A review of methods for identifying QTLs in experimental crosses, in: *Statistics in Molecular Biology and Genetics*, vol. 33,

Seiller-Moiseiwitsch, F. (Ed.), vol. 33, pp. 114–142, 1999, IMS lecture notes monograph series.

BSI, British Standards Institution, *Vocabulary – nanoparticles, publicly available specification*, PAS, Scientific Research Publishing-SCRIP, Wuhan, Hubei, China, 2005.

BSI, London, p 71, Buzea, C., Blandino, I.I.P., Robbie, K., Nanomaterials and nanoparticles: Sources and toxicity. *Biointerfaces*, 2, 4, 17–172, 2007.

Buchanan, B.B., Gruissem, W., Jones, R.L., Biochemistry and molecular biology of plants. *Plant Growth Reg.*, 35, 105–106, 2001.

Burklew, C.E., Ashlock, J., Winfrey, W.B., Zhang, B., Effects of aluminium oxide nanoparticles on the growth, development, and microRNA expression of tobacco (Nicotiana tabacum). *PLoS One*, 7, 5, e34783, 2012.

Cakmak, I., Chen, W.Y.T.S., Lackner, M., Enrichment of cereal grains with zinc: Agronomic or genetic biofortification? *Plant Soil*, 302, 1–2, 1–17, 2008.

Chen, W.Y.T.S., Lackner, M. *Handbook of Climate Change Mitigation and Adaptation*, Springer International Publishing, Switzerland, 2017.

Cakmak, I., Yilmaz, A., Kalayci, M., Ekiz, H., Torun, B., Ereno, B., Braun, H.J., Zinc deficiency as a critical problem in wheat production in Central Anatolia. *Plant Soil*, 180, 2, 165–172, 1996.

Calanca, P.P., Effects of abiotic stress in crop production, in: *Quantification of Climate Variability, Adaptation and Mitigation for Agricultural Sustainability*, M. Ahmed and C.O. Stockle (Eds.), pp. 165–180, Springer, Cham, 2017.

Calanca, P.P., Effects of abiotic stress in crop production, in: *Quantification of Climate Variability, Adaptation and Mitigation for Agricultural Sustainability*, M. Ahmed and C.O. Stockle (Eds.), pp. 165–180, Springer International Publishing, Switzerland, 2017.

Cheng, H.N., Klasson, K.T., Asakura, T., Wu, Q., Nanotechnology in agriculture, in: *Nanotechnology: Delivering on the Promise*, vol. 2, H.N. Cheng, L. Doemeny, C.L. Geraci, D.G. Schmidt (Eds.), pp. 233–242, ACS, Washington, DC, 2016.

Chen, W.Y., Suzuki, T., Lackner, M. (Eds.), *Handbook of Climate Change Mitigation and Adaptation*, Springer International Publishing, Midtown Manhattan, New York City, 2017, https://doi.org/10.1007/978-3-319-14409-2.

Chhipa, H. and Joshi, P., Nanofertilisers, nanopesticides and nanosensors in agriculture, in: *Nanoscience in Food and Agriculture 1, Sustainable Agriculture Reviews*, vol. 20, S. Ranjan, N. Dasgupta, E. Lichtfouse (Eds.), p. 247282, Springer, Cham, 2016, https://doi.org/10.1007/978-3-319-39303-2.

Chhipa, H. and Joshi, P., Nanofertilisers, nanopesticides and nanosensors in agriculture, in: *Nanoscience in Food and Agriculture 1, Sustainable Agriculture Reviews*, vol. 20, S. Ranjan, N. Dasgupta, E. Lichtfouse (Eds.), pp. 247–282, Springer International Publishing, Switzerland, 2016, https://doi.org/10.1007/ 978-3-319-39303-2.

Chichiriccò, G. and Poma, A., Penetration and toxicity of nanomaterials in higher plants. *Nanomater.*, 5, 851–873, 2015.

Chokriwal, A., Sharma, M.M., Singh, A., Biological synthesis of nanoparticles using bacteria and their applications. *Am. J. Pharm. Tech. Res.*, 4, 6, 3861, 2014.

Costa, M.N., Veigas, B., Jacob, J.M., Santos, D.S., Gomes, J., Baptista, P.V., Martins, R., Inácio, J., Fortunato, E., A low cost, safe, disposable, rapid and self-sustainable paper-based platform for diagnostic testing: Lab-on-paper. *Nanotechnology*, 25, 094006, 2014. (12pp).

Cramer, G.R., Urano, K., Delrot, S. *et al.*, Effect of abiotic stress on plants: A systems biology perspective. *BMC Plant Biol.*, 11, 163, 2011. https://doi.org/10.1186/1471-2229-11-163.

Crampton, L., *Biological vs chemical pest control: Benefits and disadvantages*, Owlcation, United State, 2017, Available at: https://owlcation.com/agriculture/Biological-vs-Chemical-Pest-Control, Accessed 29 Aug 2018.

Cushman, J.C. and Bohnert, H.J., Genomic approaches to plant stress tolerance. *Curr. Opin. Plant Biol.*, 3, 2, 117124, 2000.

Dahoumane, S.A., Jeffryes, C., Mechouet, M., Agathos, S.N., Biosynthesis of inorganic nanoparticles: A fresh look at the control of shape, size and composition. *Bioeng.*, 4, 1, 14, 2017.

Daryanto, S., Wang, L., Jacinthe, P.A., Global synthesis of drought effects on cereal, legume, tuber and root crops production: A review. *Agric. Water Manag.*, 179, 18–33, 2017.

Das, S., Yadav, A., Debnath, N., Entomotoxic efficacy of aluminium oxide, titanium dioxide and zinc oxide nanoparticles against Sitophilus oryzae (L.): A comparative analysis. *J. Stored Prod. Res.*, 83, 9296, 2019.

Davies, J.C., Nanotechnology oversight: An agenda for the new administration, in: *Project on Emerging Nanotechnologies*, Woodrow Wilson International Center for Scholars, Washington, DC, 2008.

de la Rosa, G., García-Castañeda, C., Vázquez-Núñez, E. *et al.*, Physiological and biochemical response of plants to engineered NMs: Implications on future design. *Plant Physiol. Biochem.*, 110, 226–235, 2017.

De Oliveira, J.L., Campos, E.V.R., Bakshi, M., Abhilash, P.C., Fraceto, L.F., Application of nanotechnology for the encapsulation of botanical insecticides for sustainable agriculture: Prospects and promises. *Biotechnol. Adv.*, 32, 1550–1561, 2014.

Derosa, M., Monreal, C., Schnitzer, M. *et al.*, Nanotechnology in fertilizers. *Nat. Nanotechnol*, 5, 91, 2010.

Dhankher, O.P., Li, Y., Rosen, B.P. *et al.*, Engineering tolerance and hyperaccumulation of arsenic in plants by combining arsenate reductase and γ-glutamylcysteine synthetase expression. *Nat. Biotechnol.*, 20, 11, 1140–1145, 2002.

Didunyemi, M.O., Adetuyi, B.O., Oyebanjo, O.O., Morinda lucida attenuates acetaminophen-induced oxidative damage and hepatotoxicity in rats. *J. Biomed. Sci.*, 8, 1–7, 2019. https://www.jbiomeds.com/biomedical- sciences/morinda-lucida-attenuates-acetaminopheninduced-oxidative-damage-and-hepatotoxicity-in-rats.php?aid=24482.

Didunyemi, M.O., Adetuyi, B.O., Oyewale, I.A., Inhibition of lipid peroxidation and *in-vitro* antioxidant capacity of aqueous, acetone and methanol leaf extracts of green and red Acalypha wilkesiana Muell Arg. *Int. J. Biol. Med. Res.*, 11, 3, 7089–7094, 2020.

Dimetry, N.Z. and Hussein, H.M., Role of nanotechnology in agriculture with special reference to pest control. *Int. J. Pharmtech Res.*, 9, 10, 121–144, 2016.

Donfouet, H.P.P., Barczak, A., Détang-Dessendre, C., Maigné, E., Crop production and crop diversity in France: A spatial analysis. *Ecol. Eco.*, 134, 29–39, 2017.

Dou, Y., Sun, X., Guo, G., Dong, J., Lu, M., Zhang, W., Electrospun pure chitosan nanofibrous mats with high structural stability for dura mater regeneration. *Front. Bioeng. Biotechnol.*, 10, 1, 2016. Conference Abstract: 10th World Biomaterials Congress, Frontiers https://doi.org/10.3389/conf.FBIOE.2016.01.01956.

Dresselhaus, T. and Huckelhoven, R., Biotic and abiotic stress responses in crop plants. *Agron*, 8, 11, 267, 2018. https://doi.org/10.3390/agronomy8110267.

Dsouza, C.P., *Application of nanomaterials in intergarted pest management chapter 6 traning programme on advances in applications of nanotechnology*, pp. 54–63, CICR–CIRCOT, Mumbai, 2018.

Du, W., Yang, J., Peng, Q. *et al.*, Comparison study of zinc nanoparticles and zinc sulphate on wheat growth: From toxicity and zinc biofortification. *Chemosphere*, 227, 09116, 2019.

Dubchak, S., Ogar, A., Mietelski, J.W., Turnau, K., Influence of silver and titanium nanoparticles on arbuscular mycorrhiza colonization and accumulation of radiocaesium in Helianthus annuus. *Span J. Agric. Res.*, 8, s103S108, 2010.

Dubey, A. and Mailapalli, D.R., Nanofertilisers, nanopesticides, nanosensors of pest and nanotoxicity in agriculture, in: *Sustainable Agriculture Reviews*, E. Lichtfouse (Ed.), vol. 19, pp. 307–330, Dimension publications, United Kingdom, 2016.

Dwivedi, S., Saquib, Q., Al-Khedhairy, A.A., Musarrat, J., Understanding the role of nanomaterials in agriculture, in: *Microbial Inoculants in Sustainable Agricultural Productivity*, D.P. Singh, H.B. Singh, R. Prabha (Eds.), p. 271288, Springer, New Delhi, India, 2016.

Eichert, T., Kurtz, A., Steiner, U., Goldbach, H.E., Size exclusion limits and lateral heterogeneity of the stomatal foliar uptake pathway for aqueous solutes and water-suspended nanoparticles. *Physiol. Plant*, 134, 1, 151160, 2008. https://doi.org/10.1111/j.1399-3054.2008.01135.x.

Ekhtiyari, R. and Moraghebi, F., The study of the effects of nano silver technology on salinity tolerance of cumin seed (Cuminum cyminum L). *Plant Ecosyst.*, 7, 25, 99–107, 2011.

El-bendary, H.M. and El-Helaly, A.A., First record nanotechnology in agricultural: Silicananoparticles a potential new insecticide for pest control. *Appl. Sci. Rep.*, 4, 241–246, 2013.

Farombi, E.O., Abolaji, A.O., Adetuyi, B.O., Awosanya, O.O., Fabusoro, M., Neuroprotective role of 6-Gingerol-rich fraction of Zingiber officinale

(Ginger) against acrylonitrile-induced neurotoxicity in male Wistar rats. *J. Basic Clin. Physiol. Pharmacol. (JBCPP)*, 30, 3, 1–11, 2018, https://doi.org/10.1515/jbcpp-2018-0114.

Feynman, R., There is plenty of room at the bottom. An invitation to enter new field of physics. *Annual Meeting of American Physical Society at Caltech*, Engineering and Science, pp. 22–36, 1960.

Flowers, T.J., Improving crop salt tolerance. *J. Exp. Bot.*, 55, 396, 307–319, 2004.

García-Gómez, C., Obrador, A., González, D. *et al.*, Comparative study of the phytotoxicity of ZnO nanoparticles and Zn accumulation in nine crops grown in a calcareous soil and an acidic soil. *Sci. Total Environ.*, 644, 770–780, 2018. https://doi.org/10.1016/j.scitotenv.2018.06.356.

Gatoo, M.A., Naseem, S., Arfat, M.Y. *et al.*, Physicochemical properties of nanomaterials: Implication in associated toxic manifestations. *BioMed. Res. Int.*, 8, Article ID 498420, 2014, 8, 2014. https:// doi.org/10.1155/2014/498420.

Ghanbarzadeh, B., Babazadeh, A., Hamishehkar, H., Nano-phytosome as a potential food-grade delivery system. *Food Biosci.*, 15, 126–135, 2016.

Ghidan, A.Y. and Al Antary, T.M., Applications of nanotechnology in agriculture, in: *Applications of Nanobiotechnology*, IntechOpen, London, United Kingdom, 2019.

Gil-Díaz, M., Pinilla, P., Alonso, J., Lobo, M.C., Viability of a nanoremediation process in single or multi-metal(loid) contaminated soils. *J. Haz. Mat.*, 321, 812–819, 2017.

Gilligan, T.M. and Passoa, S.C., *LepIntercept – an identification resource for intercepted Lepidoptera larvae*, Identification technology program (ITP, Fort Collins, CO, 2014, http://idtools. org/id/leps/lepintercept/morphology.html.

Godfray, H.C.J., Beddington, J.R., Crute, I.R. *et al.*, Food security: The challenge of feeding 9 billion people. *Science*, 327, 5967, 812–818, 2010.

Gogoi, R., Dureja, P., Singh, P.K., Nanoformulations a safer and effective option for agrochemicals. *Indian Farming*, 59, 7–12, 2009.

Gohari, G., Mohammadi, A., Akbari, A. *et al.*, Titanium dioxide nanoparticle (TiO2 NPs) promote growth and ameliorate salinity stress effects on essential oil profile and biochemical attributes of Dracocephalum moldavica. *Sci. Rep.*, 10, 912, 2020. https://doi.org/10.1038/s41598-020-57794-1.

Gomes, H.I., Fan, G., Ottosen, L.M. *et al.*, Nanoremediation coupled to electrokinetics for PCB removal from soil, in: *Electrokinetics Across Disciplines and Continents*, A.B. Ribeiro, E.P. Mateus, N. Couto (Eds.), pp. 331–350, Springer Publication, 2016, https://doi.org/10.1007/978-3-319-20179-5_17.

González-Melendi, P., Fernández-Pacheco, R., Coronado, M.J., Corredor, E., Testillano, P.S., Risueño, M.C., Pérez-de-Luque, A., Nanoparticles as smart treatment-delivery systems in plants: Assessment of different techniques of microscopy for their visualization in plant tissues. *Ann. Bot.*, 101, 187–195, 2008.

Gouma, P.I., Simon, S.R., Stanacevic, M., Nano-sensing and catalysis technologies for managing food-water-energy (FEW) resources in farming. *Mater. Today Chem.*, 1, 2, 40–45, 2016.

Greaves, J.A., Improving suboptimal temperature tolerance in maize- The search for variation. *J. Exp. Bot.*, 47, 307–323, 1996.

Grover, A., Kapoor, A., Lakshmi, O.S. *et al.*, Understanding molecular alphabets of the plant abiotic stress responses. *Curr. Sci. (Special Section: Plant Mol. Biol)*, 80, 206–216, 2001.

Grover, M., Ali, S.Z., Sandhya, V. *et al.*, Role of microorganisms in adaptation of agriculture crops to abiotic stresses. *World J. Microbiol. Biotechnol.*, 27, 1231–1240, 2011.

Guerra, F.D., Attia, M.F., Whitehead, D.C., Alexis, F., Nanotechnology for environmental remediation: Materials and applications. *Molecules*, 23, 7, 1760, 2018.

Gupta, A., Singh, S.K., Singh, M.K. *et al.*, Plant growth–Promoting rhizobacteria and their functional role in salinity stress management. *Abatement Env. Poll.*, 1, 151–160, 2020.

Gupta, P., Raghuvanshi, S., Tyagi, A.K., Assessment of the efficiency of various gene promoters via biolistics in leaf and regenerating seed callus of millets, Eleusine coracana and Echinochloa crusgalli. *Plant Biotechnol.*, 18, 4, 275–282, 2001.

Gupta, R. and Xie, H., Nanoparticles in daily life: Applications, toxicity and regulations. *J. Env. Patholo., Toxicolo. Oncolo.: Off. Organ Int. Soc. Environ. Toxicol. Cancer*, 37, 3, 209–230, 2018.

Gupta, P.K. and Rustogi, S., Molecular markers from the transcribed/expressed region of the genome in higher plants. *Funct. Integr. Genomics*, 4, 139–162, 2004.

Hakeem, K.R., *Crop production and global environmental issues*, Springer International Publishing, Switzerland, 2015, https://doi.org/10.1007/978-3319-23162-4.

Hasanuzzaman, M., Nahar, K., Alam, M.M. *et al.*, Physiological, biochemical, and molecular mechanisms of heat stress tolerance in plants. *Int. J. Mol. Sci.*, 14, 5, 9643–9684, 2013. https://doi.org/10. 3390/ijms14059643.

Hatami, M. and Ghorbanpour, M., Effect of nanosilver on physiological performance of pelargonium plants exposed to dark storage. *J. Horti. Res.*, 21, 1, 15–20, 2013.

Hatami, M., Kariman, K., Ghorbanpour, M., Engineered nanomaterial-mediated changes in the metabolism of terrestrial plants. *Sci. Total Environ.*, 571, 275–291, 2016.

Hattori, Y.K., Nagai, M., Ashikari, Rice growth adapting to deep water. *Curr. Opin. Plant Biol.*, 14, 100–105, 2011.

He, C., Liu, Z., Xu, M. *et al.*, Urban expansion brought stress to food security in China: Evidence from decreased cropland net primary productivity. *Sci. Tot. Env.*, 576, 660–670, 2017.

He, M., He, C.Q., Ding, N.Z., Abiotic stresses: General defenses of land plants and chances for engineering multistress tolerance. *Front. Plant Sci.*, 9, 1771, 2018. https://doi.org/10.3389/fpls.2018. 01771.

Hedayati, M., Sharma, P., Katyal, D., Fagerlund, F., Transport and retention of carbon-based engineered and natural nanoparticles through saturated porous media. *J. Nanopart. Res.*, 18, 57, 2016. https://doi.org/10.1007/s11051-016-3365-6.

Hirt, H. and Shinozaki, K., Plant responses to abiotic stress. *Topics Curr. Gene*, 4, 1–8, 2004.

Hong, F., Yang, F., Liu, C. *et al.*, Influence of nano-TiO2 on the chloroplast aging of spinach under light. *Biol. Trace Elem. Res.*, 104, 249–260, 2005.

Husen, A. and Siddiqi, K.S., Phytosynthesis of nanoparticles: Concept, controversy and application. *Nanoscale Res. Lett.*, 9, 229–252, 2014.

IFOAM, *IFOAM position paper on "the use of nanotechnologies and nanomaterials in organic agriculture"*, Organic International, Bonn, Germany, 2011, Available at: http://www.ifoam.org/press/positions/.

IFOAM, PositionPaperNanotech2011_Approved.pdf. Position, 1, 1–4, 2011, Accessed 29 Aug 2018.

Ingale, A.G. and Chaudhari, A.N., Biogenic synthesis of nanoparticles and potential applications: An eco-friendly approach. *J. Nanomed. Nanotechol.*, 4, 165, 1–7, 2013.

IPCC Climate Change, *Synthesis report*, Intergovernmental Panel on Climate Change, Switzerland, 2014, Available from: http://www.ipcc.ch/report/ar5/syr/.

Iqbal, M., Raja, N.I., Hussain, M. *et al.*, Effect of silver nanoparticles on growth of wheat under heat stress. *IJST A Sci.*, 43, 387–395, 2019.

Iqbal, M.S., Abid, M., Ansari, M.I., Agricultural crops contamination with xenobiotics and its impact on human health, in: *Perspective in Ecotoxicology*, vol. Ch 13, pp. 171–181, 2018.

Iqbal, M.S., Singh, A.K., Singh, S.P., Ansari, M.I., Nanoparticles and plant interaction with respect to stress response, in: *Nanomaterials and Environmental Biotechnology, Nanotechnology in the Life Sciences*, Bhushan, I., Singh, V.K., Tripathi, D. (Eds.), pp. 1–15, 2020, doi.org/https://doi.org/10.1007/978-3- 030-34544-01.

Iravani, S., Green synthesis of metal nanoparticles using plants. *Green Chem.*, 13, 2638–2650, 2011.

Jaberzadeh, A., Moaveni, P., Moghadam, H.R.T., Zahedi, H., Influence of bulk and nanoparticles titanium foliar application on some agronomic traits, seed gluten and starch contents of wheat subjected to water deficit stress. *Not. Bot. Horti. Agrobot.*, 41, 201–207, 2013.

James-Okoro, P.O., Iheagwam, F.N., Sholeye, M.I., Umoren, I.A., Adetuyi, B.O., Ogundipe, A.E., Braimah, A.A., Adekunbi, T.S., Ogunlana, O.E., Ogunlana, O.O., Phytochemical and *in vitro* antioxidant assessment of yoyo bitters. *World News Nat. Sci.*, 37, 1–17, 2021.

Janmohammadi, M., Sabaghnia, N., Ahadnezhad, A., Impact of silicon dioxide nanoparticles on seedling early growth of lentil (Lens culinaris Medik.) genotypes with various origins. *Poljoprivreda i Sumarstvo*, 61, 3, 19, 2015.

Jeevanandam, J., Barhoum, A., Chan, Y.S. *et al.*, Review on nanoparticles and nanostructured materials: History, sources, toxicity and regulations. *Beilstein J. Nanotechnol.*, 9, 1050–1074, 2018. https:// doi.org/10.3762/bjnano.9.98.

Jianhui, Y., Kelong, H., Yuelong, W., Suqin, L., Study on anti-pollution nanopreparation of dimethomorph and its performance. *Chin. Sci. Bull.*, 50, 2, 108–112, 2005.

Kah, M. and Hofmann, T., Nanopesticide research: Current trends and future priorities. *Env. Int.*, 63C, 224–235, 2013. https://doi.org/10.1016/j.envint.2013.11.015.

Kamaraj, C., Rajakumar, G., Rahuman, A.A., Velayutham, K., Bagavan, A., Zahir, A.A., Elango, G., Feeding deterrent activity of synthesized silver nanoparticles using Manilkara zapota leaf extract against the house fly Musca domestica (Diptera: Muscidae). *Parasitol. Res.*, 111, 2439–2448, 2012.

Karlstad University, *Nanoscience lecture*, Karlstad University Publisher, Sweden, 2018, Lecture 1, http://www3.kau.se/kurstorg/files/n/C10B9C4709ddc16A23JOTTJF6D8D/NanoI-Lecture1%202012.pdf.

Khan, A.L., Waqas, M., Asaf, S. *et al.*, Plant growth-promoting endophyte Sphingomonassp LK11 alleviates salinity stress in Solanum pimpinellifolium. *Env. Exp. Bot.*, 133, 58–69, 2017.

Khan, I., Mohd, F., Pratichi, S., Padma, T., Nanotechnology for environmental remediation. *Res. J. Pharma. Biol. Chem. Sci.*, 5, 3, 1916, 2014.

Khan, M.A. and Akhtar, M.S., Agricultural adaptation and climate change policy for crop production in Africa, in: *Crop Production and Global Environmental*, K.R. Hakeem (Ed.), Springer Publication, Switzerland, 2015, https://doi.org/10.1007/978-3-319-23162-4_18.

Khan, M.R. and Rizvi, T.F., Nanotechnology: Scope and application in plant disease management. *Plant Pathol. J.*, 13, 3, 214–231, 2014.

Kim, S.W., Jung, J.H., Lamsal, K. *et al.*, Antifungal effects of silver nanoparticles (AgNPs) against various plant pathogenic fungi. *Mycobio.*, 40, 1, 53–58, 2012.

Kole, C. and Banerjee, J., Plant nanotechnology: An overview on concepts, strategies, and tools, in: *Plant Nanotechnology: Principles and Practices*, C. Kole, K.D. Sakthi, V.K. Mariya (Eds.), pp. 1–14, Springer International Publishing, Switzerland, 2016, https://doi.org/10.1007/978-3-319-42154-4.

Komarneni, S., Nanocomposites. *J. Mater. Chem.*, 2, 1219–1230, 1992.

Kou, T.J., Yu, W.W., Lam, S.K. *et al.*, Differential root responses in two cultivars of winter wheat (Triticum aestivum L.) to elevated ozone concentration under fully open-air field conditions. *J. Agron. Crop Sci.*, 204, 325–332, 2018.

Koul, O., Microbial biopesticides: Opportunities and challenges. *CAB Rev. Perspect. Agric. Vet. Sci. Nutr. Nat. Resour.*, 6, 056, 1–26, 2011.

Koul, O., Plant biodiversity as a resource for natural products for insect pest management, in: *Biodiversity and Insect Pests Key Issues for Sustainable Management*, G.M. Gurr, S.D. Wratten, W.E. Snyder, D.M.Y. Read (Eds.), pp. 85–105, Wiley, West Sussex, 2012.

Kumar, V. and Yadav, S.K., Plant-mediated synthesis of silver and gold nanoparticles and their applications. *J. Chem. Technol. Biotechnol.*, 84, 2, 151–157, 2009.

Kumar, A., Sengar, R.S., Singh, A., Dixit, R., Singh, R., Biotechnological tools for enhancing abiotic stress tolerance in plant, in: *Eco-Friendly Agro-Biological Techniques for Enhancing Crop Productivity*, Sengar R., Singh A. (Eds.), Springer, Singapore, 2018.

Kwak, S.Y., Wong, M.H., Lew, T.T.S. *et al.*, Nanosensor technology applied to living plant systems. *Annu. Rev. Anal. Chem. (Palo Alto Calif)*, 10, 1, 113–140, 2017.

Lade, B.D., Gogle, D.P., Nandeshwar, S.B., Nano bio pesticide to constraint plant destructive pests. *J. Nanomed. Res.*, 6, 3, 1–9, 2017.

Lade, B.D. and Patil, A.S., Silver nano fabrication using leaf disc of Passiflora foetida Linn. *Appl. Nanosci*, 7, 5, 181–192, 2017.

Lal, B. and Khanna, S., Selection of salt-tolerant Rhizobium isolates of Acacia nilotica. *World J. Microbiol. Biotechnol.*, 10, 6, 637–639, 1994.

Lamsal, K., Kim, S.W., Jung, J.H. *et al.*, Application of silver nanoparticles for the control of Colletotrichum species *in vitro* and pepper anthracnose disease in field. *Mycobiol.*, 39, 3, 194–199, 2011.

Lao, S.B., Zhang, Z.X., Xu, H.H., Jiang, G.B., Novel amphiphilic chitosan derivatives: Synthesis, characterization and micellar solubilization of rotenone. *Carbohydr. Polym.*, 82, 1136–1142, 2010.

Larrouturou, L., *The nanosciences*, p. 42, Centre national De La Recherché Science, France, 2005.

Lee, D.J., Senseman, S.A., Sciumbato, A.S. *et al.*, The effect of titanium dioxide alumina beads on the photocatalytic degradation of picloram in water. *J. Agric. Food Chem.*, 51, 9, 2659–2664, 2003.

Levitt, J., Responses of plants to environmental stress, Volume 1, in: *Chilling, Freezing, and High Temperature Stresses*, Academic Press, Cambridge, United State, 1980.

Li, N., Qian, W.J., Wang, L. *et al.*, Isolation and expression features of hexose kinase genes under various abiotic stresses in the tea plant (Camellia sinensis). *J. Plant Physiol.*, 209, 95–104, 2017.

Li, Z.Z., Chen, J.F., Liu, F., Liu, A.Q., Wang, Q. *et al.*, Study of UV-shielding properties of novel porous hollow silica nanoparticle carriers for avermectin. *Pest. Manag. Sci.*, 63, 241–246, 2007.

Liang, Y., Sun, W., Zhu, Y.G., Christie, P., Mechanisms of silicon-mediated alleviation of abiotic stresses in higher plants: A review. *Environ. Pollut.*, 147, 2, 422428, 2007. https://doi.org/10.1016/j.env pol.2006.06.008.

Liu, S., Yuan, L., Yue, X., Zheng, Z., Tang, Z., Recent advances in nanosensors for organophosphate pesticide detection. *Adv. Powder Technol.*, 19, 419–441, 2008.

Love, J.C., Estroff, L.A., Kriebel, J.K. *et al.*, Self-assembled monolayers of thiolates on metals as a form of nanotechnology. *Chem. Rev.*, 105, 4, 11031169, 2005. https://doi.org/10.1021/cr0300789.

Lu, W., Senapati, D., Wang, S., Tovmachenko, O., Singh, A.K., Yu, H., Ray, P.C., Effect of surface coating on the toxicity of silver nanomaterials on human skin keratinocytes. *Chem. Phys. Lett.*, 487, 92–96, 2010.

Luque, A.P. and Rubiales, D., Nanotechnology for parasitic plant control. *Pest. Manag. Sci. J.*, 65, 540–545, 2009.

Lyons, K. and Scrinis, G., Under the regulatory radar? Nanotechnologies and their impacts for rural Australia, in: *Tracking Rural Change: Community, Policy and Technology in Australia New Zealand and Europe*, F. Merlan and D. Raftery (Eds.), pp. 151–171, Australian National University E Press, Canberra, 2009.

Ma, J.F., Role of silicon in enhancing the resistance of plants to biotic and abiotic stresses. *Soil Sci. Plant Nutr.*, 50, 11–18, 2004.

Mahmoodzadeh, H., Nabavi, M., Kashefi, H., Effect of nanoscale titanium dioxide particles on the germination and growth of canola (Brassica napus). *J. Ornam. Hortic. Plants*, 3, 25–32, 2013.

Mani, P.K. and Mondal, S., Agri-nanotechniques for plant availability of nutrients, in: *Plant Nanotechnology: Principles and Practices*, C. Kole, K.D. Sakthi, V.K. Mariya (Eds.), pp. 263–303, Springer International Publishing, Switzerland, 2016, https://doi.org/10.1007/978-3-319-42154-411.

Manjunatha, S.B., Biradar, D.P., Aladakatti, Y.R., Nanotechnology and its applications in agriculture: A review. *J. Agric. Sci.*, 29, 1, 1–13, 2016.

Manyi-Loh, C., Mamphweli, S., Meyer, E., Okoh, A., Antibiotic use in agriculture and its consequential resistance in environmental sources: Potential public health implications. *Molecules*, 23, 4, 795, 2018. https://doi.org/10.3390/molecules23040795.

Meena, R.S., Gogoi, N., Kumar, S., Alarming issues on agricultural crop production and environmental stresses. *J. Clean Prod.*, 142, 4, 3357–3359, 2017.

Monica, R.C. and Cremonini, R., Nanoparticles and higher plants. *Caryologia*, 62, 2, 161–165, 2009.

Moon, J.W., Phelps, T.J., Fitzgerald, C.L. *et al.*, Manufacturing demonstration of microbially mediated zinc sulfide nanoparticles in pilot-plant scale reactors. *Appl. Microbiol. Biotechnol.*, 100, 7921–7931, 2016. https://doi.org/10.1007/s00253-016-7556-y.

Morejon, B., Pilaquinga, F., Domenech, F., Ganchala, D., Debut, A., Neira, M., Larvicidal activity of silver nanoparticles synthesized using extracts of Ambrosia arborescens (Asteraceae) to control Aedes aegypti L. (Diptera: Culicidae). *J. Nanotechnol.*, 2018, 1–8, 2018.

Mudalkar, S., Sreeharsha, R.V., Reddy, A.R., Involvement of glyoxalases and glutathione reductase in conferring abiotic stress tolerance to Jatropha curcas L. *Env. Exp. Bot.*, 134, 141–150, 2017.

Mukhopadhyay, S.S. and Kaur, N., Nanotechnology in soil-plant system, in: *Plant Nanotechnology: Principles and Practices*, C. Kole, K.D. Sakthi, V.K. Mariya (Eds.), pp. 329–348, Springer International Publishing, Switzerland, 2016, https://doi.org/10.1007/978-3-319-42154413.

Munir, N., Hasnain, M., Waqif, H., Adetuyi, B.O., Egbuna, C., Olisah, M.C., Gelling agents, micro and nanogels in food system applications, in: *Application of Nanotechnology in Food Science, Processing and Packaging*, pp. 153–167, 2022.

Nadeem, M.A., Nawaz, M.A., Shahid, M.Q., Doğan, Y., Comertpay, G., Yıldız, M., Baloch, F.S., DNA molecular markers in plant breeding: Current status and recent advancements in genomic selection and genome editing. *Biotechnol. Biotechnol. Equip.*, 32, 2, 261–285, 2018.

Nair, R., Varghese, S.H., Nair, B.G., Maekawa, T., Yoshida, Y., Kumar, D.S., Nanoparticulate material delivery to plants. *J. Plant Sci.*, 179, 154–163, 2010.

Narayanan, K.B. and Sakthivel, N., Biological synthesis of metal nanoparticles by microbes. *Adv. Coll. Interf. Sci.*, 156, 1–13, 2010.

Naturland, *Naturland standards for organic aquaculture*, Naturland International, Germany, 2011, Available at: http://www.naturland. de/fileadmin/MDB/documents/Richtlinien_englisch/Naturland-standards_Aquaculture.pdf, Accessed 1 Mar 2012

Nazir, A., Itrat, N., Shahid, A., Mushtaq, Z., Abdulrahman, S.A., Egbuna, C., Adetuyi, B.O., Khan, J., Uche, C.Z., Toloyai, P.Y., Orange peel as a source of nutraceuticals, in: *Food and Agricultural Byproducts as Important Source of Valuable Nutraceuticals*, 1st ed, p. 400, Springer, Berlin, 2022.

Niculaes, C., Abramov, A., Hannemann, L., Frey, M., Plant protection by benzoxazinoids—Recent insights into biosynthesis and function. *Agronomy*, 8, 8, 143, 2018.

Nuruzzaman, M., Rahman, M.M., Liu, Y., Naidu, R., Nanoencapsulation, Nanoguard for pesticides: A new window for safe application. *J. Agric. Food Chem.*, 64, 1447–1483, 2016.

Oancea, S., Padureanu, S., Oancea, A.V., Growth dynamics of corn plants during anionic clays action. *Lucrari Stiintifice Seria Agronomie*, 51, 212–217, 2009.

Ogunlana, O.O., Adetuyi, B.O., Adekunbi, T.S., Adegboye, B.E., Iheagwam, F.N., Ogunlana, O.E., Ruzu bitters ameliorates high-fat diet induced nonalcoholic fatty liver disease in male Wistar rats. *J. Pharm. Pharmacogn. Res.*, 9, 3, 251–26, 2021.

Ogunlana, O.O., Adetuyi, B.O., Esalomi, E.F., Rotimi, M.I., Popoola, J.O., Ogunlana, O.E., Adetuyi, O.A., Antidiabetic and antioxidant activities of the twigs of Andrograhis paniculata on Streptozotocin-induced diabetic male rats. *BioChem*, 1, 3, 238–249, 2021.

Ogunlana, O.O., Adetuyi, B.O., Rotimi, M., Esalomi, I., Adeyemi, A., Akinyemi, J., Ogunlana, O., Adetuyi, O., Adebisi, O., Okpata, E., Baty, R., Batiha, G., Hypoglycemic activities of ethanol seed extract of *Hunteria umbellate* (Hallier F.) on Streptozotocin-induced diabetic rats. *Clin. Phytoscience*, 7, 1, 1–9, 2021.

Ogunlana, O.O., Ogunlana, O.E., Adekunbi, T.S., Adetuyi, B.O., Adegboye, B.E., Iheagwam, F.N., Anti-inflammatory mechanism of Ruzu Bitters on diet-induced nonalcoholic fatty liver disease in male wistar rats.

Evid. Based Complement. Altern. Med., 2020, 1–8, 2020. https://doi.org/10.1155/2020/5246725.

Ogunlana, O.O., Ogunlana, O.E., Popoola, J.O., Adetuyi, B.O., Adekunbi, T.S., David, O.L., Adeleye, O.J., Udeogu, S.A., Adeyemi, A.O., Twigs of Andrographis paniculata (Burn. F) nees attenuates carbon Tetrachloride (CCl4) induced liver damage in wistar albino rats. *RASAYAN J. Chem.*, 14, 4, 2598–2603, 2021.

Omena, B.O., Olanrewaju, O.S., Babalola, O.O., Plant growth promoting rhizobacterial mitigation of drought stress in crop plants: Implications for sustainable agriculture. *Agronomy*, 9, 11, 712, 2019. https:// doi.org/10.3390/agronomy9110712.

Pallavi Mehta, C.M., Srivastava, R., Arora, S., Sharma, A.K., Impact assessment of silver nanoparticles on plant growth and soil bacterial diversity. *3 Biotech.*, 6, 254, 2016.

Pandey, A.C., Sanjay, S.S., Yadav, R.S., Application of ZnO nanoparticles in influencing the growth rate of Cicer arietinum. *J. Exp. Nanosci.*, 5, 488–497, 2010.

Pandey, P., Irulappan, V., Bagavathiannan, M.V., Senthil-Kumar, M., Impact of combined abiotic and biotic stresses on plant growth and avenues for crop improvement by exploiting physiomorphological traits. *Front. Plant Sci.*, 8, 537, 2017. https://doi.org/10.3389/fpls.2017.00537.

Pandey, P., Ramegowda, V., Senthil-Kumar, M., Shared and unique responses of plants to multiple individual stresses and stress combinations: Physiological and molecular mechanisms. *Front. Plant Sci.*, 6, 723, 2015. https://doi.org/10.3389/fpls.2015.00723.

Panpatte, D.G., Jhala, Y.K., Shelat, H.N., Vyas, R.V., Nanoparticles: The next generation technology for sustainable agriculture, in: *Microbial Inoculants in Sustainable Agricultural Productivity*, D. Sing, H. Sing, R. Prava (Eds.), p. 289–300, Springer, New Delhi, India, 2016, https://doi.org/10.1007/97881-322-2644-418.

Panpatte, D.G., Jhala, Y.K., Shelat, H.N., Vyas, R.V., Nanoparticles: The next generation technology for sustainable agriculture, in: *Microbial Inoculants in Sustainable Agricultural Productivity*, D. Singh, H. Singh, R. Prabha (Eds.), pp. 289–300, Springer publication, India, 2016, https://doi.org/10.1007/978- 81-322-2644-418.

Parveen, A. and Hussain, F., Salinity tolerance of three range grasses at germination and early growth stages. *Pak. J. Bot.*, 40, 2437–2441, 2008.

Patra, A.K., Adhikari, T., Bhardwaj, A.K., Enhancing crop productivity in salt-affected environments by stimulating soil biological processes and remediation using nanotechnology, in: *Innovative Saline Agriculture*, J.C. Dagar, P.C. Sharma, D.K. Sharma, A.K. Singh (Eds.), pp. 83–103, Springer, India, 2016, https://doi.org/10.1007/978-81-322-2770-04.

Patra, J.K., Das, G., Fraceto, L.F., Campos, E. *et al.*, Nano based drug delivery systems: Recent developments and future prospects. *J. Nanobiotechnol*, 16, 1, 71, 2018. https://doi.org/10.1186/s12951- 018-0392-8.

Paul B., Merly C., Kvapil P., Hans-Peter K., Status of nanoremediation and its potential for future deployment: Risk-benefit and benchmarking appraisals. *The J Env Cleanup Costs Technolo Tech, Remediation*, 28, 43 56, 2018, https://doi.org/10.1002/rem.21559.

Plascencia-Villa, G., Starr, C.R., Armstrong, L.S. *et al.*, Imaging interactions of metal oxide nanoparticles with macrophage cells by ultra-high resolution scanning electron microscopy techniques. *Integr. Biol.: Quant. Biosci. Nano Macro*, 4, 11, 1358–1366, 2012. https://doi.org/10.1039/c2ib20172k.

Potenza, C., Aleman, L., Sengupta-Gopalan, C., Targeting transgene expression in research, agricultural, and environmental applications: Promoters used in plant transformation. *In Vitro Cell. Dev. Biol. Plant*, 40, 1, 122, 2004.

Prasad, K.G., *Advances in application of nanotechnology training module chapter 5, electrospinning for production of nano materials*, pp. 48–53, organised by ICAR–CIRCOT, MS, India, Sept 24–28, 2018.

Ragaei, M. and Sabry, A.H., Nanotechnology for insect pest control. *Int. J. Sci. Environ. Technol.*, 3, 2, 528–545, 2014.

Raliya, R., Saharan, V., Dimkpa, C., Biswas, P., Nanofertilizer for precision and sustainable agriculture: Current state and future perspectives. *J. Agric. Food Chem.*, 66, 26, 64876503, 2017.

Raliya, R. and Tarafdar, J.C., ZnO nanoparticle biosynthesis and its effect on phosphorousmobilizing enzyme secretion and gum contents in Clusterbean (Cyamopsis tetragonoloba L.). *Agric. Res.*, 2, 4857, 2013.

Ran, Y., Liang, Z., Gao, C., Current and future editing reagent delivery systems for plant genome editing. *Sci. China Life Sci.*, 60, 5, 490505, 2017.

Rashidi, L. and Khosravi, K., The applications of nanotechnology in food industry. *Crit. Rev. Food Sci. Nutr.*, 51, 723–730, 2011. https://doi.org/10.1080/10408391003785417.

Rastogi, A., Tripathi, D.K., Yadav, S. *et al.*, Application of silicon nanoparticles in agriculture. *3 Biotech.*, 9, 3, 90, 2019.

Rastogi, A., Zivcak, M., Sytar, O., Kalaji, H.M., He, X., Mbarki, S., Brestic, M., Impact of metal and metal oxide nanoparticles on plant: A critical review. *Front. Chem.*, 5, 78, 2017. https://doi.org/10.3389/fchem.2017.00078.

Raza, A., Ali, R., Sundas, S.M. *et al.*, Impact of climate change on crops adaptation and strategies to tackle its outcome: A review. *Plants*, 8, 2, 34, 2019. doi.org/https://doi.org/10.3390/plants8020034.

Reddy, P.V.L., Hernandez-Viezcas, J.A., Peralta-Videa, J.R., Gardea-Torresdey, J.L., Lessons learned: Are engineered nanomaterials toxic to terrestrial plants? *Sci. Total Environ.*, 568, 470–479, 2016.

Rejeb, I.B., Pastor, V., Mauch-Mani, B., Plant responses to simultaneous biotic and abiotic stress: Molecular mechanisms. *Plants*, 3, 4, 458–475, 2014. https://doi.org/10.3390/plants3040458.

Remedios, C., Rosario, F., Bastos, V., Environmental nanoparticles interactions with plants: Morphological, physiological and genotoxic aspects. *J. Bot.*, Article ID 751686, 8, 2012. https://doi.org/ 10.1155/2012/751686.

Rizwan, M., Ali, S., Qayyum, M.F. *et al.*, Effect of metal and metal oxide nanoparticles on growth and physiology of globally important food crops: A critical review. *J. Hazard. Mater.*, 322 A, 2–16, 2017.

Rossi, L., Fedenia, L.N., Sharifan, H. *et al.*, Effects of foliar application of zinc sulfate and zinc nanoparticles in coffee (Coffea arabica L.) plants. *Plant Physiol. Biochem.*, 135, 160–166, 2019.

Rossini, M.A., Maddonni, G.A., Otegui, M.E., Multiple abiotic stresses on maize grain yield determination: Additive vs. multiplicative effects. *Field Crops Res.*, 198, 280–289, 2016.

Sabir, S., Arshad, M., Chaudhari, S.K., Zinc oxide nanoparticle for revolutionizing agriculture: Synthesis and applications. *Sci. World J.*, Article ID 925494, 8, 2014.

Sangam, S., Jayasree, D., Reddy, K.J., Chari, P.V.B., Sreenivasulu, N., Kishor, P.B., Salt tolerance in plants-transgenic approaches. *J. Plant Biotechnol.*, 7, 1, 115, 2005.

Sasson, Y., Levy-Ruso, G., Toledano, O., Ishaaya, I., Nanosuspensions: Emerging novel agrochemical formulations, in: *Insecticides Design Using Advanced Technologies*, I. Ishaaya, R. Nauen, A.R. Horowitz (Eds.), pp. 1–32, Springer-Verlag, Netherlands, Dordrecht, 2007.

Schmidt, H., Jonschker, G., Goedicke, S., Mennig, M., The sol-gel process as a basic technology for nanoparticle-dispersed inorganic-organic composites. *J. Solgel Sci. Technol.*, 19, 39–51, 2000.

Schwerdtfeger, P., Gold goes nano-from small clusters to low-dimensional assemblies. *Angew. Chem. Int. Ed.*, 42, 1892–1895, 2003.

Scrinis, G. and Lyons, K., Nanotechnology and the techno-corporate agri-food paradigm, in: *Food Security Nutrition and Sustainability*, G. Lawrence, K. Lyons, T. Wallington (Eds.), pp. 252–270, Earthscan, London, 2010.

Seghatoleslami, M. and Forutani, R., Yield and water use efficiency of sunflower as affected by nano ZnO and water stress. *J. Adv. Agric. Technol*, 2, 1, 34–37, 2015.

Sehgal, D., Bhat, V., Raina, S.N., Applicability of DNA markers for genome diagnostics of grain legumes, in: *Handbook of New Technology for Genetic Improvement of Grain Legumes*, P.B. Kirti (Ed.), p. 497557, CRC Press, New York, 2008.

Servin, A.D. and White, J.C., Nanotechnology in agriculture: Next steps for understanding engineered nanoparticle exposure and risk. *NanoImpact*, 1, 9–12, 2016. https://doi.org/10.1016/j.impact. 2015.12.002.

Shah, V. and Belozerova, I., Influence of metal nanoparticles on the soil microbial community and germination of lettuce seeds. *Water Air Soil Pollut.*, 197, 143–148, 2009.

Shallan, M.A., Hassan, H.M., Namich, A.A.M., Ibrahim, A.A., Biochemical and physiological effects of TiO2 and SiO2 nanoparticles on cotton plant under drought stress. *Res. J. Pharm. Biol. Chem. Sci.*, 7, 4, 1541, 2016.

Shang, Y., Hasan, M.K., Ahammed, G.J. *et al.*, Applications of nanotechnology in plant growth and crop protection: A review. *Molecules*, 24, 14, 2558, 2019. https://doi.org/10.3390/molecules24142558.

Sharma, R., Dewanjee, S., Kole, C., *Utilization of nanoparticles for plant protection*, Springer International Publishing, Cham, 2016.

Sharma, V.K., Yngard, R.A., Lin, Y., Silver nanoparticles: Green synthesis and their antimicrobial activities. *Adv. Coll. Interf. Sci.*, 145, 12, 8396, 2009.

Shelar, G.B. and Chavan, A.M., Mycosynthesis of silver nanoparticles from Trichoderma harzianum and its impact on germination status of oil seed. *Biolife*, 3, 1, 109–113, 2015.

Shenashen, M., Derbalah, A., Hamza, A. *et al.*, Antifungal activity of fabricated mesoporous alumina nanoparticles against root rot disease of tomato caused by Fusarium oxysporium. *Pest. Manag. Sci.*, 73, 11211126, 2017.

Shi, W.J., Shi, W.W., Gao, S.Y., Lu, Y.T., Cao, Y.S., Zhou, P., Effects of nanopesticide chlorfenapyr on mice. *Environ. Toxicol. Chem.*, 92, 1901–1907, 2010.

Siddiqui, M.H., Al-Whaibi, M.H., Faisal, M., Al Sahli, A.A., Nano-silicon dioxide mitigates the adverse effects of salt stress on Cucurbitapepo L. *Environ. Toxicol. Chem.*, 33, 11, 2429–2437, 2014. https://doi.org/10.1002/etc.2697.

Singh, A., Singh, S., Prasad, S.M., Scope of nanotechnology in crop science: Profit or loss. *Res. Rev. J. Bot. Sci.*, 5, 1, 14, 2016.

Singh, J. and Lee, B.K., Influence of nano TiO2 particles on the bioaccumulation of Cd in soybean plants (Glycine max): A possible mechanism for the removal of Cd from the contaminated soil. *J. Env. Manag.*, 170, 88–96, 2016.

Singhal, P., Jan, A.T., Azam, M., Haq, Q.M.R., Plant abiotic stress: A prospective strategy of exploiting promoters as alternative to overcome the escalating burden. *Front. Life Sci.*, 9, 1, 5263, 2015.

Siva, C. and Kumar, M.S., Pesticidal activity of eco-friendly synthesized silver nanoparticles using Aristolochia indica extract against Helicoverpa armigera Hubner (Lepidoptera: Noctuidae). *Int. J. Adv. Sci. Technol. Res.*, 5, 197–226, 2015.

Sivapriyajothi, S., Kumar, P.M., Kovendan, K., Subramaniam, J., Murugan, K., Larvicidal and Pupicidal activity of synthesized silver nanoparticles using Leucas Aspera leaf extract against mosquito vectors Aedes Aegypti and Anopheles Stephensi. *J. Entomol. Acarol. Res.*, 46, 2, 1–8, 2014.

Smith, P., House, J., Bustamante, M. *et al.*, Global change pressures on soils from land use and management. *Global Change Biol.*, 22, 1008–1028, 2016. https://doi.org/10.1111/gcb.13068.

Song, U., Jun, H., Waldman, B. *et al.*, Functional analyses of nanoparticle toxicity: A comparative study of the effects of TiO2 and Ag on tomatoes (Lycopersicon esculentum). *Ecotoxicol. Environ. Saf.*, 93, 60–67, 2013. https://doi.org/10.1016/j.ecoenv.2013.03.033.

Stevens, J., Dunse, K., Fox, J., Evans, S., Anderson, M., Biotechnological approaches for the control of insect pests in crop plants. *Pestic Adv. Chem. Bot. Pestic*, 1, 1–41, 2012. https://doi.org/10.5772/46233.

Taniguchi, N., On the basic concept of nanotechnology, in: *Proceedings of the International Congress on Production Engineering, JSPE*, Tokyo, 1974.

Tarafdar, J.C., Raliya, R., Mahawar, H., Rathore, I., Development of zinc nanofertilizer to enhance crop production in pearl millet (Pennisetum americanum). *Agric. Res.*, 3, 3, 257–262, 2014. https://doi. org/10.1007/s40003-014-0113-y.

Tayal, D., Srivastava, P.S., Bansal, K.C., Transgenic crops for abiotic stress tolerance, in: *Plant Biotechnology and Molecular Markers*, pp. 346–365, Kluwer Academic Publishers, Springer, Dordrecht, 2004.

The Royal Society & The Royal Academy of Engineering, Nanoscience and nanotechnologies: Opportunities and uncertainties, in: *Nanoscience and Nanotechnologies*, pp. 1–127, The Royal Society publisher, London, 2004.

Tirani, M.M., Haghjou, M.M., Ismaili, A., Hydroponic grown tobaccoplants respond tozincoxide nanoparticles and bulk exposures by morphological, physiological and anatomical adjustments. *Funct. Plant Biol.*, 46, 360–375, 2019.

United Nations, Department of Economic and Social Affairs, Population Division, in: *World Population Prospects: The 2017 Revision, Key Findings and Advance Tables*, United Nations Publisher, New York, 2017, Working Paper No. ESA/P/WP/248.

USA National Nanotechnology Initiative, *Nanogov National Nanotechnology Initiative*, National Nanotechnology Initiatives, United State, 2018, https://www.nano.gov/nanotech-101/what/definition.

van der Laan, M., Bristow, K.L., Stirzaker, R.J., Annandale, J.G., Towards ecologically sustainable crop production: A south african perspective. *Agric. Ecosyst. Environ.*, 236, 108–119, 2017.

Vannini, C., Domingo, G., Onelli, E. *et al.*, Morphological and proteomic responses of Eruca sativa exposed to silver nanoparticles or silver nitrate. *PLoS One*, 8, 7, 1–8, 2013.

Vanti, G.L., Nargund, V.B., Basavesha, K.N. *et al.*, Synthesis of Gossypium hirsutum-derived silver nanoparticles and their antibacterial efficacy against plant pathogens. *Appl. Organomet. Chem.*, 33, e4630, 1–8, 2019.

Vardar, F. and Yanik, F., Toxic effects of Aluminium oxide (Al2O3) nanoparticles on root growth and development in Triticum aestivum. *Water Air Soil Pollut.*, 226, 296, 2015.

Venkatachalam, P., Jayaraj, M., Manikandan, R. *et al.*, Zinc oxide nanoparticles (ZnO NPs) alleviate heavy metal-induced toxicity in Leucaena leucocephala seedlings: A physiochemical analysis. *Plant Physiol. Biochem.*, 110, 59–69, 2016. https://doi.org/10.1016/j.plaphy.2016.08.022.

Vidyalakshmi, N., Thomas, R., Aswani, R. *et al.*, Comparative analysis of the effect of silver nanoparticle and silver nitrate on morphological and anatomical parameters of banana under *in vitro* conditions. *Inorg. Nano Metal Chem.*, 47, 1530–1536, 2017. https://doi.org/10.1080/24701556. 2017.1357605.

Vij, S. and Tyagi, A.K., Emerging trends in the functional genomics of the abiotic stress response in crop plants. *Plant Biotechnol. J.*, 5, 3, 361–380, 2007. https://doi.org/10.1111/j.1467-7652.2007.00239.x.

Vijayalakshmi, D., *Abiotic stress and its management in agriculture*, pp. 361–387, TNAU Agritech, Coimbatore, 2018.

Waage, J., Global developments in biological control and the implications for Europe. *EPPO Bull.*, 27, 5–13, 1997.

Wang, P., Sun, X., Jia, X., Ma, F., Apple autophagy-related protein MdATG3s afford tolerance to multiple abiotic stresses. *Plant Sci.*, 256, 53–64, 2017.

Wang, S.H., Wang, F.Y., Gao, S.C., Wang, X.G., Heavy metal accumulation in different rice cultivars as influenced by foliar application of nano-silicon. *Water Air Soil Pollut.*, 227, 228, 2016.

Wang, X.D., Ou-yang, C., Fan, Z.R. *et al.*, Effects of exogenous silicon on seed germination and antioxidant enzyme activities of Momordica charantia under salt stress. *J. Anim. Plant Sci.*, 6, 700–708, 2010.

Wani, S.H., Kumar, V., Shriram, V., Sah, S.K., Phytohormones and their metabolic engineering for abiotic stress tolerance in crop plants. *Crop J.*, 4, 3, 162–176, 2016.

Waqas, M.A., Kaya, C., Riaz, A. *et al.*, Potential mechanisms of abiotic stress tolerance in crop plants induced by thiourea. *Front. Plant Sci.*, 10, 13–36, 2019. https://doi.org/10.3389/fpls.2019.01336.

Xiao, M., Song, F., Jiao, J. *et al.*, Identification of the gene Pm47 on chromosome 7BS conferring resistance to powdery mildew in the Chinese wheat landrace Hongyanglazi. *Theor. Appl. Genet.*, 126, 13971403, 2013.

Yan, A. and Chen, Z., Impacts of silver nanoparticles on plants: A focus on the phytotoxicity and underlying mechanism. *Int. J. Mol. Sci.*, 20, 5, 1003, 2019. https://doi.org/doi.org/10.3390/ijms20051003.

Yang, F., Hong, F., You, W. *et al.*, Influence of nano-anatase TiO2 on the nitrogen metabolism of growing spinach. *Biol. Trace Elem. Res.*, 110, 2, 179–190, 2006.

Yang, F., Liu, C., Gao, F. *et al.*, The improvement of spinach growth by nano-anatase TiO2 treatment is related to nitrogen photoreduction. *Biol. Trace Elem. Res.*, 119, 77–88, 2007.

Yılmaz, E.E., Özgür, N., Bereli, D. *et al.*, Plastic antibody based surface plasmon resonance nanosensors for selective atrazine detection. *Mater. Sci. Eng.*, 73, 603–610, 2017.

Zaytseva, O. and Neumann, G., Carbon nanomaterials: Production, impact on plant development, agricultural and environmental applications. *Chem. Biol. Technol. Agric.*, 3, 17, 2016. https://doi.org/10. 1186/s40538-016-0070-8.

Zhao, C., Liu, B., Piao, S. *et al.*, Temperature increase reduces global yields of major crops in four independent estimates. *Proc. Natl. Acad. Sci. U. S. A.*, 114, 35, 9326–9331, 2017.

Zhu, Z.J., Wang, H., Yan, B. *et al.*, Effect of surface charge on the uptake and distribution of gold nanoparticles in four plant species. *Environ. Sci. Technol.*, 46, 12391–12398, 2012. https://doi. org/10.1016/j.cell.2016.08.029.

Zhu J.K., Abiotic stress signaling and responses in plants. *Cell,* 167, 2, 313–324. https://doi. org/10.1016/j.cell.2016.08.029.

Zuverza-Mena, N., Martínez-Fernández, D., Du, W. *et al.*, Exposure of engineered nanomaterials to plants: Insights into the physiological and biochemical responses-a review. *Plant Physiol. Biochem.*, 110, 236–264, 2017. https://doi. org/10.1016/j.plaphy.2016.05.03.

Index

Also of Interest

Handbook of Agricultural Biotechnology

Volume I Nanopesticides
Edited by Charles Oluwaseun Adetunji and Julius Kola Oloke
Published 2024. ISBN 978-1-119-83614-8

Volume II Nanobioherbicides
Edited by Charles Oluwaseun Adetunji and Julius Kola Oloke
Published 2024. ISBN 978-1-119-83615-5

Volume III Nanofungicides
Edited by Charles Oluwaseun Adetunji and Julius Kola Oloke
Published 2024. ISBN 978-1-119-83616-2

Volume IV Insecticides
Edited by Charles Oluwaseun Adetunji and Julius Kola Oloke
Published 2024. ISBN 978-1-119-83617-9

Volume V Nanobiofertilizers
Edited by Charles Oluwaseun Adetunji, Chukwuebuka Egbuna, Anton Ficai and Oluwatosin Ademola Ijabadeniyi
Forthcoming Fall 2024. ISBN 978-1-394-21149-4

Next-Generation Algae, Volume I Applications in Agriculture, Food and Environment
Edited by Charles Oluwaseun Adetunji, Julius Kola Oloke, Naveen Dwivedi, Sabeela Beevi Ummalyma, Shubha Dwivedi, Juliana Bunmi Adetunji and Daniel Ingo Hefft
Published 2023 ISBN 978-1-119-85727-3

Printed in the USA/Agawam, MA
September 16, 2026

910661.138